PID Controller Tuning Using the Magnitude Optimum Criterion

Konstantinos G. Papadopoulos

PID Controller Tuning Using the Magnitude Optimum Criterion

Springer

Konstantinos G. Papadopoulos
ATDD
ABB Industries
Turgi, Aargau
Switzerland

ISBN 978-3-319-07262-3 ISBN 978-3-319-07263-0 (eBook)
DOI 10.1007/978-3-319-07263-0

Library of Congress Control Number: 2014950412

Springer Cham Heidelberg New York Dordrecht London

Printed on acid-free paper

Springer is part of Springer Science+Business Media (www.springer.com)

To my mother Maria,

To the memory of my father Georgios,
(1937–1999)

To Eirini,

To the memory of my advisor Nikolaos Margaris.
(1949–2013)

. . . φύσις μηθὲν μήτε ἀτελὲς ποιεῖ μήτε μάτην . . .

Πολιτικά/Α.
Αριστοτέλης, 384–322 π.Χ.
Αρχαίος Έλληνας φιλόσοφος

Acknowledgments

This monograph is part of my research carried out at the Aristotle University of Thessaloniki, Greece within the Department of Electrical and Computer Engineering from 2005 to 2010. My deepest gratitude for his invaluable help and guidance during this time, goes undoubtedly to my supervisor, Professor and ex-Dean of the School of Engineering (Aristotle University of Thessaloniki), Nikolaos I. Margaris who unfortunately passed away on July 1, 2013. Professor Margaris along with Professor Loukas Petrou (Associate Professor of Microprocessor Systems, Department of Electrical and Computer Engineering, Aristotle University of Thessaloniki) taught me well how to combine the theory with practical problems and always keep in mind not to focus or recommend complex theoretical solutions, which most of the times prove to be infeasible in many industry applications when comes to the question of implementation.

In 2010, and after the end of this work, I joined ABB Switzerland Ltd., Department of Medium Voltage Drives where I had also worked during 2006, achieving the first real time implementation of the Model Predictive Direct Torque Control algorithm (also known as MPDTC) for the induction motor drive. Since that time and after working as a Control Software Development Engineer for the ABB's MV Drives ACS product family, I realized that it is a big mistake to loose contact with the real world, when discussing solutions around automatic control theory and design. This is a fundamental principle I gained throughout my discussions with my supervisor during the MPDTC implementation, Manfred Morari (Head of Automatic Control Laboratory, "Institut für Automatik", ETH Zürich, 2006) and my colleagues from my current department. All of their feedback proved to be really meaningful and invaluable, since I always remember myself learning for new problems within a real industrial control loop, problems for which many academic scientists think to have easily solved.

For that reason, I want deeply to thank one by one all my colleagues within our department, starting with Christian Stulz (Team Leader of Control Concepts, MV Drives), Patrick Bohren (ex-Head of Control Software, Platform Manager ACS software products), Gerald A. Scheuer (Vice President and Global R&D Manager of MV Drives), Georgios Papafotiou (Global R&D Manager of MV Drives control

Hardware and Software), Aleksy Burzanowski, Halina Burzanowska, Alexander Glueck (Application Software Development Engineer), Juan-Alberto Marrero-Sosa, Marc Rauer and Oliver Scheuss from the Control Software Development team, MV Drives, Peter Al-Hokayem and Tobias Geyer, both Principal Scientists with ABB Corporate Research Center, Davide Andreo (Product Manager, ABB Drives Systems), Drazen Dujic (ex R&D Platform Manager of ACS6000, Assistant Professor, École Polytechinique Fédéral de Lausanne), Jonas Wahlströem (ACS6000 Global Product Manager), Pieder Jörg (Business Development, ABB MV Drives), Tino Wymann (System Design, ABB MV Drives), Jonas Kley (Team Leader, Software Operations ABB MV Drives), Klaus Rütten, Jia Shen, Maged-Sameh Farrag and Mathieu Giroux (from the Software Operations team, ABB MV Drives), Daniel Siemaszko, (Research Fellow, CERN), Kristjan Ljubec (Service and Commissioning Engineer, ABB MV Drives ACS Product family), Konstantina Mermikli (Electrical and Computer Engineer, Prisma Electronics, Greece) and Alexandros Vouzas (M.Sc. student with the Automatic Control Laboratory, "Institut für Automatik", ETH Zürich).

The last many thanks go to my beloved people who for sure are first in my heart. These people are my mother Maria and my girlfriend Eirini. For my mother, the least I could do is to dedicate this book to her. For Eirini, I just want to remind her a statement coming from the wonderful people of the island of Crete, Greece.

...λένε πως αν ζητάς πολλά πότε σου δεν ϑα τα 'χεις γι' αυτό και 'γω σ' ευχαριστώ μονάχα που υπάρχεις ...

Thessaloniki, Greece
Zürich, Switzerland
July 2014

Konstantinos G. Papadopoulos

Synopsis

This book introduces a systematic controller design strategy for type-I, type-II and type-III linear single-input single-output closed loop control systems regardless of the process complexity. The main advantage of type-I, type-II, and type-III loops is their ability to track fast reference signals since their output variable achieves zero steady state position, velocity, and acceleration error at the presence of step, ramp, and parabolic reference input signals, respectively. Since such kind of loops are often met in many industry applications (electrical and chemical engineering) the proposed control law is of PID, and therefore fast and quick integration of the proposed approach can be achieved on a real time application platform.

The development of the proposed theory lies in the well-known Magnitude Optimum criterion, takes place in the frequency domain, and is carried out into two directions. The first direction of the proposed approach deals with the *direct tuning of PID regulators* and the second direction deals with the well-known term *automatic tuning of PID regulators*.[1]

For the direct tuning of PID regulators and further to the control law's proof, a general transfer function of the process model is involved and based on the type of the control loop to be designed (type-I, type-II, type-III), the three parameters of the PID controller are explicitly determined in terms of closed form expressions. These closed form expressions involve all process modeled parameters and can be applied for the control of any SISO process regardless of its complexity. Therefore, if system identification techniques are followed for the determination of the transfer function of the plant, the proposed PID controller parameters can be directly calculated.

Once this step is complete, a new approach to a common problem met in many real world applications is presented, which is associated with the automatic tuning of PID regulators. Note that for this problem, given little information about the

[1] Automatic tuning is often called *tuning on demand* or *one shot tuning*.

model of the process,[2] an algorithm regarding the automatic tuning of PID regulators is proposed.

Based on the proposed automatic tuning method, the aforementioned explicit solution introduces two advantages in the step of direct tuning. The first advantage is the *preservation of the shape of the step and frequency response* of the aforementioned proposed control law, by which it is meant that the step and frequency response of the control loop exhibits a certain performance (overshoot, settling time, etc.) when the PID controller is designed via the explicit solution. The second advantage is the fact that all three PID parameters can be expressed as a function of only one parameter via the explicit solution. Therefore, the proposed automatic tuning method tunes only one parameter of the controller (the other two are automatically tuned through the explicit solution) while trying to achieve the aforementioned performance.

For the development of both the explicit PID tuning and the automatic tuning method of the PID controller, background of linear systems theory is required, and all control loops are considered to be single-input single-output. The definition of the proposed theory covers both analog and digital design of the controller. Regarding the digital controller design, the sampling period T_s of the controller is also involved within the closed form expressions. This advantage, gives control engineers the flexibility to accurately investigate the effect of the choice of the controller's sampling time to the control loop's performance. Now, for the sake of a clear presentation of the proposed theory the material of this book is organized as follows: the whole book is split into three parts, Parts I, II, and III.

Part I consists of Chaps. 1 and 2. Chapter 1 gives an overview relevant to the evolution of PID control describing the current state of the art of PID tuning methods for type-I, type-II, and type-III control loops. These loops are described both on a theoretical and practical basis and concrete industrial examples from the field of electric motor drives are presented.

Chapter 2 presents the necessary background from the linear systems theory focusing on the definition of the type of the system itself, internal stability, and the Magnitude Optimum principle.

Part II consists of Chaps. 3–6. In Chap. 3 the conventional tuning for the PID controller via the Magnitude Optimum criterion for type-I control loops is presented. Advantages and drawbacks of this method are remarked and the revised PID control law is presented within the same chapter. The potential of the proposed method is justified through the control of several benchmark process models (process with dominant time constants, process with long time delay, nonminimum phase process, process with strong zeros) often met in many real world applications. Comparison results focus on the performance at the output of the control loop both for the revised and the conventional PID tuning in the time and frequency domain. Finally, an example from the field of electric motor drives

[2] Information coming from an open loop experiment of the process.

regarding the tuning of the PID current control loop of a grid connected converter through the proposed method is presented.

In similar fashion with Chap. 3, Chap. 4 presents the application of the Magnitude Optimum principle to type-II control loops, commonly known within the academic and industrial literature as Symmetrical Optimum tuning. Again, the conventional PID tuning via the Magnitude Optimum criterion for type-II control loops is presented so that advantages and disadvantages of the current state of the art are made clear to the reader. To cope with the remarked drawbacks, a revised PID control law is presented for type-II control loops, introducing again an explicit solution for the controller parameters. Once more a comparison section for several benchmark processes follows, both for the conventional and the revised method. The chapter closes with a practical example of a type-II control loop related to the control of actual DC link voltage in an AC/DC/AC converter arrangement often met in the field of electric motor drives.

In Chap. 5 the design of a PID type-III control loop is presented for first time over the literature. To achieve this, a similar to the conventional type-II PID design procedure is introduced which leads effortlessly to the development of the optimal PID control law for type-III control loops regardless of the process complexity. Again, a comparison between the conventional and the revised control law is performed for several benchmark process models. The chapter closes with the extension of the conventional PID type-III tuning to the design of type-IV, type-V and finally type-p control loops.

In Chap. 6, the revised control law is presented for digital control loops[3] and therefore the sampling period of the controller T_s is introduced within the explicit solution. Comparison results are presented for analog and digital design focusing on the effect of the sampling period on the control loop's performance both in the time and frequency domain.

Part III consists of Chaps. 7 and 8. In Chap. 7, the proposed automatic tuning method for type-I control loops is presented. The same principle is extended in cases where the process contains conjugate complex poles. The application of the proposed method requires (1) an open loop experiment for initializing the algorithm and (2) access to the output of the process and not to its states. Simulation examples between the explicit solution and the proposed method justify the potential of the current approach.

In Chap. 8, the contribution of the proposed theory is summarized and directions to control engineers are given so that the explicit solution and the automatic tuning algorithm are integrated within a real time application platform.

Finally, all proofs of the revised PID control law for type-I, type-II, type-III control loops (analog and digital design) are summarized in Appendices A, B and C. In Appendix A the principle of the Magnitude Optimum criterion is presented

---

[3] For type-I, type-II and type-III control loops.

and certain optimization conditions are extracted, which serve as the basis for the development of both the optimal analog and digital control law. Appendices B and C present the proof of the analog and digital control law (type-I, type-II, type-III), respectively.

Contents

Part I Introduction and Preliminaries

Part II Explicit Tuning of the PID Controller

Author Biography

Konstantinos G. Papadopoulos was born in Thessaloniki, Greece, on June 16, 1979. He received the Diploma in Engineering and Ph.D. degrees in Electrical and Computer Engineering from Aristotle University of Thessaloniki, Thessaloniki, Greece, in 2003 and 2010, respectively. During 2006, he was with the Automatic Control Laboratory, ETH Zürich, Switzerland and ABB Switzerland Ltd., Turgi, where he developed the model predictive direct torque control of the induction motor drive. From 2010 to 2013, he was with ABB Switzerland Ltd., Turgi, Switzerland, working as a Control Software Development Engineer at the Department of Medium Voltage Drives. Since 2014, he has been with Intracom–Telecom, Greece, working as a Software Development Engineer for Cloud Networking Technologies. His main research interests include loop-shaping control techniques, model-based control, development of control methods toward automatic tuning of PID regulators and applications of classical feedback control theory to task scheduling, power management, load balancing within cloud networking environments.

Notations

a_p	Unit step function (as a reference signal on the control loop)
a_v	Ramp function (as a reference signal on the control loop)
a_a	Quadratic function (as a reference signal on the control loop)
$G(s)$	Transfer function of the plant/process (real model)
$\widetilde{G}(s)$	Transfer function of the plant/process (approximate model)
$d_{\mathrm{i}}(s)$	Input disturbance signal (entering the input of the plant $G(s)$)
$d_{\mathrm{o}}(s)$	Output disturbance signal (entering the output of the plant $G(s)$)
$e(s)$	Error between the reference signal $r(s)$ and output of the control loop $y(s)$
e_{ss}	Error of $e(s)$ at steady state
$T(s)$	Closed loop transfer function (complementary sensitivity)
$F_{\mathrm{fp}}(s)$	Forward path transfer function (within the closed loop control system)
$F_{\mathrm{ol}}(s)$	Open loop transfer function (within the closed loop control system)
$C(s)$	Controller transfer function
$C_{\mathrm{ex}}(s)$	External controller transfer function (filter on the reference signal with a 2DoF (degrees of freedom controller) control scheme)
$C_{\mathrm{ZOH}}(s)$	Transfer function of the zero order hold (ZOH)
I	Integral control action
k_{h}	Gain of the feedback path (from the output $y(s)$ of the control loop back to the reference signal $r(s)$)
k_{p}	The controlled plant's dc gain at steady state
$n_{\mathrm{o}}(s)$	Noise signal at the output of the plant
$n_{\mathrm{r}}(s)$	Noise signal entering the reference signal of the control loop
PI	Proportional integral control action
PID	Proportional integral derivative control action
$r(s)$	Reference signal of the closed loop control system
$S_{n_{\mathrm{o}}}(s)$	Measurement sensitivity transfer function ($S_{n_{\mathrm{o}}}(s) = \frac{y(s)}{n_{\mathrm{o}}(s)}$, while all other inputs within the control loop are set to zero)

$S(s)$	Sensitivity of the closed loop control system, ($S(s) = \frac{y(s)}{d_o(s)}$, while all other inputs within the control loop are set to zero)
$S_i(s)$	Input sensitivity transfer function, ($S(s) = \frac{y(s)}{d_i(s)}$, while all other inputs within the control loop are set to zero)
$S_u(s)$	Sensitivity of the command signal at the presence of output disturbance $d_o(s)$
$S_G^T(s)$	Sensitivity of the closed loop control system in the presence of plant's model variations
$S_{k_h}^T(s)$	Sensitivity of the closed loop control system in the presence of variations of the feedback path
T_i, t_i	Time constant (and normalized time constant) of the integral action in the PID controller
$T_{\Sigma c}, t_{\Sigma c}$	Time constant (and normalized time constant) of the controller's parasitic dynamics
$T_{\Sigma p}, t_{\Sigma p}$	Parasitic time constant (and normalized time constant) of the plant
T_p, t_p	Time constant (and normalized time constant) within the plant transfer function (corresponds to poles of the process)
T_z, t_z	Time constant (and normalized time constant) within the plant transfer function (corresponds to zeros of the process)
T_d, t_d	Time delay constant (and normalized time delay constant) of the process $G(s)$
t_{rt}	Rise time of the step response
T_s, f_s	Sampling time, sampling frequency
t_{ss}	Settling time of the step response
$u(s)$	Command signal (output of the control action of $C(s)$)
$y(s)$	Output of the closed loop control system
$y_f(s)$	Output of the open loop transfer function $F_{ol}(s)$
ζ	Damping ratio of a plant with conjugate complex poles
T, τ	Time constant (and normalized time constant) of a plant with conjugate complex poles

Part I
Introduction and Preliminaries

Chapter 1
Overview

Abstract Since this book is dedicated to the definition of a general theory of tuning of the PID controller using the Magnitude Optimum criterion, a brief retrospect relevant to the evolution of the PID control and the Magnitude Optimum criterion is presented in this chapter. The strong effectiveness of the PID controller along with the simplicity of the criterion's principle justifies its strong application within many industry applications till date. By presenting concrete examples from the industry, the scope of the chapter is also to argue and justify why the functionality of tuning the PID controller via the Magnitude Optimum criterion has a long history along with still a much promising future.

1.1 Introduction

The proposed theory presented in this book copes with the design of the PID controller in single-input single-output control loops given also the fact that the complete knowledge of the process is often unknown, see [13, 20].

This unawareness of the process's behavior is owed to the fact that exact measurement of the states of the process itself is sometimes unfeasible. The reason for this issue is either the nature of the states within the process, or the lack of proper equipment able for accurate measurement. The unawareness of the process in this book is called "unmodelled dynamics", which as shown in the sequel, plays an important role in the whole control loop's performance. To this end, the proposed theory does not follow the classical line of well-established classical theories, which are often based on known process models. Examples of such approaches are (1) the linear state feedback control law, (2) the measures of a control loop's performance such as ITAE (integral time-weighted absolute error), ISE (integral squared error), IAE (integral absolute error), and (3) the root locus analysis.

In contrast to the aforementioned control design principles, an interesting method for designing control loops was proposed in the early 80s by Zames, entitled "Feedback and optimal sensitivity", see [40] or commonly known as H_∞ design control principles, see also [10, 14, 15, 30]. Let it be recalled that the goal of this principle is to design a closed-loop control system such that the maximum magnitude

K.G. Papadopoulos, *PID Controller Tuning Using the Magnitude Optimum Criterion*,
DOI 10.1007/978-3-319-07263-0_1

in any direction and at any frequency of the control loop's frequency response is minimized.

Tracking the literature further back in the past, it is found that the idea for minimizing the maximum magnitude of the frequency response in a control loop stems from the results found in the master thesis of Sartorius, which was published in Stuttgart in 1945, see [29]. Later on and in 1954, these results were also published by Oldenbourg and Sartorius in [23] where in this work, the authors concentrate on applying the proposed principle for the design of type-I[1] control loops. This kind of control design is commonly known within the German literature as Betragsoptimum, or BO or Magnitude Optimum, see [11].

The design of type-I control loops motivated Kessler in 1955 to apply the aforementioned principle to the design of type-II control loops.[2] This method is commonly known in the German literature as "Symmetrische Optimum", "Symmetrical Optimum", or "SO", see [17, 18] and again is dedicated to the design of type-II control loops by minimizing the maximum magnitude of the closed loop's frequency response at any frequency. For this reason, since the principle for designing a control loop, either type-I or type-II according to the aforementioned citations, is common, we refer to the proposed method in this book as "the principle of Magnitude Optimum criterion".

A common feature between the principle of the Magnitude Optimum criterion and the H_∞ control methods is the minimization of the maximum magnitude in the frequency response of the closed-loop transfer function. However, an H_∞ control method is looking for a controller, the order of which is most often not a constraint in the problem formulation of a strong mathematical optimization procedure, since the basic goal of this principle is to stick to the basic requirement, which is the minimization of the maximum magnitude of the control loop's frequency response. For this reason, many are the times where the resulting controller coming out of these methods is of much high order, the real-time implementation of which, is often under discussion.

In contrast to the high order of the controller that comes out of an H_∞ control method, the PID controller (only three terms) proves to be simple and effective among various control schemes proposed in the literature, see [1]. The aforementioned industrial applications reported in [1] raise automatically the big question, see [2]: Why does the PID controller stand so vigorously over the various more complex control methods that have been reported over the literature? What does the order of an industrial controller have to be? Of course, the answer to this question is not straightforward, since every industry application introduces its own requirements and specifications, which of course make the problem more complex.

However, since the three big requirements in any real-time environment are (1) effectiveness and efficiency, (2) simplicity of implementation, (3) and cost, this book

---

[1] At this point it has to be mentioned that type-I control loops are those loops that are able to track step reference signals with zero steady state position error, see Sect. 2.5.

[2] Type-II control loops, are loops that are able to track step and ramp reference signals with zero steady state position and velocity error, see Sect. 2.5.

concentrates on the PID control solution. According to the author's opinion, the effectiveness and simplicity of the PID controller along with the attractive property of the Magnitude Optimum criterion comprise a down-to-earth recipe for acceptable and satisfactory results in a wide variety of industry applications.

1.2 Target of the Proposed Theory

The first goal of the proposed theory is to present general tuning expressions for the PID controller, given the transfer function of any plant irrespective of its order. For this reason, the proposed theory is defined in the frequency domain and focuses on determining the P, I, and D parameters as a function of all time constants coming from the model of the controlled plant without following any model reduction techniques (First Order plus Dead Time models, etc.).

As mentioned in the synopsis of this book, this kind of approach is called "direct or explicit tuning" of the PID controller and can be applied in any single-input single-output process irrespective of its order complexity, see Part II. The same theory for the direct tuning of the PID controller is also extended to the design of higher order type control loops, type-I, type-II and for the first time to type-III, type-IV, type-V and type-p control loops.[3]

For defining the theory of the direct tuning of the PID controller, an introduction of the "conventional" tuning of the PID controller according to the Magnitude Optimum criterion is also presented for type-I, type-II control loops. This introduction helps the reader to understand the current state-of-the-art and clarify the drawbacks the current tuning method exhibits. Bearing these drawbacks in mind, the proposed "direct tuning" method also called "revised tuning" aims at exploiting the full potential of the PID control action along with the attractive properties introduced by the Magnitude Optimum principle.

The application of the revised method for the control of a large class of process models reveals one more property of the method called "the preservation of the shape of the step and frequency response" of the closed-loop transfer function. The "preservation of the shape of the step response" means that when the direct PID tuning is applied for the control of any single-input single-output process, the step response of the control loop exhibits a certain performance in terms of overshoot, rise, and settling time. To our eyes, the response with these time domain characteristics is reflected again by the Magnitude Optimum criterion, which is a loop shaping method defined in the frequency domain and with certain frequency domain characteristics.

With respect to the above, the preservation of the shape of the step and frequency response motivates us to define an algorithm able to tune automatically the PID controller, see Part III. In other words, the goal of the proposed algorithm is to tune the PID controller's parameters automatically so that the aforementioned performance of the control loop's output is observed both in the time and frequency domains. Note

[3] The higher the type of a control loop is, the faster reference signals can track.

that the automatic tuning of the PID controller is a practical problem, which often comes up on the table of control engineers within a real-time environment.

1.3 State of the Art—The Magnitude Optimum Criterion

In this section, a short discussion of the type of automatic control loops is presented. Type-I and type-II control loops are introduced and concrete examples from the field of electric drives and electric converters are presented.

1.3.1 Type-I Control Loops

As mentioned in Sect. 1.1 the Magnitude Optimum criterion was introduced in 1950 and since then has been widely used in the industry, see [7–9, 12, 19, 34, 38]. However, excluding the German bibliography, the Magnitude Optimum criterion is rarely referred today. In addition to this, the limited impact of both the Magnitude and Symmetrical design criteria is stressed in [33] and this might be due to the negative comments occasionally stated in the literature. Some of these comments are quoted below:

1. A significant disadvantage of the MO criterion is that systems designed with it can only be of type-I or type-0 [33].
2. A drawback of the MO criterion is that the system response due to any disturbance, applied at locations other than at the reference input, is not optimal [33].
3. A second disadvantage of this technique is that the controlled system will display only type-I or type-0 behavior, even with the presence of free integrators in the plant [33].
4. The above mentioned performance become unacceptable due to large sensitivity with respect to the modification of k_p (the plant gain) . . . [28].
5. A drawback with all design methods of this type is that process poles are canceled. This may lead to poor attenuation of load disturbances if the canceled poles are excited by disturbances and if they are slow compared to the dominant closed loop poles [3].
6. The method is very demanding since it requires reliable estimation of quite a large number of process parameters even when using relatively simple controller structures (like a PID controller). This is one of the main reasons why the method is not frequently used in practice . . . However, practical implementation of this method is comparatively difficult due to its quite demanding requirements, including the explicit identification of 12 process model parameters to calculate three parameters of the PID controller [35–37].
7. The MO criterion is not suitable for some processes with stronger zeros or complex poles, where unstable controller parameters may be obtained [36].

8. The MO technique may lead to poor attenuation of load disturbances. It was shown that disturbance rejection can be significantly improved by using a two-degrees-of freedom controller structure [36].

In our opinion, all the above remarks need to be revised for three reasons.

1. First, as it will be proved in the sequel, the conventional design[4] procedure via the Magnitude Optimum criterion for PID type controllers, restricts the controller's zeros to be tuned only with real zeros leading finally to poor tuning. This approach does not take into account the fact that the optimal values for the PID controller's zeros may be conjugate complex, which might result in more robust tuning than the principle of pole-zero cancellation.
2. Second, for determining the PID controller's zeros, exact pole-zero cancellation has to be achieved between the process's poles and the controller's zeros [3]. This approach disregards all other plant parameters for the optimal control law and as a result, the PID parameter tuning is poor and suboptimal.
3. Third, the conventional design procedure via the Magnitude Optimum criterion has been tested only on a limited class of simple process models [37] and not on benchmark processes.

Industrial examples of type-I control loops are found in the field of electric motor drives and grid connected converters. Specifically in grid connected converters, a type-I control loop is met in vector controlled AC/DC power converters where there is an inner loop responsible for regulating the current and an outer loop responsible for regulating the DC link voltage to be utilized by another DC/AC electric motor drive. In this case, the inner current control loop is of type-I since in its open-loop transfer function there exists only one integrator coming from the PID control action, see Sects. 2.5 and 3.6.

1.3.2 Type-II Control Loops

As mentioned in Sect. 1.1, the introduction of Magnitude Optimum criterion for the design of type-II control loops was initiated by Kessler in 1955. The basic characteristic of these control loops is the existence of two pure integrators within the open-loop transfer function, see Sect. 2.5. In this case, one integrator often comes from the PID control action and one more comes from the process itself (see also control of integrating processes). The basic advantage of these kinds of loops compared to type-I is their ability to track step and ramp reference signals with zero steady state position and velocity error.

The existence of such loops in the field of electric drives is found both in grid and motor connected converters. In grid connected AC/DC converters and for controlling the DC link voltage to be utilized by the motor connected converter, the outer control loop is type-II, since one integrator comes from the DC link voltage PI control action and another comes from the capacitor bank path ($\frac{1}{sC}$) within the DC link.

---

[4] Design via pole-zero cancellation.

In addition, as far as the motor connected drive is concerned, a type-II control loop is the speed control loop in vector controlled or direct torque controlled drives. In this case, one integrator comes from the speed PI control action and another integrator comes from the inertia ($\frac{1}{sJ}$) of the shaft of the motor, the speed of which is controlled.

Apart from the definition of the conventional PID tuning principle via the Symmetrical Optimum criterion for the control of integrating processes, no other work has been reported regarding the tuning of the PID controller through the Magnitude Optimum criterion. However, the problem for controlling such processes has been approached by many researchers after incorporating the Smith predictor, see [6, 21, 31, 32, 39, 41], the Internal Model Control (IMC) principle, see [22] or other optimization methods [16, 27].

1.3.3 Type-III Control Loops

The introduction of type-III control loops takes place for the first time in this book, and has already been introduced in the literature, see [24–26]. Their characteristic is the existence of three pure integrators within the open-loop transfer function. This characteristic gives them the ability to track even faster reference signals compared to type-I and type-II control loops. Therefore, type-III control loops can track step, ramp, and parabolic reference signals achieving zero steady state position, velocity, and acceleration error.

Since at least in the field of electric motor drives double integrating processes are not met, a type-III control loop can be designed by introducing free pure integrators to the PID control action.

1.4 Automatic Tuning of PID Controllers

As regards the problem of automatic tuning of regulators, a statement made by Åström et al. [5] gives in our opinion the most accurate description of what the goal of such a method should accomplish. For that reason, it is quoted below

> By automatic tuning (or auto tuning) we mean a method where a controller is tuned automatically on demand from a user. Typically the user will either push a button or send a command to the controller... Automatic tuning is sometimes called tuning on demand or one shot tuning...

The problem of automatic tuning of regulators can be seen in cases where the derivation of the process of the model is almost impossible. This may happen due to the nature of the states of the process or due to the lack of proper measuring equipment for identifying a model of the process, see [4]. Given these problem restrictions, an automatically tuned controller has basically to satisfy the requirements listed below.

1. An automatic controller tuning procedure has to decide the proper type of control action (P control action, PI control action or PID). There are many applications where the question arises whether the D term is to be added or omitted.
2. An automatic controller tuning procedure has to end up in such controller parameters so that robust performance is achieved by the process in terms of reference tracking and output disturbance rejection.
3. An automatic controller tuning procedure must have the ability to retune the controller's parameters in cases where the plant dynamics change in such a way that finally make the initial controller tuning unacceptable.

As mentioned in Sect. 1.1, the Magnitude Optimum criterion is again adopted for developing such a technique, see Part III of this book. For developing the proposed method, it is assumed that access to the output and not to the plant's states is possible[5] an open loop experiment from the process itself is available, which serves to initialize the proposed algorithm.

References

1. Ang KH, Chong G, Li Y (2005) PID control system analysis, design, and technology. IEEE Trans Control Syst Technol 13(4):559–576
2. Åström KJ, Hägglund T (2001) The future of PID control. Control Eng Pract 9(11):1163–1175
3. Åström KJ, Hägglund T (1995) PID controllers: theory, design and tuning, 2nd edn. Instrument Society of America, North Carolina
4. Åström KJ, Wittenmark B (1973) On self tuning regulators. Automatica 9(2):185–199
5. Åström KJ, Hägglund T, Hang CC, Ho WK (1993) Automatic tuning and adaptation for PID controllers–a survey. Control Eng Pract 1(4):699–714
6. Åström KJ, Hang CC, Lim BC (1994) A new Smith predictor for controlling a process with an integrator and long dead time. IEEE Trans Autom Control 39(2):343–345
7. Buxbaum A (1967) Berechnung von regelkreisen der antriebstechnik. Frankfurt am Main, AEG–Telefunken AG, Berlin
8. Buxbaum A, Schierau K, Straughen A (1990) Design of control systems for DC drives. Springer, Berlin
9. Courtiol B, Landau ID (1975) High speed adaptation system for controlled electrical drives. Automatica 11(2):119–127
10. Doyle CJ, Francis BA, Tannenbaum AR (2009) Feedback control theory. Dover Publications, New York
11. Föllinger O (1994) Regelungstechnik. Hüthig, Heidelberg
12. Fröhr F, Orttenburger F (1982) Introduction to electronic control engineering. Siemens, Berlin
13. Gunter S (2003) Respect the unstable. IEEE Control Syst Mag 23(4):12–25
14. Helton JW, Merino O (1998) Classical control using H_∞ methods. SIAM: Society for Industrial and Applied Mathematics, Philadelphia
15. Ho MT (2003) Synthesis of H_∞ PID controllers: a parametric approach. Automatica 39(6):1069–1075
16. Isaksson AJ, Graebe SF (1999) Analytical PID parameter expressions for higher order system. Automatica 35(6):1121–1130
17. Kessler C (1955) UG ber die Vorausberechnung optimal abgestimmter regelkreise teil III. Die optimale einstellung des reglers nach dem betragsoptimum. Regelungstechnik 3:40–49

[5] As frequently happens in many real-time applications.

18. Kessler C (1958) Das Symmetrische Optimum. Regelungstechnik, pp 395–400 and 432–426
19. Loron L (1997) Tuning of PID controllers by the non-symmetrical optimum method. Automatica 33(1):103–107
20. Margaris NI (2003) Lectures in applied automatic control (in Greek), 1st edn. Tziolas
21. Mataušek MR, Micić AD (1999) On the modified Smith predictor for controlling a process with an integrator and long dead-time. IEEE Trans Autom Control 44(8):1603–1606
22. Morari M, Zafiriou E (1989) Robust process control, 1st edn. Prentice-Hall, New Jersey
23. Oldenbourg RC, Sartorius H (1954) A uniform approach to the optimum adjustment of control loops. Trans ASME 76:1265–1279
24. Papadopoulos KG, Margaris NI (2012) Extending the symmetrical optimum criterion to the design of PID type-p control loops. J Process Control 12(1):11–25
25. Papadopoulos KG, Papastefanaki EN, Margaris NI (2011) Optimal tuning of PID controllers for type-III control loops. In: 19th Mediterranean conference on control & automation (MED). IEEE, Corfu, Greece, pp 1295–1300
26. Papadopoulos KG, Papastefanaki EN, Margaris NI (2012) Automatic tuning of PID type-III control loops via the symmetrical optimum criterion. In: International conference on industrial technology, (ICIT). IEEE, Athens, Greece, pp 881–886
27. Poulin E, Pomerleau A (1999) PI settings for integrating processes based on ultimate cycle information. IEEE Trans Control Syst Technol 7(4):509–511
28. Preitl S, Precup RE (1999) An extension of tuning relation after symmetrical optimum method for PI and PID controllers. Automatica 35(10):1731–1736
29. Sartorius H (1945) Die zweckmässige festlegung der frei wählbaren regelungskonstanten. Master thesis, Technische Hochscule, Stuttgart, Germany
30. Skogestad S, Postlethwaite I (2005) Multivariable feedback control: analysis and design. Wiley, New York
31. Smith OJM (1959) Closed control of loops with dead-time. Chem Eng Sci 53:217–219
32. Stojić MR, Matijević MS, Draganović LS (2001) A robust Smith predictor modified by internal models for integrating process with dead time. IEEE Trans Autom Control 46(8):1293–1298
33. Umland WJ, Safiuddin M (1990) Magnitude and symmetric optimum criterion for the design of linear control systems: what is it and how does it compare with the others? IEEE Trans Ind Appl 26(3):489–497
34. Voda AA, Landau ID (1995) A method for the auto-calibration of PID controllers. Automatica 31(1):41–53
35. Vrančić D (1997) Design of anti-windup and bumpless transfer protection. PhD thesis, University of Ljubljana, Faculty of Electrical Engineering, Ljubljana
36. Vrančić D, Peng YSS (1999) A new PID controller tuning method based on multiple integrations. Control Eng Pract 7(5):623–633
37. Vrančić D, Strmčnik S (1999) Practical guidelines for tuning PID controllers by using MOMI method. In: International symposium on industrial electronics, IEEE, vol 3, pp 1130–1134
38. Washburn DC (1967) Optimization of feedback control loops. Westinghouse Industrial Control Seminars
39. Watanabe K, Ito M (1981) A process model control for linear systems with delay. IEEE Trans Autom Control 26(6):1261–1268
40. Zames G (1981) Feedback and optimal sensitivity: model reference transformations, multiplicative seminorms, and approximate inverses. IEEE Trans Autom Control 26(2):301–320
41. Zhang W, Sun Y, Xu X (1998) Two degree-of-freedom Smith predictor for processes with time delay. Automatica 34(10):1279–1282

Chapter 2
Background and Preliminaries

Abstract In this chapter, fundamental definitions and terminology are given to the reader regarding the closed-loop control system. The analysis of the control loop takes place in the frequency domain and, therefore all necessary transfer functions of the control loop are presented in Sect. 2.2. The important aspect of internal stability of a control loop is presented in Sect. 2.3, whereas in Sect. 2.4 the property of robustness in a control loop is analyzed. In Sect. 2.5, a clear definition of the type of the control loop is given, since in Part II, the proposed theory is dedicated to the design of type-I, type-II, and type-III, … type-p control loops. Last but not least, in Sect. 2.6, the definitions of sensitivity and complementary sensitivity functions are presented so that the tradeoff feature in terms of controller performance that these two functions introduce is made clear to the reader. Finally, in Sect. 2.7, the principle of the Magnitude Optimum criterion is presented and certain optimization conditions are proved that comprise the basic tool for all control laws' proof throughout this book. These optimization conditions serve to maintain the magnitude of the closed-loop frequency response equal to the unity in the widest possible frequency range as the Magnitude Optimum criterion implies. In the same section, the Magnitude Optimum criterion is proved to be considered as a practical aspect of the H_∞ design control principle.

2.1 Definitions and Preliminaries

The core of a closed-loop control system is namely the plant or the process, see Fig. 2.1. The plant receives signals from the outer world, commonly known as inputs, depicted by $u(t)$ in Fig. 2.1, and acts at the same time to the outer world with its response, known as output, $y(t)$. Moreover, the whole process can also be described by its states $x(t)$, which along with the inputs $u(t)$, determine the response $y(t)$ of the plant itself.

Ideally, there are two fundamental requirements of a process in any real-time application:

1. From a plant, it is required that its output $y(t)$ must track perfectly its input $u(t)$.
2. The aforementioned output tracking of the input $u(t)$ must also be repetitive and for several different input signals $u(t)$.

K.G. Papadopoulos, *PID Controller Tuning Using the Magnitude Optimum Criterion*,
DOI 10.1007/978-3-319-07263-0_2

Fig. 2.1 The plant or process

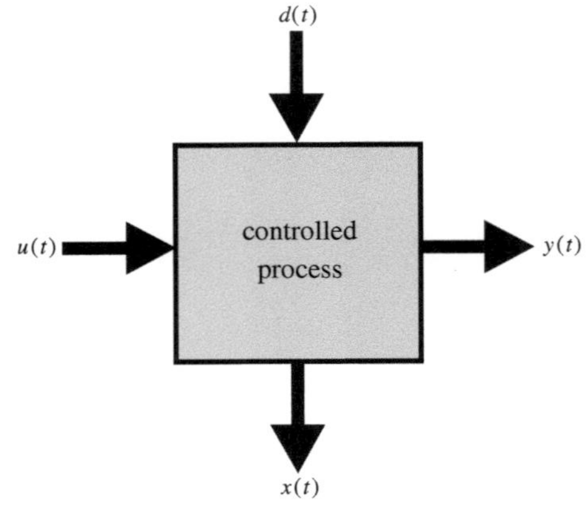

Of course, these two aforementioned requirements are practically impossible to be satisfied at the same time in real-world plants and applications, since the existence of disturbances $d(t)$ alters the behavior of the process during its operation.

In real-world problems, disturbances $d(t)$ are classified into two categories. The first category involves disturbances coming from the process itself, known as internal disturbances.[1]

The second category includes any external or exogenous disturbance that can be relevant basically to the environmental conditions the process is located at, i.e., varying loads acting as input signals to the output of the process, noise coming from the measuring equipment, etc.

With respect to the above, it is without any doubt apparent that during the plant's operation, perfect tracking of the output $y(t)$ for repetitive and different input signals $u(t)$ can only be satisfied if fast suppression of internal and external disturbances is achieved.

For achieving fast suppression of disturbance in a closed-loop control system, the solution of the well-known principle of negative feedback,[2] widely used in mechanical, chemical, and electronic engineering is adopted. The introduction of negative feedback in a control loop leads to a control system presented in Fig. 2.2 the basic elements of which are (1) the process (plant), (2) the measuring equipment, (3) the reference signal, (4) the comparator, and (5) the controller along with the actuator unit.

In such a control loop, the path that connects the reference input $r(t)$ with the output of the control loop $y(t)$ is called *forward path*. This path includes the (1) process, (2) the actuator or power part unit, (3) the controller, and the (4) comparator.

[1] I.e., rise of temperature during a motor's operation, aging of materials after a certain time (for example, copper conductors in a squirrel-cage induction motor).

[2] Negative feedback is present to the water clock invented by Ktesibios (Greek inventor and mathematician in Alexandreia, 285–222 BC) and in the steam engine governor patented by James Watt in 1788.

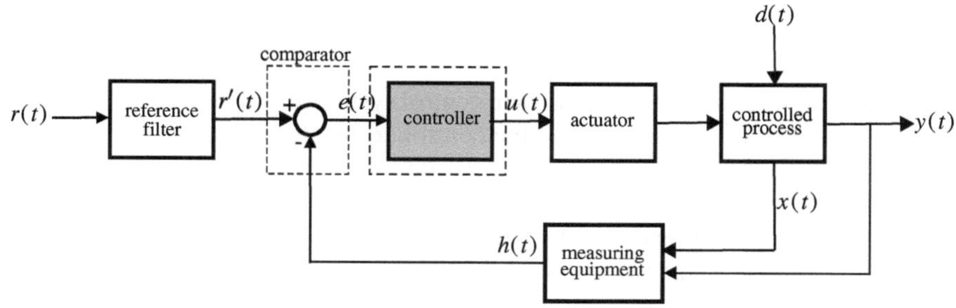

Fig. 2.2 General form of a closed-loop control system with negative feedback. The path that connects the reference input $r(t)$ with the output control loop $y(t)$ is called forward path. The path that connects the measuring equipment with the states and the output of the process is the feedback path

The path connecting the measuring equipment with the states and the output of the process is called *feedback path*. The logic for the existence of these two paths is as follows:

All information $h(t)$ that can be potentially accessed in the process either from the output $y(t)$ or the states $x(t)$, is collected by the measuring equipment and is transferred to the input of the comparator. This information $h(t)$, is then compared with a reference signal $r(t)$ that describes the desired behavior of the process. This comparison takes place within the comparator unit, the output of which is the error $e(t) = r(t) - h(t)$. This error enters the controller unit, passes through the actuator, and finally enters the input $u(t)$ of the process. The goal of $u(t)$ is to make the output of the process $y(t)$ track perfectly the reference signal $r(t)$.

With respect to the above, it becomes apparent that the aforementioned goal has to be achieved by the controller unit, which basically, given the error $e(t)$ and the presence of any disturbance $d(t)$ entering the plant, tries to calculate the proper $u(t)$ command signal such that the output $y(t)$ tracks perfectly the reference signal $r(t)$. In principle and as previously mentioned, both *perfect* disturbance rejection and *perfect* tracking of the reference at the same time cannot take place. Therefore, the design of a control action has to take always into account this compromise and deliver this command signal to the plant, which satisfies certain constraints according to the application.

In many industry applications, a control engineer sets as a first priority to design such a control unit able for fast suppression of disturbances. The reason for this is due to the nature of these signals which often enter the control loop suddenly and without any prediction. Thus, tracking of the reference signal is set as a second priority for the control action's design, since $r(t)$ does not change frequently while its value is known a priori before setting the control loop into operation.

Finally, once disturbance rejection is achieved for improving reference tracking, many are the times when the reference signal is filtered to avoid high overshoot at step changes of $r(t)$, see Fig. 2.2. This control scheme is also known in the literature as a two degree of freedom controller (2DoF).

2.2 Frequency Domain Modeling

In this section, we refer to the closed-loop control system presented in Fig. 2.3 where $G(s)$, $C(s)$ stand for the process and the controller transfer functions, respectively. Output of the control loop is defined as $y(s)$ and k_h stands for the feedback path for the output $y(s)$.

Signal $r(s)$ is the reference input to the control loop, $d_o(s)$ and $d_i(s)$ are the output and input disturbance signals, respectively, and $n_r(s)$, $n_o(s)$ are the noise signals at the reference input and the process output, respectively. Finally, k_p stands for the plant's dc gain at steady state.[3]

- **Closed-loop transfer function**

$$T(s) = \frac{y(s)}{r(s)} = \frac{F_{fp}(s)}{1 + F_{ol}(s)}, \tag{2.1}$$

where

$$F_{fp}(s) = \frac{y(s)}{e(s)} = k_p C(s) G(s), \tag{2.2}$$

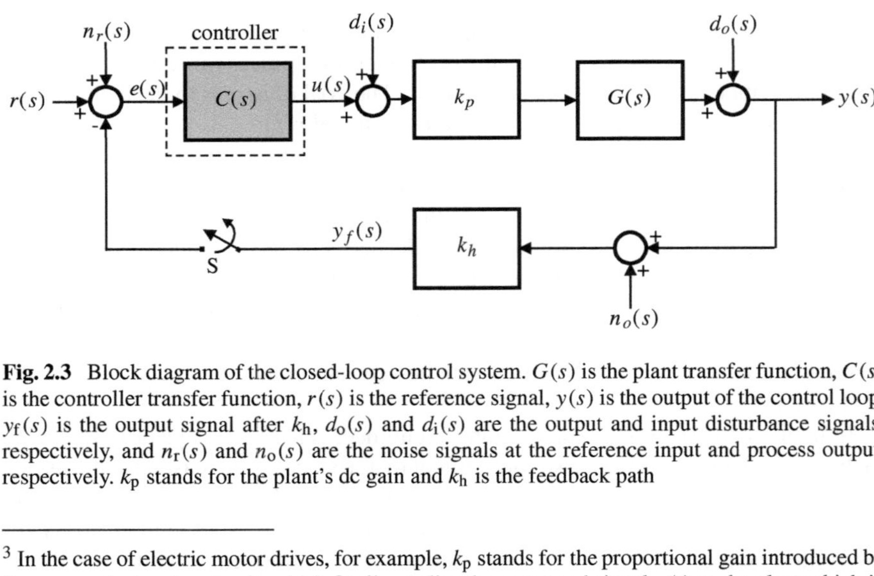

Fig. 2.3 Block diagram of the closed-loop control system. $G(s)$ is the plant transfer function, $C(s)$ is the controller transfer function, $r(s)$ is the reference signal, $y(s)$ is the output of the control loop, $y_f(s)$ is the output signal after k_h, $d_o(s)$ and $d_i(s)$ are the output and input disturbance signals, respectively, and $n_r(s)$ and $n_o(s)$ are the noise signals at the reference input and process output, respectively. k_p stands for the plant's dc gain and k_h is the feedback path

[3] In the case of electric motor drives, for example, k_p stands for the proportional gain introduced by the power electronics circuit, which finally applies the command signal $u(t)$ to the plant which in this case is the electric motor. For voltage source inverters, the command signal $u(t)$ is voltage. In the sequel it is explained that the gain introduced by the actuator has to be linear and proportional so that the command signal of the controller remains unaltered. In the specific case of electric motor drives, the power part introduces also a time delay with time constant T_d which corresponds to the time the controller decides the command $u(t)$, until the time it is finally applied by the power electronic circuit. Therefore in this case, the model of the actuator is given as $k_p e^{-sT_d}$

is the *forward path transfer function* and

$$F_{\mathrm{ol}}(s) = \frac{y_{\mathrm{f}}(s)}{e(s)} = k_{\mathrm{h}} k_{\mathrm{p}} C(s) G(s), \tag{2.3}$$

is the *open-loop transfer function*.

- **Output sensitivity or sensitivity function**

$$S(s) = \frac{y_{d_{\mathrm{o}}}(s)}{d_{\mathrm{o}}(s)} = \frac{1}{1 + F_{\mathrm{ol}}(s)}, \tag{2.4}$$

which expresses the variation of the output $y_{d_{\mathrm{o}}}(s)$, in the presence of output disturbance $d_{\mathrm{o}}(s)$.

- **Input sensitivity function**

$$S_{\mathrm{i}}(s) = \frac{y_{d_{\mathrm{i}}}(s)}{d_{\mathrm{i}}(s)} = \frac{k_{\mathrm{p}} G(s)}{1 + F_{\mathrm{ol}}(s)} = k_{\mathrm{p}} G(s) S(s), \tag{2.5}$$

which expresses the variation of the output $y_{d_{\mathrm{i}}}(s)$, in the presence of input disturbance $d_{\mathrm{i}}(s)$.

- **Control (command) signal sensitivity function**

$$S_u(s) = \frac{u(s)}{d_{\mathrm{o}}(s)} = -\frac{k_{\mathrm{h}} C(s)}{1 + F_{\mathrm{ol}}(s)} = -k_{\mathrm{h}} C(s) S(s), \tag{2.6}$$

which expresses the variation of the command signal $u(s)$ of the controller in the presence of output disturbance $d_{\mathrm{o}}(s)$.

In general, if we consider that all inputs of the control loop are acting at the same time, then after applying the theorem of superposition among (2.1), (2.4), and (2.5), it becomes apparent that the output of the control loop is determined as

$$y(s) = T(s)[r(s) + n_{\mathrm{r}}(s) - k_{\mathrm{h}} n_{\mathrm{o}}(s)] + S(s)[d_{\mathrm{o}}(s) + k_{\mathrm{p}} G(s) d_{\mathrm{i}}(s)]. \tag{2.7}$$

2.3 Internal Stability

The problem of stability in a control loop is considered of highest priority in many real-world applications. Loss of stability in an industrial plant may lead often to damage of expensive components or even to loss of human life. Therefore, control engineers are often willing and determined to spend much effort on designing stable control loops so that the aforementioned cases are avoided.

A classic reference that remains modern till date, is the paper by Gunter Stein, "Respect the Unstable", which describes accurately the importance of stability in modern control systems, see [2].

Definition 1 Any closed-loop control system is said to be internally stable if for any bounded signal entering the control loop, all other generated responses (states, output) remain bounded.

Definition 2 A linear time-invariant system (LTI) is said to be internally stable, if and only if, every transfer function from whichever input to whichever output within the control loop is stable. In other words, every transfer function from whichever input to whichever output within the control loop must introduce poles only in the left-half plane (LHP).

From the control loop structure presented in Fig. 2.3, it is seen that the difference between the reference signal $r(s)$ and the output of the control loop $y(s)$ is expressed by the error signal $e(s)$, because $e(s) = r(s) - y(s)$. Since $r(s)$ is bounded and $r(s) = e(s) + y(s)$, for checking the internal stability of the control loop, it is sufficient to track either the response of the output signal $y(s)$ or the error signal $e(s)$. Assuming a stable controller design of $C(s)$ it is apparent that $u(s)$ is also stable, since $u(s) = C(s)e(s)$. As a result, for checking the internal stability of the control loop, it is again sufficient to track either the response of the output signal $y(s)$ or the controller's command signal $u(s)$ in the presence of the bounded signal $r(s)$.

The same investigation has to take place also for the affect of the disturbance signals $d(s)$ which enter the control loop either on the input $d_i(s)$ or the output $d_o(s)$ of the process. Therefore, it is necessary to investigate the effect of the signals $d_i(s)$ or $d_o(s)$ on the response of $u(s)$, since both $d_i(s)$ and $d_o(s)$ are bounded.

For investigating the way how signals $y(s),\ u(s)$ are affected in the presence of the reference signal $r(s)$ and disturbance $d(s)$ ($d_i(s)$ or $d_i(s)$), the internal stability matrix of (2.8) is introduced

$$\begin{bmatrix} y(s) \\ u(s) \end{bmatrix} = \begin{bmatrix} T(s) & k_p G(s)S(s) \\ C(s)S(s) & -k_h T(s) \end{bmatrix} \begin{bmatrix} r(s) \\ d_i(s) \end{bmatrix}. \tag{2.8}$$

From (2.8), it is concluded that internal stability for the control loop of Fig. 2.3 is guaranteed only if each one of the transfer functions $T(s),\ S_i(s),\ S_u(s)$ is stable. For the definition of $T(s),\ S_i(s),\ S_u(s)$ see accordingly (2.1), (2.5), and (2.6). After algebraic manipulation of (2.8), it is seen that

$$y(s) = r(s)T(s) + d_i(s)k_p G(s)S(s) \tag{2.9}$$

which is valid if we set in the general expression of $y(s)$ (2.7), $n_r(s) = 0$, $n_o(s) = 0$, and $d_o(s) = 0$. Moreover, from (2.8) it is seen that

$$u(s) = r(s)C(s)S(s) - k_h d_i(s)T(s) \tag{2.10}$$

which is also valid. If $d_{\mathrm{i}}(s) = 0$ and assuming then that $r(s)$ is the only active input in the control loop, it is necessary to prove that $u(s) = r(s)C(s)S(s)$. This is proved from Fig. 2.3, since if $d_{\mathrm{i}} = d_{\mathrm{o}} = n_{\mathrm{r}}(s) = n_{\mathrm{o}}(s) = 0$ then $e(s) = r(s) - k_{\mathrm{h}}k_{\mathrm{p}}G(s)u(s)$

$$\frac{u(s)}{C(s)} = r(s) - k_{\mathrm{h}}k_{\mathrm{p}}G(s)u(s), \tag{2.11}$$

or

$$u(s) = r(s)C(s) - k_{\mathrm{h}}k_{\mathrm{p}}G(s)u(s)C(s), \tag{2.12}$$

or

$$u(s) = r(s)C(s) - u(s)k_{\mathrm{h}}k_{\mathrm{p}}G(s)C(s), \tag{2.13}$$

or finally

$$u(s)[1 + k_{\mathrm{h}}k_{\mathrm{p}}C(s)G(s)] = r(s)C(s). \tag{2.14}$$

From (2.14) and along with (2.4) it is apparent that $u(s)\frac{1}{S(s)} = r(s)C(s)$ or finally

$$u(s) = r(s)C(s)S(s). \tag{2.15}$$

In a similar fashion, it can be proved that $u(s) = -d_{\mathrm{i}}(s)k_{\mathrm{h}}T(s)$ assuming all other inputs within the control loop are set to zero.

From Fig. 2.3 it is obvious that

$$u(s) + d_{\mathrm{i}}(s) = -\frac{u(s)}{k_{\mathrm{p}}k_{\mathrm{h}}C(s)G(s)} \tag{2.16}$$

or

$$u(s) + \frac{u(s)}{k_{\mathrm{p}}k_{\mathrm{h}}C(s)G(s)} + d_{\mathrm{i}}(s) = 0. \tag{2.17}$$

From (2.17) it is seen that

$$u(s)\left[\frac{1}{k_{\mathrm{h}}}\left(\frac{1 + k_{\mathrm{p}}k_{\mathrm{h}}C(s)G(s)}{k_{\mathrm{p}}C(s)G(s)}\right)\right] + d_{\mathrm{i}}(s) = 0 \tag{2.18}$$

or finally along with the use of (2.1)

$$u(s)\frac{1}{k_{\mathrm{h}}T(s)} + d_{\mathrm{i}}(s) = 0, \tag{2.19}$$

which is equal to

$$u(s) = -k_{\mathrm{h}}d_{\mathrm{i}}(s)T(s). \tag{2.20}$$

2.4 Robustness

Robust performance is of primary importance when designing a control law. In other words, it is related to the ability of the controller to deliver the necessary command signal to the plant, which both makes the plant achieve perfect tracking of the reference along with satisfactory disturbance rejection and regardless of the changes that might take place within the process during its operation.

For measuring robustness, the functions of sensitivity and complementary sensitivity are introduced. The sensitivity function for two functions F, S is given as

$$S_G^F(s) = \frac{\mathrm{d}F/F}{\mathrm{d}G/G} = \frac{G}{F}\frac{\mathrm{d}F}{\mathrm{d}G} \tag{2.21}$$

see [3]. By applying the aforementioned definition to the sensitivity of the closed-loop transfer function T with respect to changes in the transfer function of the process G see Fig. 2.3, results in

$$S_G^T(s) = \frac{G}{T}\frac{\mathrm{d}T}{\mathrm{d}G} = \frac{1}{1+k_\mathrm{p}k_\mathrm{h}C(s)G(s)} = \frac{1}{1+F_\mathrm{ol}(s)} = S(s). \tag{2.22}$$

Further to (2.22), by applying (2.21) to the sensitivity of the closed-loop transfer function T with respect to changes in the feedback path k_h, results in

$$S_{k_\mathrm{h}}^T(s) = \frac{k_\mathrm{h}}{T}\frac{\mathrm{d}T}{\mathrm{d}k_\mathrm{h}} = -\frac{k_\mathrm{p}k_\mathrm{h}C(s)G(s)}{1+k_\mathrm{p}k_\mathrm{h}C(s)G(s)} = \frac{F_\mathrm{ol}(s)}{1+F_\mathrm{ol}(s)}. \tag{2.23}$$

If the magnitude of the open-loop transfer function $|F_\mathrm{ol}(s)|$ is fairly high compared to unity ($|F_\mathrm{ol}(s)| \gg 1$) then (2.22) and (2.23) are transformed into

$$S_G^T(s) = \frac{G}{T}\frac{\mathrm{d}T}{\mathrm{d}G} \ll 1, \tag{2.24}$$

and

$$S_{k_\mathrm{h}}^T(s) = \frac{k_\mathrm{h}}{T}\frac{\mathrm{d}T}{\mathrm{d}k_\mathrm{h}} \approx 1. \tag{2.25}$$

Equation (2.24) reveals that possible changes on the model G of the process do not affect seriously the behavior of the closed-loop transfer function T and therefore of the closed-loop control system. Moreover, from (2.25), it is concluded that any variation that takes place in the feedback path k_h, is transferred directly and without any change to the output of the closed-loop control system T.

With respect to the above, it is apparent that the sensitivity of the units located in the forward path of the closed-loop control system is directly transmitted to the feedback path. As a result, when designing a closed-loop control system, extra care must be taken by the control engineer for the sensitivity of the feedback path. After summing up together (2.1) and (2.22), it is seen that

$$k_h T(s) + S(s) = 1. \tag{2.26}$$

Note at this point that (2.26) is the fundamental equation that connects the sensitivity S with the transfer function of the closed-loop control system T, via the feedback path k_h. In case of unity feedback systems $k_h = 1$, (2.26) is rewritten as follows:

$$T(s) + S(s) = 1, \tag{2.27}$$

which is considered as one more fundamental relation in a closed-loop control system, see [4–7, 9].

2.5 Type of Control Loop

Preliminary definitions regarding the type of control loop are given in this section. According to Fig. 2.3, the error $e(s)$ is defined by $e(s) = r(s) - y(s) = (1 - T(s))\, r(s) = S(s)r(s)$. If the closed-loop transfer function $T(s) = \frac{y(s)}{r(s)}$ from reference $r(s)$ to output $y(s)$ while all other inputs in the control loop are assumed zero is defined as

$$T(s) = \frac{b_m s^m + b_{m-1}s^{m-1} + \cdots + b_1 s + b_0}{a_n s^n + a_{n-1}s^{n-1} + \cdots + a_1 s + a_0} \tag{2.28}$$

then the resulting error $e(s)$ is given as

$$e(s) = \left(\frac{a_n s^n + \cdots + c_m s^m + \cdots + c_1 s + c_0}{a_n s^n + a_{n-1}s^{n-1} + \cdots + a_1 s + a_0}\right) r(s) \tag{2.29}$$

where $c_j = (a_j - b_j)$ $(j = 0 \ldots m)$. According to the final value theorem and if $e(s)$ is stable, $e(\infty)$ is equal to

$$e(\infty) = \lim_{s\to 0} s \left(\frac{a_n s^n + \cdots + c_2 s^2 + c_1 s + c_0}{a_n s^n + a_{n-1}s^{n-1} + \cdots + a_1 s + a_0}\right) r(s). \tag{2.30}$$

If $r(s) = \frac{1}{s}$ then

$$e(\infty) = \lim_{s\to 0} \left(\frac{c_0}{a_0}\right) \tag{2.31}$$

which becomes zero when $c_0 = 0$ or when $a_0 = b_0$. Hence, sensitivity $S(s) = \frac{y(s)}{d_o(s)}$ [4] and closed-loop transfer function $T(s)$ are defined as

---

[4] $S(s)$ stands for the sensitivity of the closed-loop control system and is defined as $S(s) = \frac{y(s)}{d_o(s)}$ when $r(s) = n_r(s) = d_i(s) = n_r(s) = 0$.

$$T(s) = \frac{s^m b_m + \cdots + s^2 b_2 + s b_1 + a_0}{s^n a_n + \cdots + s^2 a_2 + s a_1 + a_0}, \tag{2.32}$$

$$S(s) = s \frac{\begin{pmatrix} a_n s^{n-1} + a_{n-1} s^{n-2} + \cdots + (a_m - b_m) s^{m-1} \\ + (a_{m-1} - b_{m-1}) s^{m-2} + s(a_2 - b_2) + a_1 - b_1 \end{pmatrix}}{s^n a_n + s^{n-1} a_{n-1} + \cdots + s^2 a_2 + s a_1 + a_0}, \tag{2.33}$$

respectively. If (2.32) and (2.33) hold by the closed-loop control system is said to be of type-I. In a similar fashion, if $r(s) = \frac{1}{s^2}$ then the velocity error is equal to

$$e(\infty) = \lim_{s \to 0} \left(\frac{a_n s^n + \cdots + c_m s^m + \cdots + c_1 s + c_0}{a_n s^n + a_{n-1} s^{n-1} + \cdots + a_1 s + a_0} \right) \frac{1}{s} \tag{2.34}$$

which becomes finite if $c_0 = 0$ or $a_0 = b_0$. As a result, the final value of the error is given as

$$\lim_{t \to \infty} e_{\mathrm{vss}}(t) = \lim_{s \to 0} \left(\frac{c_1}{a_0} \right) = \lim_{s \to 0} \left(\frac{a_1 - b_1}{a_0} \right) \tag{2.35}$$

and becomes zero when $c_1 = 0$ or when $a_1 = b_1$. In this case, the closed-loop control system is said to be of type-II.[5] Sensitivity $S(s)$ and closed-loop transfer function $T(s)$ take the following form, respectively:

$$T(s) = \frac{s^m b_m + s^{m-1} b_{m-1} + \cdots + s a_1 + a_0}{s^n a_n + s^{n-1} a_{n-1} + \cdots + s a_1 + a_0}, \tag{2.36}$$

$$S(s) = s^2 \frac{a_n s^{n-2} - \cdots - b_m s^{m-2} - b_{m-1} s^{m-3} + \cdots + a_2 - b_2}{s^n a_n + s^{n-1} a_{n-1} + \cdots + s^2 a_2 + s a_1 + a_0}. \tag{2.37}$$

According to the above analysis, a closed-loop control system is said to be of type-p when sensitivity $S(s)$ and complementary sensitivity $T(s)$ have the following form:

$$S(s) = s^p \frac{\begin{pmatrix} a_n s^{n-p} + a_{n-1} s^{n-1-p} - \cdots - b_m s^{m-p} \\ - b_{m-1} s^{m-1-p} + a_p - b_p \end{pmatrix}}{s^n a_n + s^{n-1} a_{n-1} + \cdots + s^2 a_2 + s a_1 + a_0} \tag{2.38}$$

---

[5] In grid-connected power converters and when vector control is followed for regulating the DC link voltage to be utilized by the motor connected converter, there is one inner loop for regulating the current of the power converter and one outer loop for regulating its DC link voltage. In this case, the inner current control loop is of type-I, since in its open-loop transfer function there exists only one integrator coming from the current PI control action, whereas the outer control loop is of type-II, since the open-loop transfer function introduces two integrators, one coming from the DC link voltage PI control action and another coming from the capacitor bank path ($\frac{1}{sC}$). A case of type-II control loop in the field of electric motor drives is the speed control loop in vector-controlled or direct torque-controlled drives. In this case, one integrator comes from the speed PI control action and another integrator comes from the inertia ($\frac{1}{sJ}$) of the shaft of the motor the speed of which is controlled.

and

$$T(s) = \frac{b_m s^m + \cdots + a_p s^p + a_{p-1} s^{p-1} + \cdots + a_1 s + a_0}{a_n s^n + \cdots + a_p s^p + a_{p-1} s^{p-1} + \cdots + a_1 s + a_0}, \tag{2.39}$$

respectively. Also, one could argue according, to (2.38), that type-p control loops are characterized by the order of zeros at $s = 0$ in the sensitivity function $S(s)$, see (2.33), (2.37) and (2.38). In a similar fashion, the type of the control loop is automatically defined by the closed-loop transfer function $T(s)$ when observing the terms of s^j $(j = 0 \ldots p-1)$ both in the numerator and the denominator's polynomial.

2.6 Sensitivity and Complementary Sensitivity Function

The calculation of the magnitude of (2.27) results in

$$|T(s) + S(s)| = 1. \tag{2.40}$$

Ideally, in a closed-loop control system it is necessary to have the magnitude of S sufficient small, or in other words

$$|S(s)| \ll 1, \tag{2.41}$$

so that optimal disturbance rejection is achieved. However, perfect tracking of the reference signal $r(s)$ by the output $y(s)$ of the control loop requires also that

$$|T(s)| \approx 1. \tag{2.42}$$

At this point, it would be necessary to recall that relation

$$y(s) = T(s)[r(s) + n_\mathrm{r}(s) - k_\mathrm{h} n_\mathrm{o}(s)] + S(s)[d_\mathrm{o}(s) + k_\mathrm{p} G(s) d_\mathrm{i}(s)] \tag{2.43}$$

holds by within the closed-loop control system of Fig. 2.3. From (2.43), it is apparent that if sensitivity S is large enough, any disturbance signal ($d_\mathrm{o}(s)$ or $d_\mathrm{i}(s)$) entering the control loop is amplified, and as a result the output of the control loop $y(s)$ can hardly track the reference signal $r(s)$. To this end, the main problem which a control engineer faces when designing an output feedback control loop, is that in such a system, it is impossible to have perfect tracking of the reference signal $r(s)$ along with optimal disturbance and noise rejection at the same time.

Looking further on this statement, one can claim that the aforementioned conclusion is not 100 % correct, if we consider the frequency spectrum of both the noise and disturbance signals that enter the control loop. Often in many real-time applications, the reference signal $r(s)$ along with disturbances $d_\mathrm{o}(s)$ (i.e., load disturbance in electric motor drives operation) that appear at the output of the process, are signals of low frequency. By contrast, noise signals come basically by measuring equipment and most of the time contain high-frequency components.

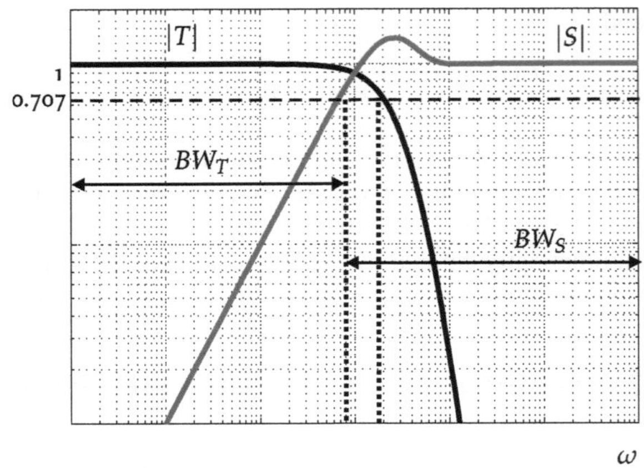

Fig. 2.4 Typical frequency response of sensitivity S and complementary T

Taking into account these facts, it becomes apparent that if the magnitude of T remains equal to unity in the widest possible frequency range, then complementary sensitivity is low enough, see Fig. 2.4 and therefore low-frequency disturbances are not amplified by low sensitivity S in the low-frequency region. As a result, satisfactory tracking of the reference can be achieved while disturbances are suppressed.

On the other hand, since noise signals appear in the higher frequency region, they cannot be amplified by the low complementary sensitivity T since it is close to zero in the high-frequency region; see Fig. 2.4. Finally, no disturbances can be amplified by the high magnitude of the complementary sensitivity S, since they do not exist in this high-frequency region.

However, it has to be pointed out that a high magnitude of T does not necessarily mean that the magnitude of S is low, since the relation (2.40) between T, S is a relation between vectors, see also Fig. 2.5. The aforementioned statement is true only in the

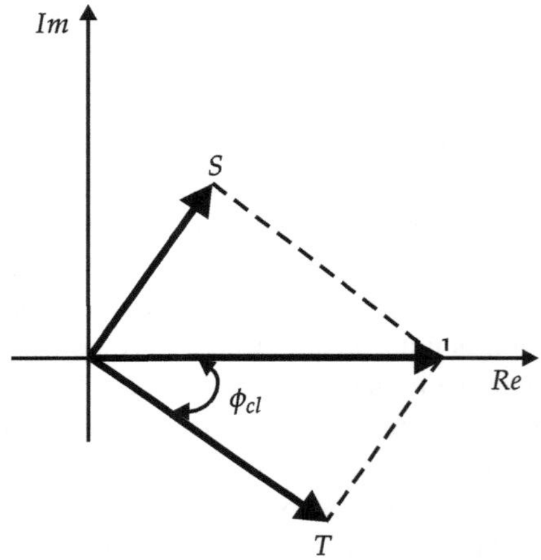

Fig. 2.5 Geometric interpretation of $|T(s) + S(s)| = 1$ in the complex plane

case where the angle ϕ_{cl} of T is very low. As a result, it becomes apparent that optimal disturbance rejection along with perfect reference tracking can be achieved only when

$$T(j\omega) \approx 1\angle 0^{\circ}. \tag{2.44}$$

Since practically this kind of design cannot be achieved, control engineers have to design control loops such that the frequency response of the closed-loop control system does not exhibit any resonance all over the low- and high-frequency regions.

2.7 The Magnitude Optimum Design Criterion

Further to the requirements defined by (2.42) and (2.44) in a closed-loop control system, in this section the principle of the Magnitude Optimum criterion is introduced. The target of the Magnitude Optimum (Betragsoptimum) criterion is to maintain the amplitude $|T(j\omega)|$ of the closed-loop frequency response equal to unity in the widest possible frequency range. This target can be mathematically expressed by

$$|T(j\omega)| \simeq 1. \tag{2.45}$$

The aforementioned equation can be considered as a practical implementation of the H_∞ controller design principle, see [8], since as mentioned in Chap. 1, the H_∞ design principle tries to optimize the amplitude of the closed-loop transfer function regardless of the resulting order of the controller. For this reason, most often times, the order of the controller of such a solution is so high that it makes its practical implementation unattractive or even sometimes unfeasible.

Back to Fig. 2.2 again, it is assumed that the transfer function of the closed-loop control system is given as

$$T(s) = \frac{b_m s^m + b_{m-1}s^{m-1} + \cdots + b_2 s^2 + b_1 s + b_0}{a_n s^n + a_{n-1}s^{n-1} + \cdots + a_2 s^2 + a_1 s + a_0} \quad (n \geq m). \tag{2.46}$$

The H_∞ controller design principle can be mathematically described as

$$H_\infty = \min\left[\lim_{n\to\infty} \sqrt[n]{\int_0^\infty |T(j\omega)|^n \, \mathrm{d}\omega}\right]. \tag{2.47}$$

A typical frequency response of $|T(j\omega)|$ involving its maximum $T_{\max}$ at a certain resonance frequency is presented in Fig. 2.6a. In this case, a good approximation of the area of the frequency response $|T(j\omega)|$ is given as

$$E = T_{\max}\Delta\omega. \tag{2.48}$$

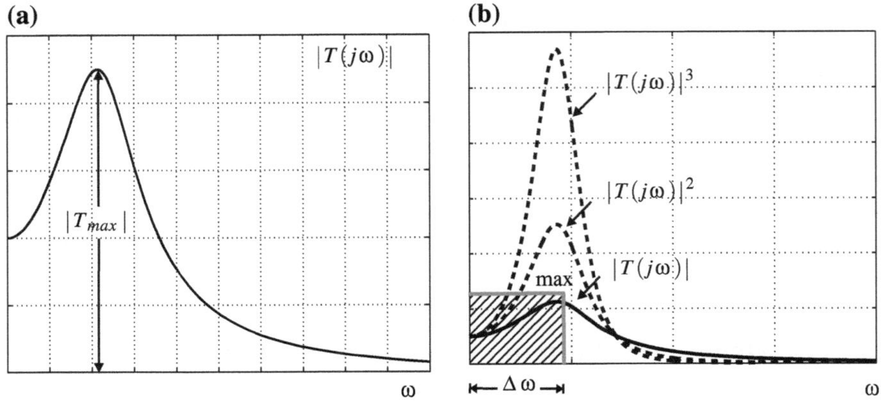

Fig. 2.6 **a** Frequency response of $|T(j\omega)|$ with resonant peak $T_{\max}$. **b** Frequency response of $|T(j\omega)|^n$ for various values of parameter n, $n = 1, 2, \ldots$

In a similar way, for calculating the surface of $|T\ (j\omega)|^n$, we can rewrite according to

$$E = T^n_{\max} \Delta\omega. \tag{2.49}$$

For calculating the surface of $|T(j\omega)|^n$, we can also write

$$E = \int_0^\infty |T(j\omega)|^n \, \mathrm{d}\omega. \tag{2.50}$$

Note that (2.49) is equal to (2.50) in case where n becomes sufficiently high and the term $\Delta\omega$ becomes sufficiently small. Strictly speaking, (2.49) is equal to (2.50) when $n \to \infty$ and $\Delta\omega \to 0$. For this reason, after taking the lim of both (2.49) and (2.50) when $n \to \infty$, we can rewrite

$$\lim_{n\to\infty} \int_0^\infty |T(j\omega)|^n \, \mathrm{d}\omega = \lim_{n\to\infty} (\Delta\omega T^n_{\max}). \tag{2.51}$$

The algebraic manipulation of (2.51) results in

$$\lim_{n\to\infty} \sqrt[n]{\int_0^\infty |T(j\omega)|^n \mathrm{d}\omega} = \lim_{n\to\infty} \left(T_{\max} \sqrt[n]{\Delta\omega} \right). \tag{2.52}$$

From (2.52) it is obvious that, if $n \to \infty$ then

$$\lim_{n\to\infty} \sqrt[n]{\Delta\omega} = 1 \quad \forall \Delta\omega. \tag{2.53}$$

Substituting (2.53) into (2.52) results in

$$\lim_{n\to\infty} \sqrt[n]{\int_0^\infty |T(j\omega)|^n \, d\omega} = \lim_{n\to\infty} (T_{\max}). \tag{2.54}$$

Therefore, for minimizing (2.54), we can rewrite

$$H_\infty = \min\left(\lim_{n\to\infty} \sqrt[n]{\int_0^\infty |T(j\omega)|^n \, d\omega} \right) = \min\left(\lim_{n\to\infty} (T_{\max}) \right). \tag{2.55}$$

For this, we have to invent a systematic approach that satisfies the condition

$$H_\infty = \min(T_{\max}). \tag{2.56}$$

Such a systematic strict mathematical approach of optimizing (2.56) can be found in [10] the final result of which is graphically depicted in Fig. 2.7.

Since in this book, the goal is to present tuning rules for the PID controller which is often met and applicable in the majority of industrial applications see [1], a less strict mathematical optimization is presented for forcing the magnitude of the closed-loop frequency response $|T(j\omega)|$ equal to unity in the widest possible frequency range.

From (2.46) it becomes apparent that if we substitute $s = j\omega$ results in

$$|T(j\omega)| = \left| \frac{N(j\omega)}{D(j\omega)} \right|. \tag{2.57}$$

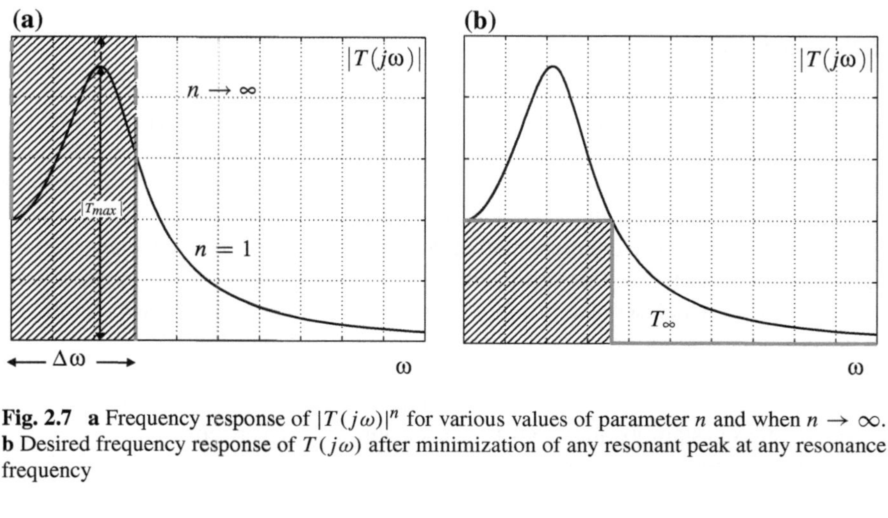

Fig. 2.7 **a** Frequency response of $|T(j\omega)|^n$ for various values of parameter n and when $n \to \infty$. **b** Desired frequency response of $T(j\omega)$ after minimization of any resonant peak at any resonance frequency

Calculating $|N(j\omega)|^2$, $|D(j\omega)|^2$, results in

$$|N(j\omega)|^2 \simeq \cdots + B_8\omega^{16} + B_7\omega^{14} + B_6\omega^{12} + B_5\omega^{10} \quad (2.58)$$

$$+ B_4\omega^8 + B_3\omega^6 + B_2\omega^4 + B_1\omega^2 + B_0 \quad (2.59)$$

$$|D(j\omega)|^2 \simeq \cdots + A_8\omega^{16} + A_7\omega^{14} + A_6\omega^{12} + A_5\omega^{10} \quad (2.60)$$

$$+ A_4\omega^8 + A_3\omega^6 + A_2\omega^4 + A_1\omega^2 + A_0\,, \quad (2.61)$$

respectively. For forcing $|T(j\omega)| \approx 1$ in the widest possible frequency range

$$A_i = B_j \quad \forall i, j \quad (i = 0, n)\ (j = 0, m) \quad (2.62)$$

must hold by. In Appendix A it is proved that for setting $A_i = B_j$, $\forall i, j$ $(i = 0, n)$ $(j = 0, m)$ results finally in

$$a_0 = b_0 \quad (2.63)$$

$$a_1^2 - 2a_2a_0 = b_1^2 - 2b_2b_0 \quad (2.64)$$

$$a_2^2 - 2a_3a_1 + 2a_4a_0 = b_2^2 - 2b_3b_1 + 2b_4b_0 \quad (2.65)$$

$$a_3^2 + 2a_1a_5 - 2a_6a_0 - 2a_4a_2 = b_3^2 + 2b_1b_5 - 2b_6b_0 - 2b_4b_2 \quad (2.66)$$

$$\begin{pmatrix} a_4^2 + 2a_0a_8 + 2a_6a_2 - 2a_1a_7 \\ -2a_3a_5 \end{pmatrix} = \begin{pmatrix} b_4^2 + 2b_0b_8 + 2b_6b_2 - 2b_1b_7 \\ -2b_3b_5 \end{pmatrix} \quad (2.67)$$

Equations (2.63)–(2.67) comprise the basis for the proof of every optimal control law for every type of control loop presented in the sequel within this book.

2.8 Summary

In this chapter, preliminary definitions of the operation of closed-loop control systems were presented in Sect. 2.1. It was shown how the problem of perfect reference tracking is in conflict with any kind of disturbance entering the control loop from the outer world. To justify this statement, in Sect. 2.2, the closed-loop control system was presented in a more concrete mathematical modeling by the frequency domain approach. With respect to this approach, basic transfer functions of the control loop were presented, which serve as proof of the proposed PID control law, which follows in the next chapters for any type-I, type-II, and type-III control loops.

Given the aforementioned necessary definitions regarding the transfer functions involved within a closed-loop control system, in Sects. 2.3 and 2.4 the important aspect of internal stability and robustness in a control loop were covered. In Sect. 2.5, a mathematical approach was presented relevant to the type of feedback control loop. This section aims at giving the reader quick hints on how to easily identify, given the closed-loop transfer function of a control system, its exact type (type-I, type-II, and type-III).

In Sect. 2.6 it was shown why it is important to keep the magnitude of the closed-loop control system equal to unity in the widest possible frequency range ($|T(j\omega)|$), since under certain circumstances this principle leads to satisfactory disturbance rejection both at the input and output of the process. This section is also the connecting ring to the principle of the Magnitude Optimum criterion which is finally presented in Sect. 2.7. The principle of the Magnitude Optimum criterion is considered as a practical aspect of the H_{∞} and is used to deploy the proposed PID control laws presented in the following chapters. Finally, certain optimization conditions are derived in Sect. 2.7 which serve as the basis for the explicit definition of the PID control action irrespective of the process complexity.

References

1. Ang KH, Chong G, Li Y (2005) PID control system analysis, design, and technology. IEEE Trans Control Syst Technol 13(4):559–576
2. Gunter S (2003) Respect the unstable. IEEE Control Syst Mag 23(4):12–25
3. Horowitz I (1963) Synthesis of feedback systems. Academic Press, London
4. Margaris NI (2003) Lectures in applied automatic control (in Greek), 1st edn. Tziolas, Greece
5. Middleton RH (1991) Trade-offs in linear control system design. Automatica 27(2):281–292
6. Morari M, Zafiriou E (1989) Robust process control, 1st edn. Prentice-Hall, New Jersey
7. Petridis V (2001) Automatic control systems, part B (in Greek), 2nd edn. Ziti, Greece
8. Voda AA, Landau ID (1995) A method for the auto-calibration of PID controllers. Automatica 31(1):41–53
9. Voronov AA (1985) Basic principles of automatic control theory—special linear and nonlinear systems. MIR Publishers, Moscow
10. Zames G, Francis BA (1983) Feedback, minimax sensitivity, and optimal robustness. IEEE Trans Autom Control 28(5):585–600

Part II
Explicit Tuning of the PID Controller

Chapter 3
Type-I Control Loops

Abstract In this chapter, the tuning of the PID controller via the Magnitude Optimum criterion for type-I control loops is presented. Initially, the revision of the conventional Magnitude Optimum design criterion for tuning the PID type controller's parameters is presented in Sect. 3.2, which serves as a basis for the reader to understand the current state of the art, see Sects. 3.2.1–3.2.4. This revision reveals three fundamental drawbacks, which are summarized in Sect. 3.2.5 and prove to restrict the PID controller's optimal tuning in terms of robustness and disturbance rejection at the output of the plant. Sorting out these drawbacks in the beginning, one can argue that: (1) with the conventional PID tuning and for determining the PID controller's zeros, exact pole-zero cancellation has to be achieved between the processes' poles and the controller's zeros. (2) To this end, the conventional PID tuning via the Magnitude Optimum criterion restricts the controller's zeros to be tuned only with real zeros. (3) Last but not the least, the conventional design procedure via the Magnitude Optimum criterion has been tested only to a limited class of simple process models. To overcome the aforementioned drawbacks, a revised PID type control law is then proposed in Sect. 3.3. For the development of the control law a general transfer function process model is employed in the frequency domain. The final control law consists of analytical expressions that involve all modeled process parameters. The resulting control law can be applied directly to any linear single input single output stable process regardless of its complexity. A summary of the explicit solution is presented in Sect. 3.3 and the analytical proof of the control law is presented in Appendix B.1. For evaluating the proposed theory, an extensive simulation test batch between the conventional and the revised PID tuning is performed in Sect. 3.4 for various benchmark processes. Throughout this evaluation, the validity of several literature comments related to the Magnitude Optimum criterion is discussed in Sects. 3.4.6 and 3.4.7. Finally, it is shown that the performance of the proposed control law compared to the conventional PID design procedure achieves satisfactory results both in the time and the frequency domain, in terms of robustness and disturbance rejection.

K.G. Papadopoulos, *PID Controller Tuning Using the Magnitude Optimum Criterion*,
DOI 10.1007/978-3-319-07263-0_3

3.1 Introduction

The principle of the Magnitude Optimum criterion, introduced by Sartorius and Oldenbourg [21, 22] is based on the idea of designing a controller, which renders the magnitude of the closed loop frequency response as close as possible to unity, in the widest possible frequency range. Oldenbourg and Sartorius applied the Magnitude Optimum criterion in type-I systems with stable real poles. In succession, Kessler suggested the Symmetrical Optimum criterion [15, 16]. The name of this criterion comes from the symmetry exhibited by the open loop frequency response. In reality, the Symmetrical Optimum criterion is not something different, but the application of the Magnitude Optimum criterion to type-II control systems.

The design of control systems both with the Magnitude Optimum and Symmetrical Optimum criteria of Oldenbourg–Sartorius and Kessler, respectively, presents at least two important advantages according to [2, 24]: firstly they do not require the complete plant model see [24] and secondly, the setpoint response of the closed loop system is excellent [2]. Further to these statements, the Symmetrical Optimum criterion is more known because of its successful application in the control of electric motor drives, see [6, 9, 17, 25]. However, excluding the German bibliography [8, 10, 11, 18], the Magnitude Optimum criterion is rarely referred today. In addition to this, the limited impact of both the Magnitude and Symmetrical design criteria is stressed in [24], and it might be owed to the negative comments that occasionally have been stated in the literature. Some of these comments are presented in Sect. 1.3.1 all of which, need, in our opinion, to be revised for three reasons.

1. Firstly, as it is proved in the sequel, the conventional design[1] procedure via the Magnitude Optimum criterion for PID type controllers, restricts the controller's zeros to be tuned only with real zeros leading finally to poor tuning. This approach, does not take into account the fact that the optimal values for the PID controller's zeros may be conjugate complex, which might result to more robust tuning than the principle of pole-zero cancellation.
2. Secondly, for determining the PID controller's zeros, exact pole-zero cancellation has to be achieved between the process's poles and the controller's zeros [2]. This approach disregards all other plant parameters for the optimal control law and as a result, the PID parameter tuning is poor and suboptimal.
3. Thirdly, the conventional design procedure via the Magnitude Optimum criterion has been tested only to a limited class of simple process models [26, 27], and not to benchmark processes as it is carried out in Sect. 3.4 of this chapter.

Based on the above and for the sake of a clear presentation of the proposed theory, this chapter is organized as follows. In Sect. 3.2, the conventional tuning method of PID controllers via the Magnitude Optimum criterion is presented, so that all drawbacks are made clear, see Sect. 3.2.5.

---

[1] Design via pole-zero cancellation.

Taking into account the aforementioned drawbacks, in Sect. 3.3, the revised PID type control law is developed. For the control law's proof, a general transfer function of the process model is employed consisting of n poles m zeros plus time delay d. In Sect. 3.4, the conventional and the revised PID control law are compared via simulation examples for benchmark processes met in many industry applications.

The comparison focuses on the performance of the control law in terms of disturbance rejection and reference tracking. Finally, after the verification of the proposed control law, the validity of the several negative comments toward the Magnitude Optimum criterion presented in this section, is investigated in Sects. 3.4.6, 3.4.7 and 3.4.10.

3.2 Conventional PID Tuning Via the Magnitude Optimum Criterion

For presenting the conventional PID tuning via the Magnitude Optimum criterion, the closed loop system of Fig. 3.1 is considered. The involved signals in the frequency domain $r(s)$, $e(s)$, $u(s)$, $y(s)$, $d_o(s)$ and $d_i(s)$ stand for the reference input, the control error, the input and output of the plant, the output and the input disturbances respectively. In addition, a real process met in many industry applications can be described by

$$G(s) = \frac{1}{\left(1 + sT_{p1}\right)\left(1 + sT_{p2}\right)\cdots\left(1 + sT_{pn}\right)}, \tag{3.1}$$

for which $T_{p1} > T_{p2} > \cdots > T_{pn}$ is also considered. Note that k_p stands for the plant's DC gain at steady state. Supposing that no information about the real process is available, it is conceived as a first order one [2, 9, 13], defined by the approximation

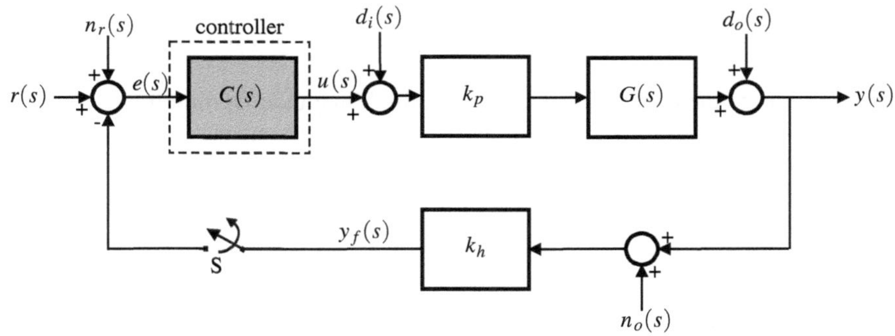

Fig. 3.1 Block diagram of the closed-loop control system. $G(s)$ is the plant transfer function, $C(s)$ is the controller transfer function, $r(s)$ is the reference signal, $d_o(s)$ and $d_i(s)$ are the output and input disturbance signals, respectively, and $n_r(s)$, $n_o(s)$ are the noise signals at the reference input and process output, respectively. k_p stands for the plant's DC gain and k_h is the feedback path

$$\widetilde{G}(s) = \frac{1}{1 + sT_{\Sigma_\mathrm{p}}}, \tag{3.2}$$

where $T_{\Sigma_\mathrm{p}} = \sum_{i=1}^{n} T_{\mathrm{p}_i}$ is the equivalent sum time constant of the plant.

3.2.1 I Control

When the information about the plant is limited, the control that can consciously be applied is limited to integral control action, so that the system exhibits at least zero steady state position error. By applying integral action given by

$$C(s) = \frac{1}{sT_{\mathrm{i_I}}\left(1 + sT_{\Sigma_\mathrm{c}}\right)} \tag{3.3}$$

to the approximate plant (3.2), the resulting closed loop transfer function takes the form of

$$\begin{aligned}\widetilde{T}(s) &= \frac{k_\mathrm{p}}{sT_{\mathrm{i_I}}(1 + sT_{\Sigma_\mathrm{c}})(1 + sT_{\Sigma_\mathrm{p}}) + k_\mathrm{h}k_\mathrm{p}} \\ &\approx \frac{k_\mathrm{p}}{s^2 T_{\mathrm{i_I}} T_\Sigma + sT_{\mathrm{i_I}} + k_\mathrm{p}k_\mathrm{h}},\end{aligned} \tag{3.4}$$

for which

$$T_{\Sigma_\mathrm{c}} T_{\Sigma_\mathrm{p}} \approx 0 \quad \text{and} \quad T_\Sigma = T_{\Sigma_\mathrm{c}} + T_{\Sigma_\mathrm{p}} \tag{3.5}$$

has been considered. Note that T_{Σ_c} stands for the controller's unmodeled dynamics, which are involved between the output of the controller and the input signal to the plant.

According to the conventional design via the Magnitude Optimum criterion, the integration time constant $T_{\mathrm{i_I}}$ of the controller and the parameter k_h in the feedback path are determined so that the amplitude of the closed loop transfer function $T(s)$ is forced equal to one $|\widetilde{T}(j\omega)| \simeq 1$ in the wider possible frequency range. The magnitude of (3.4) is given by

$$|\widetilde{T}(j\omega)| \approx \sqrt{\frac{k_\mathrm{p}^2}{T_{\mathrm{i_I}}^2 T_\Sigma^2 \omega^4 + \left(T_{\mathrm{i_I}} - 2k_\mathrm{p}k_\mathrm{h}T_\Sigma\right) T_{\mathrm{i_I}}\omega^2 + k_\mathrm{p}^2 k_\mathrm{h}^2}}. \tag{3.6}$$

Condition $|\widetilde{T}(j\omega)| \simeq 1$ is satisfied if $T_{\mathrm{i_I}} - 2k_\mathrm{p}k_\mathrm{h}T_\Sigma = 0$ and $k_\mathrm{h} = 1$, or finally

$$k_\mathrm{h} = 1 \quad \text{and} \quad T_{\mathrm{i_I}} = 2k_\mathrm{p}k_\mathrm{h}T_\Sigma. \tag{3.7}$$

In that case, the magnitude of (3.6) takes the form

$$|\widetilde{T}\,(j\omega)| \approx \sqrt{\frac{1}{4T_{\Sigma}^{4}\omega^{4}+1}}, \tag{3.8}$$

which is close to unity in the low frequency region, if $\omega \to 0$. If condition $k_{\mathrm{h}} = 1$ is fulfilled, it is implied that the closed loop system has zero steady state position error. At this point, and after substituting (3.7) into (3.4), results in

$$\widetilde{T}(s) = \frac{1}{2T_{\Sigma}^{2}s^{2}+2T_{\Sigma}s+1}. \tag{3.9}$$

Normalizing the time by setting $s' = sT_{\Sigma}$ leads to

$$\widetilde{T}(s') = \frac{1}{2s'^{2}+2s'+1}. \tag{3.10}$$

3.2.2 Preservation of the Shape of the Step and Frequency Response

If now, the same control law defined in (3.7) is applied to the real plant (3.1), the resulting closed loop transfer function is given by

$$T(s') = \frac{1}{\left(\begin{array}{l}\frac{2}{T_{\Sigma}^{n}}\prod_{j=1}^{n}T_{\mathrm{p}_j}s'^{n+1}+\cdots+\frac{2}{T_{\Sigma}^{2}}\sum_{i\neq j=1}^{n}T_{\mathrm{p}_i}T_{\mathrm{p}_j}s'^{3}\\ +2s'^{2}+2s'+1\end{array}\right)} \tag{3.11}$$

where $s' = sT_{\Sigma}$. Comparing (3.9) with (3.11), it becomes apparent that in the approximate design, the terms of order higher than s'^{2} are being neglected in the denominator polynomial.

However, these terms have negligible effect on the dynamic behavior of the control loop, because their coefficients are small (they are divided by a power of the closed loop sum time constant T_{Σ} of higher order). Therefore, the two systems exhibit almost the same dynamic behavior. The accuracy of the approximation depends on the distribution of the plant time constants $T_{\mathrm{p}_j},\ j = 1, 2, \ldots, n$. In cases where ratio $\rho = \frac{T_{\mathrm{p}1}}{T_{\Sigma}} \to 0$, the accuracy is especially satisfactory both in the time and frequency domain, Fig. 3.2.

Figure 3.2a presents the step response of the exact and approximate closed loop system to the reference input $r(s)$ and to the output disturbance $d_{\mathrm{o}}(s)$, for two extreme distributions of the plant time constants ($\rho = 0.3$ and $\rho = 0.9$). The coincidence of the two responses is especially satisfactory, despite the fact that the determination of

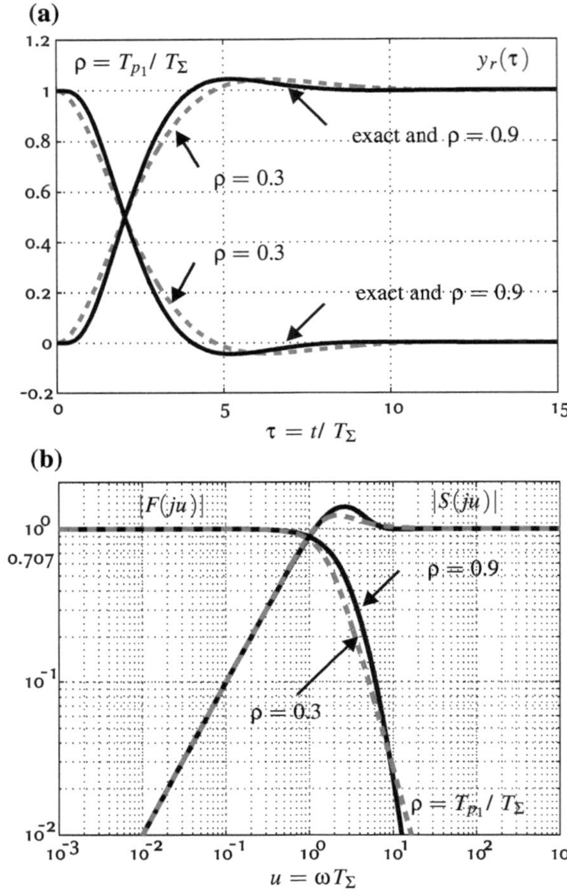

Fig. 3.2 **a** Step response of the control loop. **b** Closed loop frequency response. Comparison of the exact and the approximate control systems with integral control according to the conventional design via the Magnitude Optimum criterion

parameters T_{i_I} and k_h was based on a rough plant model. Figure 3.2b, presents the closed loop transfer and output sensitivity frequency responses of the exact and the approximate systems, for the two extreme distributions of the plant time constants ($\rho = 0.3$ and $\rho = 0.9$).

With respect to the above analysis, it is concluded that by using a rough model of the plant and applying only integral control through the conventional design method via the Magnitude Optimum criterion, a closed loop system with satisfactory response results. The features of these response are listed below.

- Mean rise time $t_r = 4.40T_\Sigma$ ($4.7T_\Sigma$ for $\rho \geq 0.9$ and $4.1T_\Sigma$ for $\rho = 0.3$).
- Mean settling time $t_{ss} = 7.86T_\Sigma$ ($8.40T_\Sigma$ for $\rho \geq 0.9$ and $7.32T_\Sigma$ for $\rho = 0.3$).
- Mean overshoot 4.47 % (4.32 % for $\rho \geq 0.9$ and 4.62 % for $\rho = 0.3$).
- Gain margin $\alpha_m = 205$ db.
- Phase margin $\phi_m = 65.27°$.

3.2.3 PI Control

In cases now where the dominant time constant T_{p_1} of the plant (conventional design method via the Magnitude Optimum criterion) is evaluated, an approximate transfer function of (3.1) is defined by

$$\widetilde{G}(s) = \frac{1}{(1+sT_{\mathrm{p}_1})(1+sT_{\Sigma_{1\mathrm{p}}})}, \tag{3.12}$$

where

$$T_{\Sigma_{1\mathrm{p}}} = \sum_{i=2}^{n} T_{\mathrm{p}_i} \tag{3.13}$$

stands for the parasitic time constant of the plant. Since the plant has a dominant time constant, by imposing PI control through the controller

$$C(s) = \frac{1+sT_{\mathrm{n}}}{sT_{\mathrm{i_{PI}}}(1+sT_{\Sigma_{\mathrm{c}}})}, \tag{3.14}$$

the following closed loop transfer function results

$$\widetilde{T}(s) = \frac{k_{\mathrm{p}}(1+sT_{\mathrm{n}})}{sT_{\mathrm{i_{PI}}}(1+sT_{\mathrm{p}_1})(1+sT_{\Sigma_1}) + k_{\mathrm{h}}k_{\mathrm{p}}(1+sT_{\mathrm{n}})}. \tag{3.15}$$

Note again that for the derivation of (3.15), $T_{\Sigma_{\mathrm{c}}} \ll T_{\Sigma_{1\mathrm{p}}}$ and $T_{\Sigma_1} = T_{\Sigma_{1\mathrm{p}}} + T_{\Sigma_{\mathrm{c}}} = T_{\Sigma} - T_{\mathrm{p}_1}$ has been set.

According to the conventional Magnitude Optimum criterion design, for determining the zero T_{n} of the PI controller, pole-zero cancellation between the process's dominant time constant T_{p_1} and the controller's zero T_{n} has to take place. To this end

$$T_{\mathrm{n}} = T_{\mathrm{p}_1} \tag{3.16}$$

is set in (3.15). This results in

$$\widetilde{T}(s) = \frac{k_{\mathrm{p}}}{sT_{\mathrm{i_{PI}}}(1+sT_{\Sigma_1}) + k_{\mathrm{h}}k_{\mathrm{p}}} \tag{3.17}$$

which yields

$$\widetilde{T}(s) = \frac{k_{\mathrm{p}}}{s^2 T_{\mathrm{i_{PI}}} T_{\Sigma_1} + sT_{\mathrm{i_{PI}}} + k_{\mathrm{h}}k_{\mathrm{p}}}. \tag{3.18}$$

In a similar fashion with (3.4) condition $|\widetilde{T}(j\omega)| \simeq 1$ is now satisfied if

$$k_h = 1 \quad \text{and} \quad T_{iPI} = 2k_p k_h T_{\Sigma_1} \tag{3.19}$$

is set. Note that $T_{\Sigma_1} = T_\Sigma - T_{p_1}$ and therefore the final PI control law is equal to

$$k_h = 1, \tag{3.20}$$

$$T_n = T_{p_1}, \tag{3.21}$$

$$\begin{aligned} T_{iPI} &= 2k_p k_h T_{\Sigma_1} \\ &= 2k_p k_h (T_\Sigma - T_{p_1}) \end{aligned} \tag{3.22}$$

$$= 2k_p k_h (T_\Sigma - T_n). \tag{3.23}$$

Let it be noted that for the derivation of the control law, (3.20)–(3.22), exact pole-zero cancellation has been assumed (conventional design method via the Magnitude Optimum criterion) (3.21). Substituting (3.20)–(3.22) into (3.15) results in

$$\widetilde{T}(s) = \frac{1}{2T_{\Sigma_1}^2 s^2 + 2T_{\Sigma_1} s + 1}. \tag{3.24}$$

Setting again $s' = sT_{\Sigma_1}$ leads to

$$\widetilde{T}(s') = \frac{1}{2s'^2 + 2s' + 1}. \tag{3.25}$$

Comparing (3.25) with (3.10), it is concluded that with the application of PI control via the conventional design of the Magnitude Optimum criterion, a closed loop system with time and frequency response of the *same shape* results.

However, the response of (3.25) is faster than of (3.10), because the time scale ($T_{\Sigma_1} < T_\Sigma$) is smaller. In other words, the compensation of the dominant time constant T_{p_1} has left the *shape of the system time and frequency responses unaltered* and produced only a change both in the time and frequency scale, respectively.

3.2.4 PID Control

In cases now when two dominant time constants T_{p_1}, T_{p_2} of the plant are evaluated, the transfer function of the real process (3.1) can be approximated by

$$\widetilde{G}(s) = \frac{1}{(1 + sT_{p_1})(1 + sT_{p_2})(1 + sT_{\Sigma_{2p}})}, \tag{3.26}$$

where

$$T_{\Sigma_{2p}} = \sum_{i=3}^{n} T_{\Sigma_{p_i}} \tag{3.27}$$

represents the parasitic time constant of the plant. Since the plant has now two dominant time constants, the PID control law defined by

$$C(s) = \frac{(1 + sT_n)(1 + sT_v)}{sT_{iPID}\left(1 + sT_{\Sigma_c}\right)} \tag{3.28}$$

is imposed to (3.28). Assuming that $T_{\Sigma_c} \ll T_{\Sigma_{2_p}}$ and $T_{\Sigma_2} = T_{\Sigma_{2_p}} + T_{\Sigma_c} = T_{\Sigma} - T_{p_1} - T_{p_2}$, the transfer function of the closed loop control system is equal to

$$\widetilde{T}(s) = \frac{k_p(1 + sT_n)(1 + sT_v)}{sT_{iPID}(1 + sT_{p_1})(1 + sT_{p_2})(1 + sT_{\Sigma_2}) + k_h k_p(1 + sT_n)(1 + sT_v)}. \tag{3.29}$$

According to the conventional Magnitude Optimum criterion design, for determining the zeros T_n, T_v of the PID controller, pole-zero cancellation between the process's dominant time constants T_{p_1}, T_{p_2} and the controller' zero T_n, T_v has to take place. To this end

$$T_n = T_{p_1} \tag{3.30}$$

$$T_v = T_{p_2} \tag{3.31}$$

is set in (3.29), which results in

$$\widetilde{T}(s) = \frac{k_p}{sT_{iPID}(1 + sT_{\Sigma_2}) + k_h k_p} \tag{3.32}$$

or

$$\widetilde{T}(s) = \frac{k_p}{s^2 T_{iPID} T_{\Sigma_2} + sT_{iPID} + k_h k_p}. \tag{3.33}$$

In a similar fashion with (3.4), condition $|\widetilde{T}(j\omega)| \simeq 1$ is now satisfied when

$$k_h = 1 \quad \text{and} \quad T_{iPID} = 2k_p k_h T_{\Sigma_2}. \tag{3.34}$$

Note that $T_{\Sigma_2} = T_{\Sigma} - T_{p_1} - T_{p_2}$, and therefore the final PID control law is equal to

$$k_h = 1, \tag{3.35}$$

$$T_n = T_{p_1}, \tag{3.36}$$

$$T_v = T_{p_2}, \tag{3.37}$$

$$T_{iPID} = 2k_p k_h T_{\Sigma_2} \tag{3.38}$$

$$= 2k_p k_h (T_{\Sigma} - T_{p_1} - T_{p_2})$$

$$= 2k_p k_h (T_{\Sigma} - T_n - T_v). \tag{3.39}$$

After substituting the control law given by (3.35)–(3.38) into (3.29), results in

$$\widetilde{T}(s) = \frac{1}{2T_{\Sigma_2}^2 s^2 + 2T_{\Sigma_2} s + 1}. \tag{3.40}$$

Normalizing the time by setting $s' = sT_{\Sigma_2}$ leads to

$$\widetilde{T}(s') = \frac{1}{2s'^2 + 2s' + 1}. \tag{3.41}$$

Comparing (3.41) with (3.25) and (3.10) it is concluded that with the application of PID control, a closed loop system with time and frequency responses of the same shape results, but with even smaller time scale ($T_{\Sigma_2} < T_{\Sigma_1} < T_{\Sigma}$) and consequently even faster (Fig. 3.3).

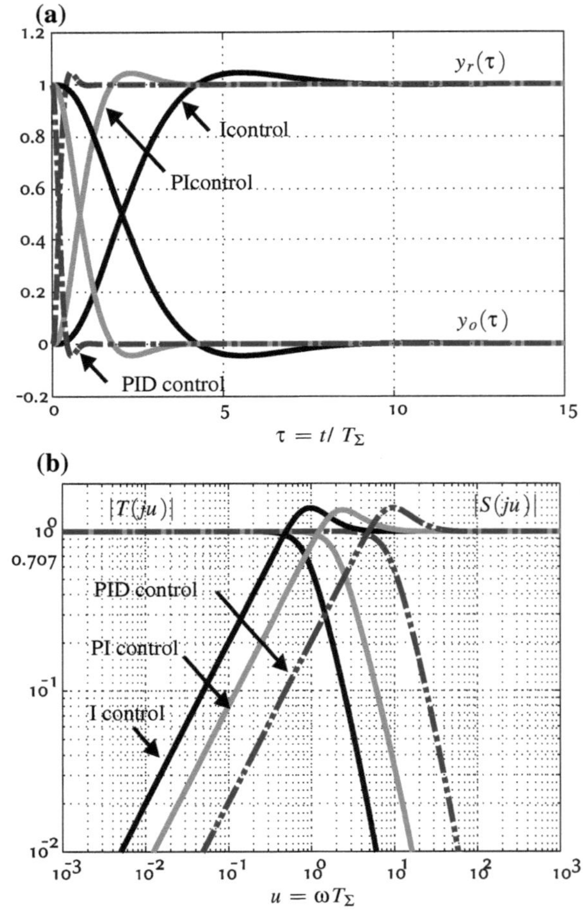

Fig. 3.3 **a** Step response. **b** Frequency response. Comparison study of the step and frequency response of the closed loop control system defined by (3.10), (3.25) and (3.41), respectively. The approximate process $\widetilde{G}(s)$ is controlled by I, PI, PID control action through the conventional tuning

3.2.5 Drawbacks of the Conventional Tuning Method

According to the above, from the conventional design procedure via the Magnitude Optimum criterion it is apparent that

- for the PID controller's tuning real zeros are always considered, see (3.21) and (3.35)–(3.36).
- For tuning the PI, PID type controller zeros, exact pole-zero cancellation has to be achieved.
- Since this type of tuning disregards any other fundamental dynamics of the process, the resulting PID tuning is considered suboptimal.

In the next section, these three restrictions are thoroughly revised and the development of the proposed theory takes place.

3.2.6 Why PID Control?

In Sect. 3.2 it was shown that the Magnitude Optimum criterion exploits the power and the advantages in terms of implementation, the PID controller offers. These advantages stem from the fact that the order of a large variety of processes within many industry applications is not high enough, or at least their model can be approximated by a second or third order process model.

To justify this conclusion let the plant transfer function be defined by

$$G(s) = \frac{k_{\mathrm{p}}}{(1+sT_{\mathrm{p}1})(1+sT_{\mathrm{p}2})\cdots(1+sT_{\mathrm{p}_{n-2}})\cdots(1+sT_{\mathrm{p}_n})}. \tag{3.42}$$

After some algebraic manipulation it is easily seen that

$$G(s) = \frac{k_{\mathrm{p}}}{\left(\begin{array}{l} s^n \prod_{i=1}^{n} T_{\mathrm{p}_i} + \cdots + s^3 \sum_{i\neq j\neq k=1}^{n} T_{\mathrm{p}_i}T_{\mathrm{p}_j}T_{\mathrm{p}_k} + s^2 \sum_{i\neq j=1}^{n} T_{\mathrm{p}_i}T_{\mathrm{p}_j} \\ +s\sum_{i=1}^{n} T_{\mathrm{p}_i} + 1 \end{array}\right)}. \tag{3.43}$$

By making the substitution

$$s' = s\sum_{j=1}^{n} T_{\mathrm{p}_j} = sT_{\Sigma} \tag{3.44}$$

transfer function defined by (3.43) is rewritten as follows

$$G(s') = \frac{k_p}{\left(\begin{array}{l} s'^n \frac{1}{T_\Sigma^n} \prod_{i=1}^n T_{p_j} + \cdots + s'^3 \frac{1}{T_\Sigma^3} \sum_{j \neq k \neq l=1}^n T_{p_j} T_{p_k} T_{p_l} \\ + s'^2 \frac{1}{T_\Sigma^2} \sum_{j \neq k=1}^n T_{p_j} T_{p_k} + s' + 1 \end{array}\right)}. \tag{3.45}$$

From (3.45) it is apparent that the higher order terms of s' are divided by T_Σ^n where $n = 0, 1, 2 \ldots$ At this point, let it be noted that in principle, the value of the sum time constant T_Σ is relatively high.

With respect to the above, it can be concluded that higher order type systems, can under certain circumstances, be approximated by low order systems. The error of this approximation lies in the distribution of the time constants of the process itself. Obviously, the worst case takes place in case the time constants of the plant are equally distributed. For example, in case of a process with five equal dominant time constants given by

$$G(s) = \frac{k_p}{(1+s)^5} = \frac{k_p}{1 + 5s + 10s^2 + 10s^3 + 5s^4 + s^5} \tag{3.46}$$

it is concluded, according to the aforementioned analysis, that can be rewritten in the form of

$$G(s') = \frac{k_p}{\frac{1}{3125}s'^5 + \frac{5}{625}s'^4 + \frac{10}{125}s'^3 + \frac{10}{25}s'^2 + s' + 1} \tag{3.47}$$

or

$$G(s') = \frac{k_p}{3.2 \times 10^{-4}s'^5 + 8 \times 10^{-3}s'^4 + 8 \times 10^{-2}s'^3 + 4 \times 10^{-1}s'^2 + s' + 1} \tag{3.48}$$

which can be easily controlled by a PID controller tuning according to the method described in Sect. 3.2.

3.3 Revised PID Tuning Via the Magnitude Optimum Criterion

For deriving the revised PID type control law, a general transfer function of the process model consisting of $(n - 1)$ poles, m zeros plus a time delay constant T_d is adopted, see (3.49). Zeros of the plant may lie both in the left or right imaginary half plane. The plant transfer function may also contain second order oscillatory terms in the denominator, described by polynomials of the form $1 + 2\zeta Ts + s^2T^2$, where $\zeta \in (0, 1], \in \Re$ and $T > 0, \in \Re$. Hence, the plant transfer function can be described in general by

$$G(s) = \frac{s^m \beta_m + s^{m-1}\beta_{m-1} + \cdots + s^2\beta_2 + s\beta_1 + 1}{s^{n-1}\alpha_{n-1} + \cdots + s^3\alpha_3 + s^2\alpha_2 + s\alpha_1 + 1} e^{-sT_\mathrm{d}} \tag{3.49}$$

where $n - 1 > m$. The proposed PID-type controller is given by the flexible form

$$C(s) = \frac{1 + sX + s^2Y}{sT_\mathrm{i}(1 + sT_{\mathrm{p}n})} \tag{3.50}$$

allowing its zeros to become conjugate complex. Time constant $T_{\mathrm{p}n}$ stands for the unmodeled controller dynamics coming from the controller's implementation.

According to Fig. 3.1, the closed loop transfer function $T(s)$ is given by

$$T(s) = \frac{k_\mathrm{p} C(s) G_p(s)}{1 + k_\mathrm{h} k_\mathrm{p} C(s) G_p(s)} = \frac{N(s)}{D(s)} = \frac{N(s)}{D_1(s) + k_\mathrm{h} N(s)}. \tag{3.51}$$

Polynomials $N(s),\ D_1(s)$ are equal to

$$N(s) = k_\mathrm{p}(1 + sX + s^2Y)\sum_{i=0}^{m}(s^i\beta_i), \tag{3.52}$$

$$D_1(s) = sT_\mathrm{i} e^{sT_\mathrm{d}} \sum_{j=0}^{n}(s^j\alpha_j) \tag{3.53}$$

where $\alpha_0 = \beta_0 = 1$ according to (3.49). Normalizing $N(s)$, $D_1(s)$ by making the substitution $s' = sc_1$ results in

$$N(s') = k_\mathrm{p}(1 + s'x + s'^2 y)\sum_{i=0}^{m}(s'^i z_i) \tag{3.54}$$

$$D_1(s') = s' t_\mathrm{i} e^{s'd} \sum_{j=0}^{n}(s'^j r_j) \tag{3.55}$$

respectively. The corresponding normalized terms involved in the control loop are given by

$$x = \frac{X}{c_1},\quad y = \frac{Y}{c_1^2},\quad t_i = \frac{T_\mathrm{i}}{c_1},\quad d = \frac{T_\mathrm{d}}{c_1},$$

$$r_i = \frac{\alpha_i}{c_1^i},\quad \forall\, i = 1, 2, \ldots, n,\quad z_j = \frac{\beta_j}{c_1^j},\quad \forall\, j = 1, 2, \ldots, m.$$

The normalized time delay constant d is substituted with the "all pole" series approximation

$$e^{s'd} = \sum_{k=0}^{n} \tfrac{1}{k!} s'^k d^k = 1 + s'd + \tfrac{1}{2!} s'^2 d^2 + \tfrac{1}{3!} s'^3 d^3 + \tfrac{1}{4!} s'^4 d^4 + \tfrac{1}{5!} s'^5 d^5 + \cdots \tag{3.56}$$

By substituting (3.50) into (3.55), $D_1\left(s'\right)$ becomes

$$D_1\left(s'\right) = \sum_{i=1}^{k} (t_i s'^i q_{(i-1)}), \quad q_0 = 1, \tag{3.57}$$

where

$$q_k = \sum_{i=0}^{k} r_{(k-i)} \left(\frac{1}{i!} d^i\right), \quad k = 0, 1, 2, \ldots, n, \quad r_0 = 1 \tag{3.58}$$

or

$$\begin{bmatrix} q_0 \\ q_1 \\ q_2 \\ q_3 \\ q_4 \\ q_5 \\ \vdots \end{bmatrix} = \begin{bmatrix} 1 \\ r_1 + d \\ r_2 + r_1 d + \frac{1}{2!} d^2 \\ r_3 + r_2 d + \frac{1}{2!} d^2 r_1 + \frac{1}{3!} d^3 \\ r_4 + r_3 d + \frac{1}{2!} d^2 r_2 + \frac{1}{3!} d^3 r_1 + \frac{1}{4!} d^4 \\ r_5 + r_4 d + \frac{1}{2!} d^2 r_3 + \frac{1}{3!} d^3 r_2 + \frac{1}{4!} d^4 r_1 + \frac{1}{5!} d^5 \\ \vdots \end{bmatrix}. \tag{3.59}$$

Polynomials $N(s')$, $D(s') = N(s') + k_h D_1(s')$ are then finally defined by

$$N(s') = \sum_{i=0}^{n} \left[s'^i k_p (z_{(i)} + z_{(i-1)} x + z_{(i-2)} y) \right], \tag{3.60}$$

$$D\left(s'\right) = \sum_{j=0}^{k} s'^j \left[t_i q_{(j-1)} + \left(k_p k_h \left(z_{(j)} + z_{(j-1)} x + z_{(j-2)} y \right) \right) \right] \tag{3.61}$$

where $z_{(-2)} = z_{(-1)} = 0$, $z_0 = 1$. Therefore, the resulting closed loop transfer function is given by (3.62)

$$\begin{aligned} T(s') &= \tfrac{N(s')}{D(s')} \\ &= \frac{\sum_{i=0}^{n} \left[s'^i k_p \left(z_{(i)} + z_{(i-1)} x + z_{(i-2)} y \right) \right]}{\sum_{j=0}^{k} s'^j \left[t_i q_{(j-1)} + \left(k_p k_h \left(z_{(j)} + z_{(j-1)} x + z_{(j-2)} y \right) \right) \right]}. \end{aligned} \tag{3.62}$$

The problem to be solved now for determining the optimal control law is as follows: given measured the parameters of the process $k_{\mathrm{p}},\ z_i,\ q_j$, calculate controller parameters $t_{\mathrm{i}},\ x,\ y,\ k_{\mathrm{h}}$ as a function of $k_{\mathrm{p}},\ z_i,\ q_j$. For doing this, the principle of the Magnitude Optimum criterion is adopted, which is presented in Appendix B.1. There, a general closed loop transfer function is formulated the magnitude of which is forced to be equal to the unity in the widest possible frequency range, $|T(j\omega)| \simeq 1$. Once this is completed, a set of optimization conditions[2] are derived, which comprise the basis for proving the proposed optimal control law.

In Appendix B.1 the optimal control law is proved to be equal to

$$\begin{bmatrix} t_{\mathrm{i}} \\ x \\ y \end{bmatrix} = \begin{bmatrix} 1 & 2k_{\mathrm{p}}k_{\mathrm{h}} & 0 \\ 0 & 1 & -a_{12} \\ 0 & 1 & a_{22} \end{bmatrix}^{-1} \begin{bmatrix} 2k_{\mathrm{p}}k_{\mathrm{h}}\left(q_1 - z_1\right) \\ b_1 \\ b_2 \end{bmatrix} \tag{3.63}$$

where parameters $a_{12},\ a_{22}$ and $b_1,\ b_2$ are equal to

$$a_{12} = \frac{q_1 - z_1}{\left(q_1 - z_1\right) q_1 - \left(q_2 - z_2\right)}, \tag{3.64}$$

$$b_{11} = \frac{\left(q_1^2 - 2q_2\right)\left(q_1 - z_1\right) + q_1 z_2 - q_2 z_1 + q_3 - z_3}{\left(q_1 - z_1\right) q_1 - \left(q_2 - z_2\right)} \tag{3.65}$$

$$a_{22} = \frac{q_1 z_2 - q_2 z_1 + q_3 - z_3}{q_2^2 - 2q_1 q_3 - q_2 z_2 + q_1 z_3 + q_3 z_1 + q_4 - z_4} \tag{3.66}$$

and

$$b_{22} = \frac{Q_0 Q_1 + Q_2}{Q_3} \tag{3.67}$$

and

$$Q_0 = q_2^2 - 2q_1 q_3 + 2q_4 \tag{3.68}$$

$$Q_1 = q_1 - z_1 \tag{3.69}$$

$$Q_2 = q_2 z_3 - q_3 z_2 - q_1 z_4 + q_4 z_1 - q_5 + z_5 \tag{3.70}$$

$$Q_3 = q_2^2 - 2q_1 q_3 - q_2 z_2 + q_1 z_3 + q_3 z_1 + q_4 - z_4. \tag{3.71}$$

Finally, the corresponding I, PI control law can be easily derived in Table 3.1. It is necessary to mention that the new integrator's time constant is equal to $t_{\mathrm{i}} = 2k_{\mathrm{p}}k_{\mathrm{h}}(q_1 - z_1 - x)$ or finally

---

[2] These optimization conditions are between the numerator and the denominator of the closed loop transfer function.

Table 3.1 Optimal control law for type-I control loops

Controller	k_h	t_i	x	y
I	1	$2k_p(r_1 + d - z_1)$	–	–
PI	1	$2k_p(r_1 + d - z_1 - x)$	b_{11}	–
PID	1	$2k_p(r_1 + d - z_1 - x)$	$\frac{a_{11} \quad b_{22} + a_{22} b_{11}}{a_{11} + a_{22}}$	$\frac{b_{22} - \quad b_{11}}{a_{11} + a_{22}}$

$$T_i = 2k_p k_h \left(\underbrace{\sum_{i=1}^{n} T_{p_i} + T_d}_{} - \underbrace{\sum_{i=1}^{m} T_{z_i} - X}_{} \right). \tag{3.72}$$

Another conclusion which is derived from (3.72) is that the integrator's time constant is equal to the sum o the poles of $F_{ol}(s)$ minus the sum of zeros of $F_{ol}(s)$. As a result, necessary condition for the control loop to be controllable is T_i or

$$\sum_{i=1}^{n} T_{p_i} + T_d > \sum_{i=1}^{m} T_{z_i} - X. \tag{3.73}$$

3.4 Performance Comparison Between Conventional and Revised PID Tuning

In this section, a comparison between the conventional Magnitude Optimum design criterion and the revised control law is carried out. Several benchmark processes met over the industry have been chosen. In all cases, we compare both the performance in terms of tracking the reference signal and robustness of the final control loop as far as disturbance rejection is concerned.

Comparison takes place both in the time and frequency domain. Controller's unmodeled dynamics have been chosen equal to $t_{sc} = 0.1$ and all time constants have been normalized by $s' = sT_{p_1}$.

3.4.1 *Plant with One and Two Dominant Time Constants*

Consider the processes defined by

$$G_1(s') = \frac{1}{\prod_{j=1}^{5} (1 + a^{j-1} s')}, \tag{3.74a}$$

$$G_2(s') = \frac{1}{(1+s')^2 \prod_{j=2}^{4}(1+a^j s')} \tag{3.74b}$$

where $a = 0.1$. For controlling G_1, the resulting PI controllers via the conventional and the revised design procedure are given by,

$$C_{\text{con}}(s') = \frac{1+s'}{0.42s'(1+s't_{\text{sc}})}, \tag{3.75a}$$

$$C_{\text{rev}}(s') = \frac{1+1.0035s'}{0.415s'(1+s't_{\text{sc}})}. \tag{3.75b}$$

For controlling G_2, the respective PID controllers are given by

$$C_{\text{con}}(s') = \frac{(1+s')(1+s')}{0.22s'(1+s't_{\text{sc}})}, \quad (t_n = 1,\ t_v = 1), \tag{3.76a}$$

$$C_{\text{rev}}(s') = \frac{1+2s'+1.0024s'^2}{0.218s'(1+s't_{\text{sc}})} \tag{3.76b}$$

$$= \frac{[1+s'(1+0.0224\text{i})][1+s'(1-0.0224\text{i})]}{0.218s'(1+s't_{\text{sc}})}. \tag{3.76c}$$

From (3.75a), (3.75b), (3.76a), (3.76b) and Fig. 3.4 it is apparent that the two control loops, both for PI and PID control law exhibit almost the same behavior regarding reference tracking and output disturbance rejection.

Let it be noted that the revised PID control law has led to a PID controller consisting of conjugate complex zeros with a very close to zero imaginary part. In both cases (PI, PID control) a step disturbance is applied in the input $d_{\text{i}}(s)$ and the output $d_{\text{o}}(s)$ of the process. Disturbance rejection remains the same for both tuning methods (conventional and revised).

3.4.2 Plant with Five Dominant Time Constants

Consider the process defined by

$$G(s') = \frac{1}{(1+s')^5}. \tag{3.77}$$

The conventional and revised PID controllers are given by

$$C_{\text{con}}(s') = \frac{(1+s')(1+s')}{6.2s'(1+s't_{\text{sc}})}, \tag{3.78a}$$

$$C_{\text{rev}}(s') = \frac{1+3.42s'+3.26s'^2}{3.34s'(1+s't_{\text{sc}})} \tag{3.78b}$$

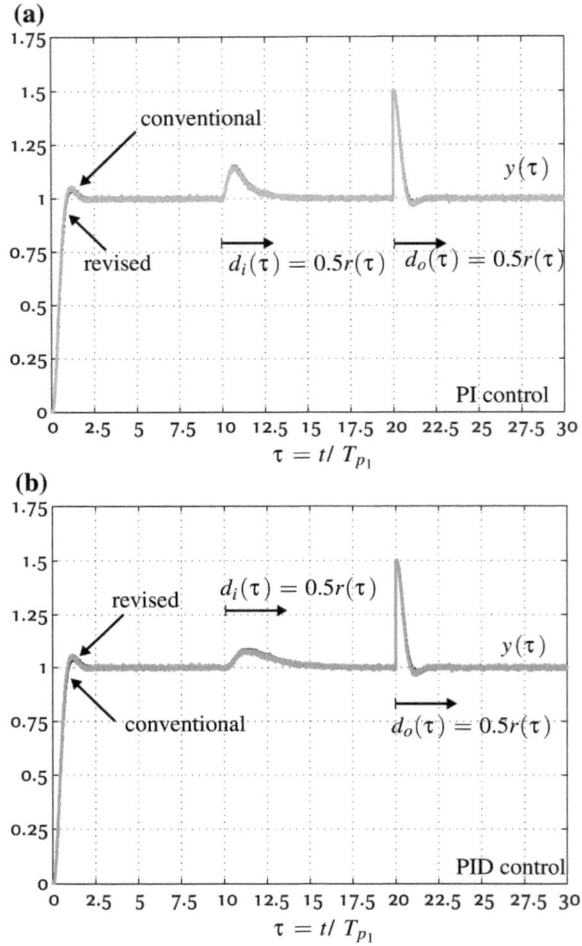

Fig. 3.4 **a** Control of a process with one dominant time constant defined by (3.74a) for $a = 0.1$, PI control. **b** Control of a process with two dominant time constants defined by (3.74b) for $a = 0.1$, PID control. Comparison between the conventional and the revised Magnitude Optimum criterion. Input and output disturbance $d_{\mathrm{i}}(s)$ and $d_{\mathrm{o}}(s)$ are applied at $t = 10\tau$ and $t = 20\tau$ respectively

$$= \frac{\left[1 + s'(1.7 + 0.57\mathrm{i})\right]\left[1 + s'(1.7 - 0.57\mathrm{i})\right]}{3.34s'(1 + s't_{\mathrm{sc}})}. \tag{3.78c}$$

From Fig. 3.5a it is apparent that disturbance rejection has been improved since $t_{\mathrm{ss}} = 110\tau + 21.9\tau$ to $t_{\mathrm{ss}} = 110\tau + 10.6\tau$ (51.6 % decrease) when the PID controller is tuned via the revised method.

Robustness of the control loop has been increased, since in the frequency domain, $|S_{\mathrm{rev}}(ju)| < |S_{\mathrm{con}}(ju)|$ holds by in the lower frequency region, Fig. 3.5b. The cost of this improvement is paid in the overshoot of the output where there has been an increase from 4.65 to 8.07 %. Once more, the revised PID type controller involves conjugate complex zeros in its transfer function.

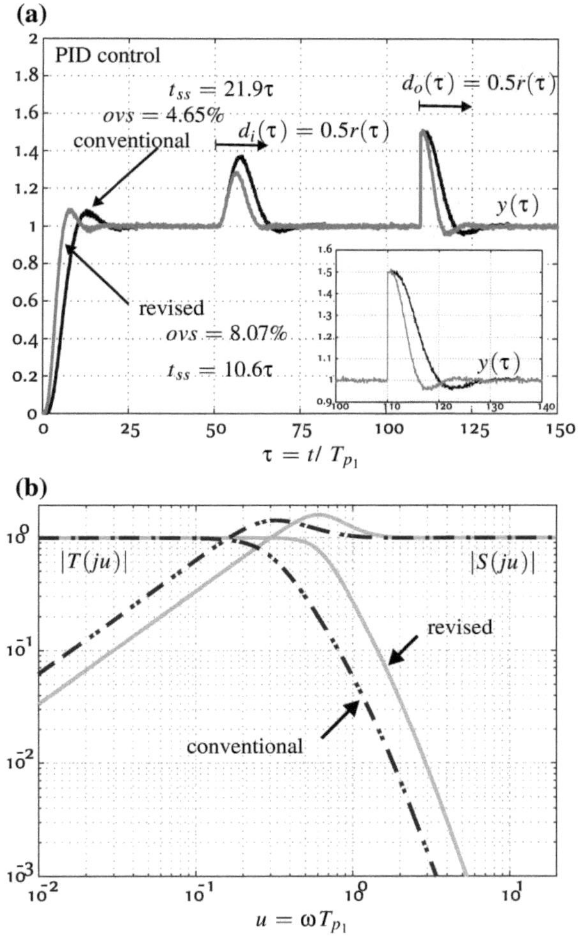

Fig. 3.5 **a** Step response of the control loop. Input and output disturbance $d_{\mathrm{i}}(s)$ and $d_{\mathrm{o}}(s)$ are applied at $t = 50\tau$ and $t = 110\tau$ respectively. **b** Frequency response of complementary sensitivity $T(s) = \frac{y(s)}{r(s)}$ and sensitivity $S(s) = \frac{y(s)}{d_{\mathrm{o}}(s)}$. Control of a process with five dominant time constants defined by (3.77). Comparison between the conventional and the revised Magnitude Optimum criterion

3.4.3 A Pure Time Delay Process

Consider the plant with time delay four times larger than its dominant time constant

$$G(s') = \frac{1}{(1+s')^5}\mathrm{e}^{-4s'}. \tag{3.79}$$

The conventional and the revised PID tuning via the Magnitude Optimum criterion has led to

$$C_{\mathrm{con}}(s') = \frac{(1+s')(1+s')}{14.2s'(1+s't_{\mathrm{sc}})}, \tag{3.80a}$$

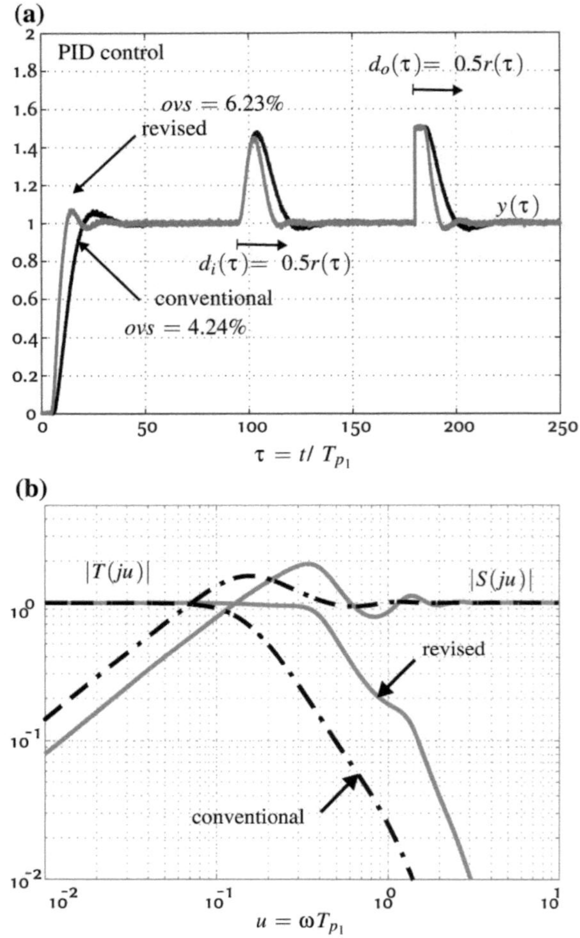

Fig. 3.6 **a** Step response of the control loop. Input and output disturbance $d_i(s)$ and $d_o(s)$ are applied at $t = 90\tau$ and $t = 180\tau$, respectively. **b** Frequency response of complementary sensitivity $T(s) = \frac{y(s)}{r(s)}$ and sensitivity $S(s) = \frac{y(s)}{d_o(s)}$. Control of a process with long time delay defined by (3.79). Comparison between the conventional and the revised Magnitude Optimum criterion

$$C_{rev}(s') = \frac{1 + 5.08s' + 9.22s'^2}{8.02s'(1 + s't_{sc})} \tag{3.80b}$$

$$= \frac{\left[1 + s'(2.5 + 1.66\mathrm{i})\right]\left[1 + s'(2.5 - 1.66\mathrm{i})\right]}{8.02s'(1 + s't_{sc})} \tag{3.80c}$$

respectively. The revised PID controller involves conjugate complex zeros while disturbance rejection has been improved ($t_{ss} = 180\tau + 44.6\tau \rightarrow 180\tau + 23.4\tau$) up to (47.5 % decrease) compared to the standard design, Fig. 3.6a. Let it be noted that $|T_{rev}(ju)| > |T_{con}(ju)|$ holds for a wider band in the lower frequency region as well, Fig. 3.6b.

3.4.4 A Nonminimum Phase Process

Consider the nonminimum phase process defined by

$$G(s') = \frac{(1-0.7s')(1-0.9s')}{(1+s')^5}. \tag{3.81}$$

The respective PID controllers via the conventional and the revised method are defined by

$$C_{\mathrm{con}}(s') = \frac{(1+s')(1+s')}{9.4s'(1+s't_{\mathrm{sc}})}, \tag{3.82a}$$

$$C_{\mathrm{rev}}(s') = \frac{1+3.77s'+4.04s'^2}{5.85s'(1+s't_{\mathrm{sc}})} \tag{3.82b}$$

$$= \frac{\left[1+s'(1.88+0.7\mathrm{i})\right]\left[1+s'(1.88-0.7\mathrm{i})\right]}{5.85s'(1+s't_{\mathrm{sc}})}. \tag{3.82c}$$

The resulting step and frequency responses in terms of disturbance rejection show an improvement of up to 50.6 % as far as t_{ss} is concerned, Fig. 3.7a. Robustness of the control loop has also been improved since output sensitivity $|S_{\mathrm{rev}}(ju)| < |S_{\mathrm{con}}(ju)|$ holds by, in the lower frequency region, Fig. 3.7b.

3.4.5 A Process with Large Zeros

Let us now consider the process defined by

$$G(s') = \frac{(1+1.2s')(1+1.6s')}{(1+s')(1+0.9s')(1+0.8s')(1+0.2s')(1+0.1s')}. \tag{3.83}$$

In that case, there is a loss of controllability both for the revised PI and PID type control law, Fig. 3.8. This is due to the fact that the integral gain becomes negative since large zeros are involved in the process (3.83). This is justified by taking into account that the revised definition of the integral gain is given by

$$T_{\mathrm{i}} = 2k_{\mathrm{p}}k_{\mathrm{h}}\left(\sum_{i=1}^{n} T_{\mathrm{p}_i} + T_{\mathrm{d}} - \sum_{i=1}^{m} T_{\mathrm{z}_i} - X\right), \tag{3.84}$$

from which it is apparent that T_{i} becomes negative ($T_{\mathrm{i}} < 0$) when

$$\sum_{i=1}^{n}(T_{\mathrm{p}_i}) + T_{\mathrm{d}} < \sum_{i=1}^{m}(T_{\mathrm{z}_i}) + X$$

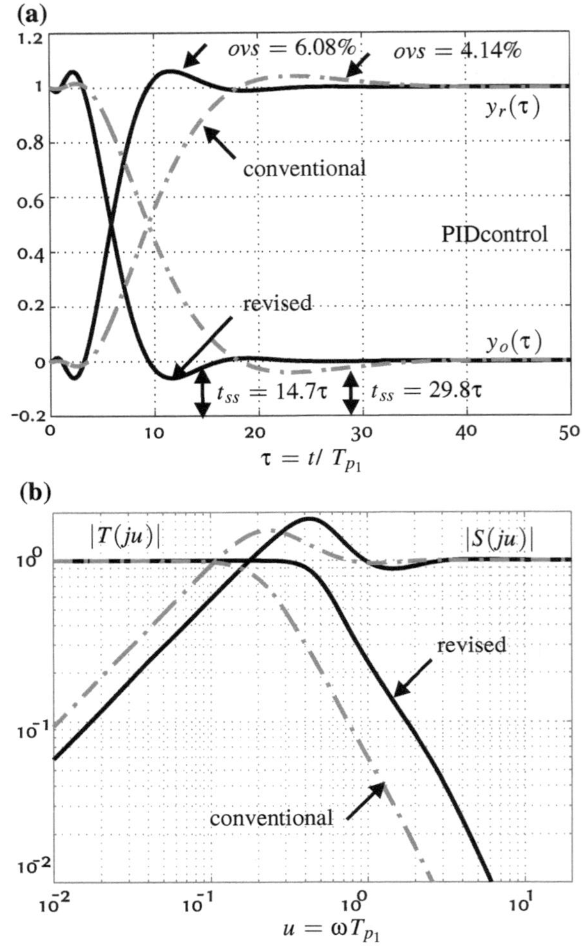

Fig. 3.7 **a** Step response of the closed loop transfer function $T(s) = \frac{y(s)}{r(s)}$ and output disturbance rejection $S(s) = \frac{y(s)}{d_o(s)}$. **b** Frequency response of the closed loop transfer function $T(s) = \frac{y(s)}{r(s)}$ and output disturbance rejection $S(s) = \frac{y(s)}{d_o(s)}$. Control of a nonminimum phase process defined by (3.81). Comparison between the conventional and the revised Magnitude Optimum criterion

More specifically in the case of PI control the integral gain is $t_i = -3.4286$ and in the case of PID control law the integral gain is $t_i = -2.7379$. Note that only I control leads to a stable but still oscillatory control loop, Fig. 3.8. In order to overcome that obstacle, PI and PID control are turned into PI-lag and PID-lag, respectively, by adding a lag time constant t_x in the initial PI, PID controller so that T_i becomes positive again, $T_i > 0$. By choosing a lag time constant $t_x = 5$ ($T_x = 5T_{p_1}$) results in

$$C_{\text{revPI}}(s') = \frac{1 + s'x}{6.5714s'(1 + s't_{sc})} \frac{1}{(1 + 5s')},$$
$$C_{\text{revPID}}(s') = \frac{1 + s'x + s'^2 y}{7.2621s'(1 + s't_{sc})} \frac{1}{(1 + 5s')}. \tag{3.85}$$

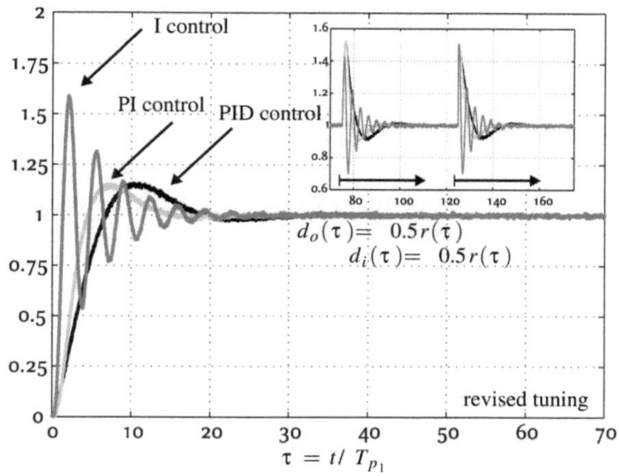

Fig. 3.8 Control of a process with large zeros defined by (3.83). PID type tuning via the Magnitude Optimum criterion. The revised tuning for PI and PID control leads to unstable response because of the negative integral gain. I control leads to stable but oscillatory response. PI, PID control are turned into PI-lag, PID-lag control so that the control loop becomes again controllable

The new integral gain is now defined by

$$T_{\mathrm{i}} = 2k_{\mathrm{p}}k_{\mathrm{h}}\left(\sum_{i=1}^{n} T_{\mathrm{p}_i} + T_{\mathrm{d}} + T_x - \sum_{i=1}^{m} T_{z_i} - X\right) \tag{3.86}$$

while the optimal solutions for X, Y remain the same. The resulting step responses are shown in Fig. 3.8. In conclusion, the revised design procedure can overcome the obstacle of large zeros in a process by turning the PI or PID control law into PI or PID-lag respectively.

3.4.6 Comments on Pole-Zero Cancellation

Let us now consider a simple process of the form G and the PI controller defined by

$$G(s) = \frac{1}{\prod_{i=1}^{5}\left(1 + sT_{\mathrm{p}_i}\right)}, \quad C(s) = \frac{1 + sX}{sT_{\mathrm{i}}\left(1 + sT_{\mathrm{p}6}\right)} \tag{3.87}$$

respectively [3]. Normalizing all time constants by setting $s' = sT_{\mathrm{p}_1}$, (3.87) is rewritten as follows

$$G(s') = \frac{1}{\prod_{i=1}^{5}\left(1 + s't_{p_i}\right)}, \quad C(s') = \frac{1 + s'x}{s't_i\left(1 + s't_{p6}\right)} \tag{3.88}$$

where $t_{p_i} = \frac{T_{\mathrm{p}_i}}{T_{\mathrm{p}_1}}$, $i = 1, \ldots, 5$, $t_{\mathrm{i}} = \frac{T_{\mathrm{i}}}{T_{\mathrm{p}_1}}$ and $x = \frac{X}{T_{\mathrm{p}_1}}$. Substituting again by $t_{p_j} = a^{(j-1)}$, $j = 1, \ldots, 5$ into $G(s')$ we obtain

$$G\left(s'\right) = \frac{1}{\prod_{j=1}^{5}\left(1 + s'a^{(j-1)}\right)}, \quad C\left(s'\right) = \frac{1 + s'x}{s't_{\mathrm{i}}\left(1 + s't_{p6}\right)}. \tag{3.89}$$

If $a < 0.3$ the resulting process consists of a relatively large time constant whereas if $a > 0.8$ the process consists of relatively equivalent dominant time constants. The optimal PI control law proved in Sect. 3.3 results in $x = b_{11}$. Since the class of processes $\forall a$ does not contain any zeros, $z_i = 0$, $i = 1, 2, \ldots, m$, it is concluded from (3.63) that,

$$x = \frac{(q_1^2 - 2q_2)q_1 + q_3}{q_1^2 - q_2} \tag{3.90}$$

where the q_i coefficients are defined in (3.59). Rolling back the q_i coefficients, results in

$$q_1 = \sum_{i=0}^{1} r_{(1-i)}\left(\frac{1}{i!}d^i\right) = r_1 + r_0 d, \tag{3.91}$$

$$q_2 = \sum_{i=0}^{2} r_{(1-i)}\left(\frac{1}{i!}d^i\right) = r_{(2)} + r_1 d + \frac{1}{2!}d^2 r_0, \tag{3.92}$$

$$q_3 = \sum_{i=0}^{3} r_{(3-i)}\left(\frac{1}{i!}d^i\right) = r_3 + r_2 d + \frac{1}{2!}d^2 r_1 + \frac{1}{3!}d^3 r_0. \tag{3.93}$$

Since no delay exists at the output of the plant, $d = 0$, $q_1 = r_1$, $q_2 = r_2$, $q_3 = r_3$. Finally, the optimal x component of the PI control law is equal to

$$x = \frac{(r_1^2 - 2r_2)r_1 + r_3}{r_1^2 - r_2} \tag{3.94}$$

where the r_j coefficients are defined by

$$r_1 = \sum_{i=1}^{6} t_{p_i}, \quad r_2 = \sum_{i \neq j = 1}^{6} t_{p_i} t_{p_j}, \quad r_3 = \sum_{i \neq j \neq k = 1}^{6} t_{p_i} t_{p_j} t_{p_k}.$$

From Fig. 3.9 it is evident that the revised design method via the Magnitude Optimum criterion, also drives the optimal PI controller parameter x to pole-zero cancellation [2] only in case when the process contains one dominant time constant, Sect. 3.4.1. The same result can be proved also for the PID controller, Sect. 3.4.1.

Hence, in cases where the process contains only one or two dominant time constants the revised PI, PID control law leads to pole-zero cancellation, respectively. In any other case, neither the PI nor the PID controller tuning through the revised

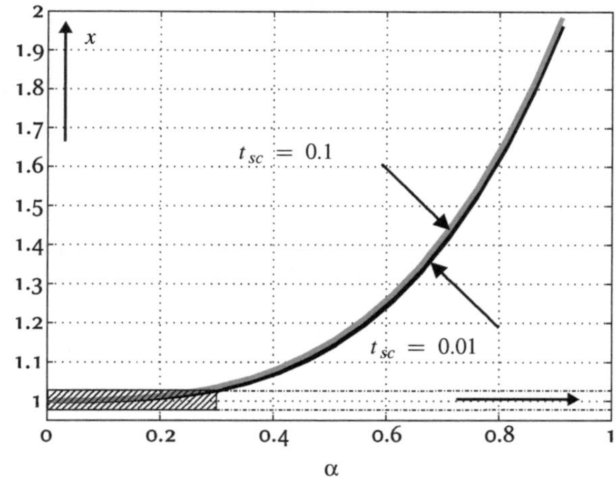

Fig. 3.9 The revised PI control law leads to pole-zero cancellation in cases where the process contains only one dominant time constant, if $a < 0.3$ then $x = \frac{X}{T_{\mathrm{p}1}} \simeq 1$, region of compensation

Magnitude Optimum design criterion do lead to pole-zero cancellation. This is also evident from the examples presented in Sect. 3.4 where the proposed PID controller consists of conjugate complex zeros, not leading finally to pole-zero cancellation.

3.4.7 Comments on Disturbances Rejection

Since in this section comments related to the conventional design method via the Magnitude Optimum criterion are investigated, the process defined by (3.1) and the PID controller defined by (3.28) are considered. The output sensitivity function $S_{\mathrm{o}}(s)$, Fig. 3.1, is then given by

$$S_{\mathrm{o}}(s) = \frac{y(s)}{d_{\mathrm{o}}(s)} = \frac{1}{1 + k_{\mathrm{p}}k_{\mathrm{h}}C(s)G(s)} \tag{3.95}$$

whereas input sensitivity is equal to

$$S_{\mathrm{i}}(s) = \frac{y(s)}{d_{\mathrm{i}}(s)} = k_{\mathrm{p}}G(s)S_{\mathrm{o}}(s) \tag{3.96}$$

and control signal sensitivity is given by whereas input sensitivity is equal to

$$S_u(s) = \frac{u(s)}{d_{\mathrm{o}}(s)} = -k_{\mathrm{h}}S_{\mathrm{o}}(s)C(s). \tag{3.97}$$

Substituting (3.1), (3.28) and (3.35)–(3.38) into S_{o} results in

$$S_{\mathrm{o}}(s) = \frac{y_{d_{\mathrm{o}}}(s)}{d_{\mathrm{o}}(s)} = \frac{N_{y_{d_{\mathrm{o}}}}(s)}{D_{y_{d_{\mathrm{o}}}}(s)}, \tag{3.98}$$

respectively, where

$$N_{y_{d_o}}(s) = 2s\left(\sum_{i=1}^{n} T_{p_i} - T_n - T_v\right)\prod_{i=1}^{n}(1 + sT_{p_i}), \tag{3.99}$$

$$D_{y_{d_o}}(s) = \left[2s\left(\sum_{i=1}^{n} T_{p_i} - T_n - T_v\right)\prod_{i=3}^{n}(1 + sT_{p_i}) + 1\right] \times (1 + sT_{p_1})(1 + sT_{p_2}). \tag{3.100}$$

Additionally, we have

$$S_i(s) = \frac{y_{d_i}(s)}{d_i(s)} = \frac{N_{y_{d_i}}(s)}{D_{y_{d_i}}(s)} \tag{3.101}$$

where

$$N_{y_{d_i}}(s) = 2k_p s\left(\sum_{i=1}^{n} T_{p_i} - T_n - T_v\right), \tag{3.102}$$

$$D_{y_{d_i}}(s) = \left[2s\left(\sum_{i=1}^{n} T_{p_i} - T_n - T_v\right)\prod_{i=3}^{n}(1 + sT_{p_i}) + 1\right] \times (1 + sT_{p_1})(1 + sT_{p_2}). \tag{3.103}$$

Finally,

$$S_u(s) = \frac{u_{d_o}(s)}{d_o(s)} = \frac{N_{u_{d_o}}(s)}{D_{u_{d_o}}(s)} \tag{3.104}$$

$$N_{u_{d_o}}(s) = -(1 + sT_{p_1})(1 + sT_{p_2})\prod_{i=1}^{n}(1 + sT_{p_i}) \tag{3.105}$$

$$D_{u_{d_o}}(s) = k_p\left[2s\left(\sum_{i=1}^{n} T_{p_i} - T_n - T_v\right)\prod_{i=3}^{n}(1 + sT_{p_i}) + 1\right] \times (1 + sT_{p_1})(1 + sT_{p_2}). \tag{3.106}$$

According to [2], pole-zero cancellation may lead to poor rejection of load and input disturbances [12], if the compensated modes are excited by disturbances, especially if they are slow compared to the dominant closed-loop poles. K.J. Åström and T. Hägglund discovered the above drawbacks of the pole-zero cancellation by examining the tuning method of [13] and extended their conclusion to other methods such as the Internal Model Control [20], and the Magnitude Optimum design criterion [21, 22]. K.J. Åström and T. Hägglund attribute the poor rejection of load disturbance on the loss of the system controllability for the specific modes.

In (3.98), along with (3.99), and (3.100) it is observed that indeed, there is a pole-zero cancellation for the compensated time constants and the loss of controllability is possibly justified. On the contrary, as observed in (3.101), in the case of input disturbances a pole-zero cancellation does not occur. Therefore, in this case the loss of controllability is not justified. For the verification of the correctness of this belief, let us examine the sensitivity functions of the closed loop system, by imposing disturbances of the form (3.107),

$$d_{\mathrm{i}}(s) = d_{\mathrm{o}}(s) = \frac{T_{\mathrm{p}j}}{1 + sT_{\mathrm{p}j}}, \quad j = 1, 2, \ldots, n. \tag{3.107}$$

3.4.8 Rejection of Output Disturbances

Substituting (3.107) into (3.98), (3.99) and (3.100) respectively, results in

$$y_{d_{\mathrm{o}}}(s) = \frac{N_{y_{d_{\mathrm{o}}}}}{D_{y_{d_{\mathrm{o}}}}} \frac{T_{\mathrm{p}j}}{(1 + sT_{\mathrm{p}j})} = \frac{N_{1d_{\mathrm{o}}}(s)}{D_{1d_{\mathrm{o}}}(s)} \tag{3.108}$$

where

$$N_{1d_o}(s) = 2sT_{\mathrm{p}j}\left(\sum_{i=1}^{n} T_{\mathrm{p}i} - T_{\mathrm{n}} - T_{\mathrm{v}}\right)\left[\prod_{i=1}^{n}(1 + sT_{\mathrm{p}i})\right] \tag{3.109}$$

$$D_{1d_{\mathrm{o}}}(s) = \left[2s\left(\sum_{i=1}^{n} T_{\mathrm{p}i} - T_{\mathrm{n}} - T_{\mathrm{v}}\right)\left[\prod_{i=1}^{n}(1 + sT_{\mathrm{p}i})\right] + 1\right](1 + sT_{\mathrm{p}j}) \tag{3.110}$$

while

$$u_{d_{\mathrm{o}}}(s) = \frac{N_{u_{d_{\mathrm{o}}}}}{D_{u_{d_{\mathrm{o}}}}} \frac{T_{\mathrm{p}j}}{(1 + sT_{\mathrm{p}j})} = \frac{N_{2d_{\mathrm{o}}}(s)}{D_{2d_{\mathrm{o}}}(s)} \tag{3.111}$$

and

$$N_{2d_o}(s) = -T_{\mathrm{p}j}(1 + sT_{\mathrm{p}1})(1 + sT_{\mathrm{p}2})\prod_{i=1}^{n}(1 + sT_{\mathrm{p}i}) \tag{3.112}$$

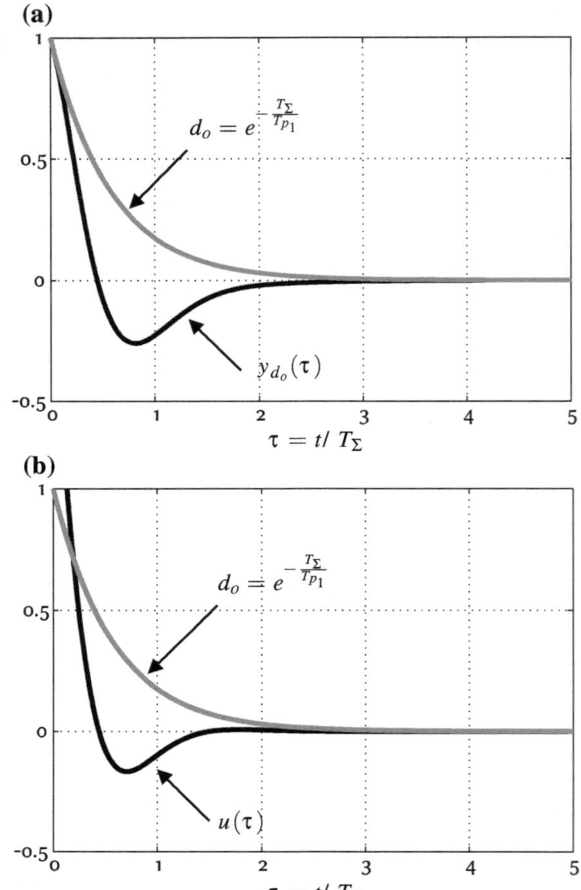

Fig. 3.10 **a** $y_{d_o}(\tau)$ response to output disturbances. **b** $u(\tau)$ response to output disturbances. The disturbance excites a canceled slow mode

$$D_{2d_o}(s) = \left\{ \begin{array}{l} k_p[2s\left(\sum_{i=1}^{n} T_{p_i} - T_n - T_v\right)\prod_{i=3}^{n}(1+sT_{p_i})+1] \\ \times(1+sT_{p_1})(1+sT_{p_2}) \end{array} \right\}(1+sT_{p_j}). \quad (3.113)$$

Figs. 3.10a, b, and 3.11a, b present the responses $y_{d_o}(\tau)$ and $u(\tau)$ to the output disturbance (3.107). In Fig. 3.10a, b the disturbance excites the compensated dominant time constant T_{p_1}.

Let it be noted that in these cases no poor rejection of the output disturbance or loss of controllability is observed, respectively. In Fig. 3.11a, b the disturbance excites the relatively small uncompensated time constant T_{p_3}. It is observed that the system behavior remains the same, as in the case of the compensated time constant. On the contrary, poor rejection of the load disturbance is observed, when the integration time constant has not been correctly tuned, as shown in Figs. 3.12a, b, and 3.13a, b.

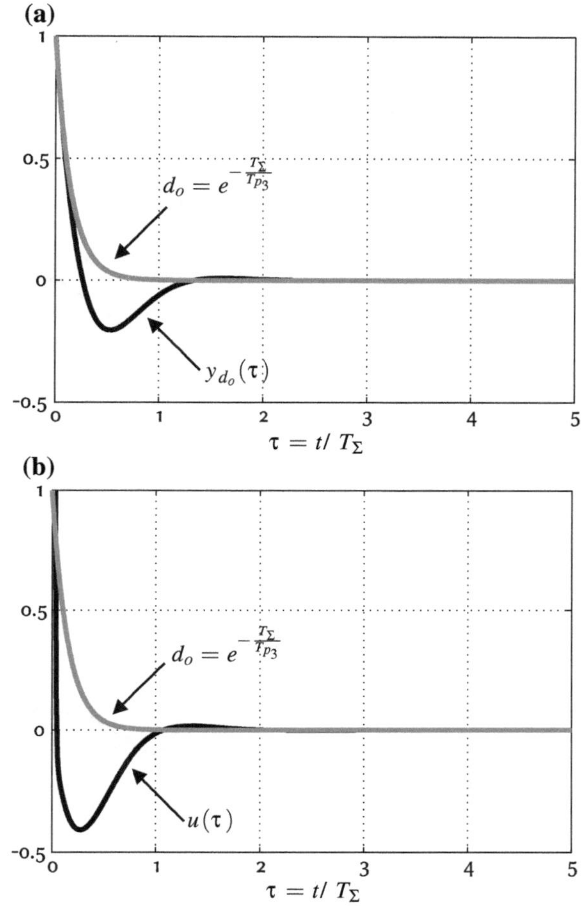

Fig. 3.11 **a** $y_{d_o}(\tau)$ response to output disturbances. **b** $u(\tau)$ response to output disturbances. The disturbance excites an uncanceled fast mode

As observed in Figs. 3.12b, and 3.13b if we violate the optimal control law through incorrect tuning of the integration time constant, the control input $u(\tau)$ is kept almost constant and consequently the system appears as uncontrollable. Let it be noted that this behavior appears independently of the excitation of a compensated or uncompensated mode. These results seem to agree with the statement of K.J. Åström and T. Hägglund, that "the attenuation of load disturbance is improved considerably by reducing the integral time of the controller" [2].

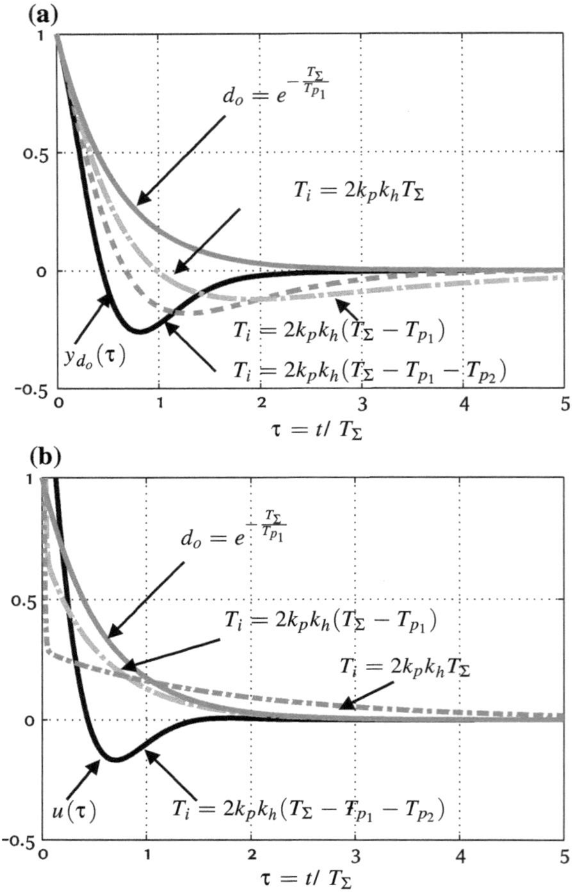

Fig. 3.12 **a** $y_{d_o}(\tau)$ response to output disturbances. **b** $u(\tau)$ response to output disturbances. The rejection of the output disturbances becomes poor when the integration time constant is not adjusted correctly. The disturbance excites a canceled slow mode

3.4.9 Rejection of Input Disturbances

The same applies also in the case of input disturbance rejection. Substituting (3.107) and (3.101) takes the form of

$$y_{d_i}(s) = \frac{N_{y_{d_i}}}{D_{y_{d_i}}} \frac{T_{p_j}}{(1 + sT_{p_j})} = \frac{N_{3d_i}(s)}{D_{3d_i}(s)} \quad (3.114)$$

where

$$N_{3d_i}(s) = 2k_p T_{p_j} (\sum_{i=1}^{n} T_{p_i} - T_n - T_v)s \quad (3.115)$$

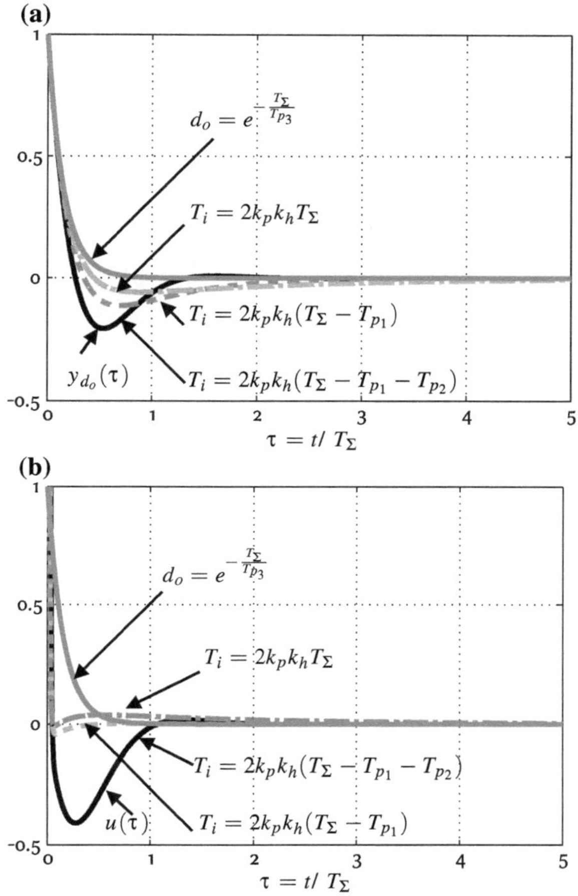

Fig. 3.13 **a** $y_{d_o}(\tau)$ response to output disturbances. **b** $u(\tau)$ response to output disturbances. The rejection of the output disturbances becomes poor when the integration time constant is not adjusted correctly. The disturbance excites an uncanceled fast mode

$$D_{3d_i}(s) = \left[2s\left(\sum_{i=1}^{n} T_{p_i} - T_n - T_v\right)\prod_{i=3}^{n}(1 + sT_{p_i}) + 1\right] \times (1 + sT_{p_1})(1 + sT_{p_2})(1 + sT_{p_j}) \tag{3.116}$$

As shown in Fig. 3.14a, b the rejection of the input disturbance is not poor, whether the disturbance excites the compensated dominant time constant T_{p_1}, or the uncompensated time constant T_{p_3} respectively.

On the contrary, poor attenuation of the disturbance occurs again, when the integration time constant has not been properly tuned, as shown in Fig. 3.15a, b. Again the attenuation of load disturbances is improved considerably by reducing the integral time of the controller [2]. From the analysis presented in that section, it is evident that the pole-zero cancellation does not lead to poor disturbances rejection. On the contrary, poor disturbance rejection is caused by incorrect tuning of the integration

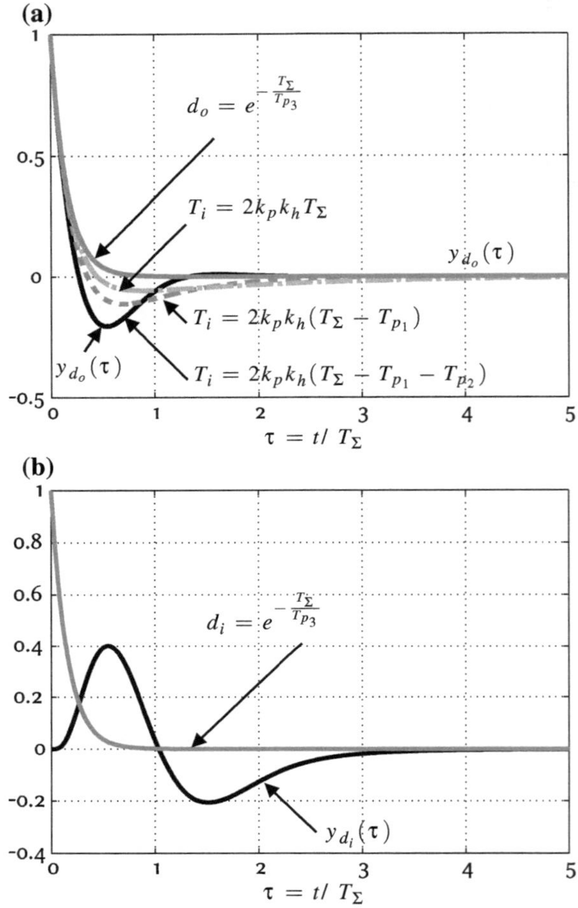

Fig. 3.14 **a** The disturbance excites a canceled slow mode. **b** The disturbance excites an uncanceled fast mode. Response to input disturbance

time constant of the PID controller and not by the pole-zero cancellation tuning method.

3.4.10 Robustness to Model Uncertainties

During the operation of the control system it is possible that some of the system parameters vary, see [1]. In this section, the effect of the parameters variation on the dynamics of the closed loop system is examined. For that reason, at the time of the tuning the PID controller is assumed to have the following form

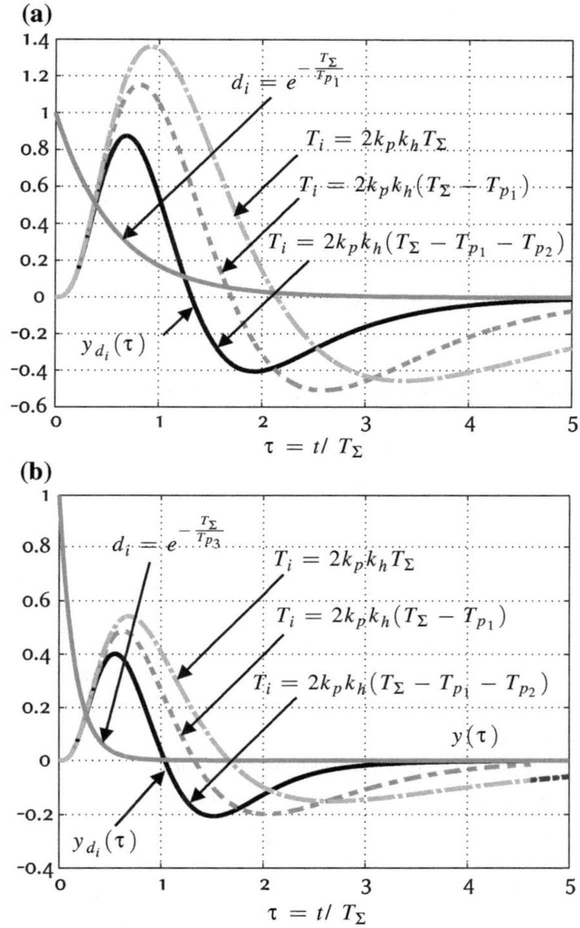

Fig. 3.15 **a** The disturbance excites a canceled slow mode. **b** The disturbance excites an uncanceled fast mode. The rejection of the input disturbances becomes poor when the integration time constant is not adjusted correctly

$$C(s) = \frac{(1 + sT_{n0})(1 + sT_{v0})}{2k_{p0}k_{h0}s(\sum_{i=1}^{n} T_{p_i} - T_{n0} - T_{v0})(1 + sT_{p_n})}. \tag{3.117}$$

where T_{n0}, T_{v0}, $T_{\Sigma 2_0} = \sum_{i=1}^{n} T_{p_i} - T_{n0} - T_{v0}$, k_{p0}, k_{h0}, are the nominal values of the system parameters, see (3.35)–(3.38). Using (3.117), the closed loop transfer function (3.29) obtains the form,

$$T(s) = \frac{k_p(1 + sT_{n0})(1 + sT_{v0})}{\begin{bmatrix} 2k_{p0}k_{h0}T_{\Sigma 2_0}s(1 + sT_{p1})(1 + sT_{p2})(1 + sT_{\Sigma 2}) \\ + k_p k_h(1 + sT_{n0})(1 + sT_{v0}) \end{bmatrix}}. \tag{3.118}$$

3.4.10.1 Variations of Feedback Path

Assuming that parameter k_{h} varies, so that $k_{\mathrm{h}} = (1 + a)k_{\mathrm{h}_0}$ while all other parameters retain their nominal values (3.118) becomes

$$T(s) = \frac{1}{2T_{\Sigma_{2_0}}^2 s^2 + 2T_{\Sigma_{2_0}} s + 1 + a}. \tag{3.119}$$

From (3.119), it is apparent that a variation of the parameter k_{h} manifests with a steady state position error.

3.4.10.2 Variations of Plant's DC Gain

Assuming that only parameter k_{p} varies, so that $k_{\mathrm{p}} = (1+b)k_{\mathrm{p}0}$, while all other parameters retain their nominal values, the closed loop transfer function (3.29) becomes

$$T(s) = \frac{1 + b}{2T_{\Sigma_{2_0}}^2 s^2 + 2T_{\Sigma_{2_0}} s + 1 + b}. \tag{3.120}$$

Figure 3.16, shows the step responses of the nominal ($b = 0$) and modified system. It is apparent that variations of parameter k_{p} cause variations on the overshoot, but for variation up to 20 %, the settling time remains practically unchanged. Moreover, from Fig. 3.16 it becomes obvious that a variation less than 10 % in k_{p} does not have a significant effect on the system response.

Therefore, system response cannot be considered unacceptable while modifications to parameter k_{p} take place. Let it be noted that in vector controlled induction motor drives and when carrier based modulation methods are adopted, k_{p} stands for the pulse width modulator gain when the modulator is modulating in its linear region. Variations of k_{p} can take place in cases when modulation enters the so called non-linear region.

3.4.10.3 Variations of the Plant's Dominant Time Constant

Assuming that only the dominant time constant $T_{\mathrm{p}1}$ varies, so that $T_{\mathrm{p}1} = (1 + c)T_{\mathrm{p}1_0}$, the closed loop transfer function (3.29) takes the form

$$T(s) = \frac{sb_1 + b_0}{a_3 s^3 + a_2 s^2 + a_1 s + a_0} \tag{3.121}$$

where $a_0 = b_0 = 1$ and

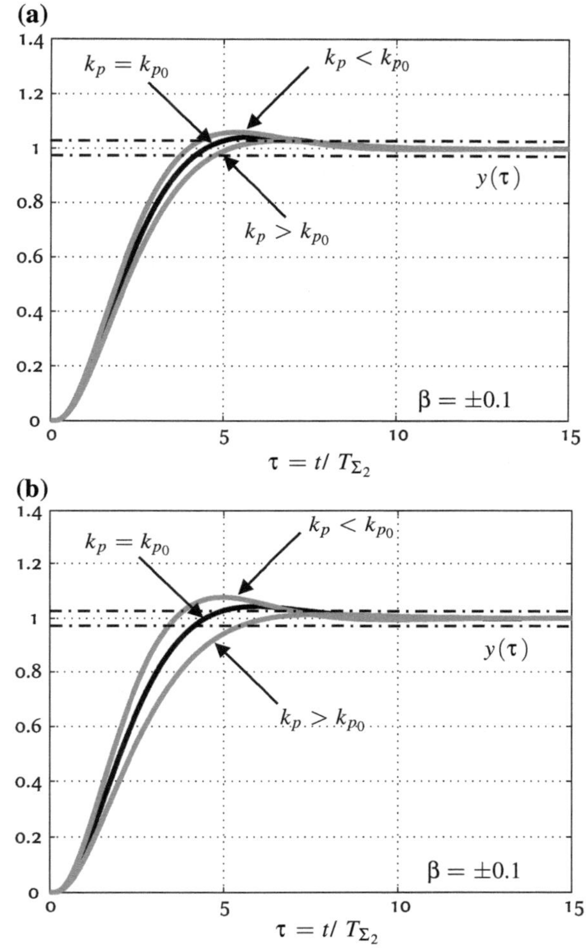

Fig. 3.16 **a** 10 % variation of the process DC gain. **b** 20 % variation of the process DC gain. Effect of variation of parameter k_p. PID controller remains tuned with the nominal values while variations occur in the process's DC gain

$$a_3 = 2(1+c)\frac{T_{p1_0}}{T_{\Sigma 2_0}}T^3_{\Sigma 2_0}, \quad a_2 = 2\left[1+\frac{T_{p1_0}}{T_{\Sigma 2_0}}(1+c)\right]T^2_{\Sigma 2_0}$$

$$a_1 = \left(2+\frac{T_{p1_0}}{T_{\Sigma 2_0}}\right)T_{\Sigma 2_0}, \quad b_1 = \frac{T_{p1_0}}{T_{\Sigma 2_0}}T_{\Sigma 2_0}.$$

From the step responses of the nominal ($c = 0$) and modified system, presented in Fig. 3.17a, it is concluded that the variation of the dominant time constant T_{p_1} manifests with a variation of the overshoot and mainly of the settling time. Moreover, as shown in Fig. 3.17b, variations of the dominant time constant T_{p_1} less than 10 % have no effect on the system response.

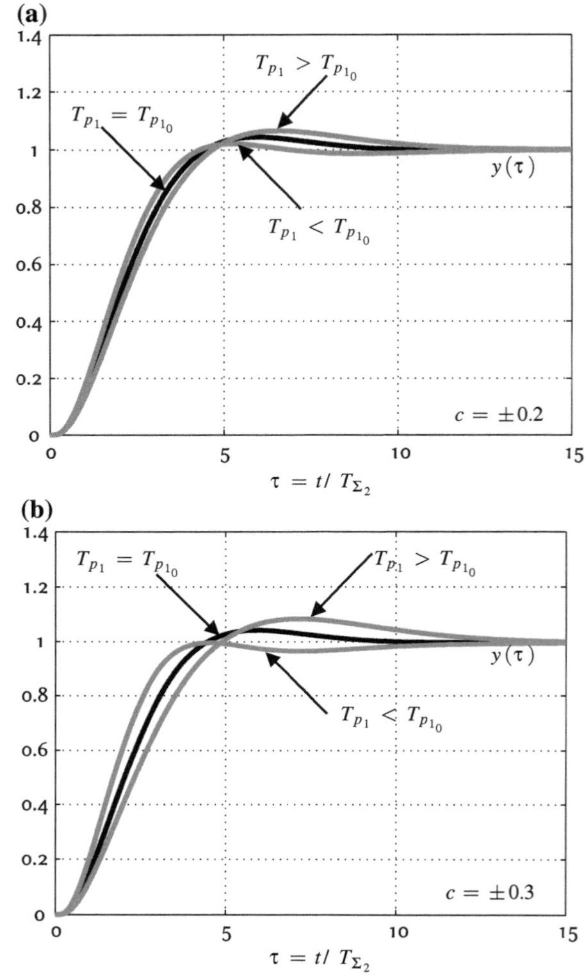

Fig. 3.17 **a** Dominant time constant T_{p_1} varies up to $\pm 20\,\%$. **b** Dominant time constant T_{p_1} varies up to $\pm 30\,\%$. Effect of variation of parameter T_{pl} affect the rise and settling time of the optimal closed loop control system

3.5 Performance Comparison Between Revised PID Tuning and Other Methods

In this section a performance comparison analysis is presented between the revised PID tuning rules via the Magnitude Optimum criterion and two methods commonly used over many industry applications; the Internal Model Control principle (IMC) and the Ziegler–Nichols step response method. The analysis focuses on the time domain and the response of the control loop to reference changes, input, and output disturbance rejection is observed. Within all examples, the three aforementioned

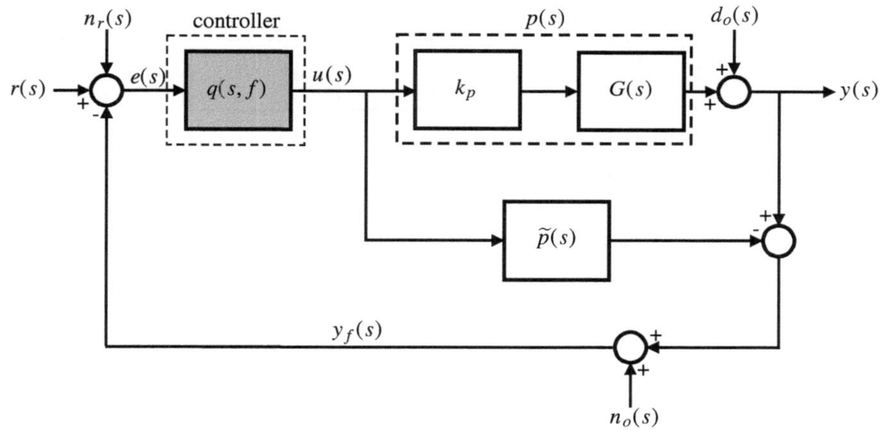

Fig. 3.18 The internal model control (IMC) principle. $G(s)$ is the plant transfer function, $C(s)$ is the controller transfer function, $r(s)$ is the reference signal, $d_\mathrm{o}(s)$ and $d_\mathrm{i}(s)$ are the output and input disturbance signals respectively and $n_\mathrm{r}(s)$, $n_\mathrm{o}(s)$ are the noise signals at the reference input and process output respectively. k_p stands for the plant's DC gain and k_h is the feedback path. $\widetilde{p}(s)$ is the approximated model of $k_\mathrm{p}G(s)$ coming out of an open loop experiment, measurements etc

methods are used for regulating the same process and three curves are presented in each figure, in Sect. 3.5.3.

3.5.1 Internal Model Control

The principle of Internal Model Control is presented in Fig. 3.18. Note that $p(s)$ stands for the real process and $\widetilde{p}(s)$ stands for an approximate model of the process. From Fig. 3.18 it is easily proved that the structure of Fig. 3.18 can be transformed to the one presented in Fig. 3.19. Therefore, based on Fig. 3.19 the following transfer functions can be defined.

- **Controller transfer function**

$$C(s) = \frac{u(s)}{e(s)} = \frac{q(s)}{1 - q(s)\widetilde{p}(s)}. \tag{3.122}$$

- **Closed loop transfer function**

$$T(s) = \frac{y(s)}{r(s)} = \frac{p(s)C(s)}{1 + p(s)C(s)} = \frac{p(s)q(s)}{1 + q(s)\left(p(s) - \widetilde{p}(s)\right)}. \tag{3.123}$$

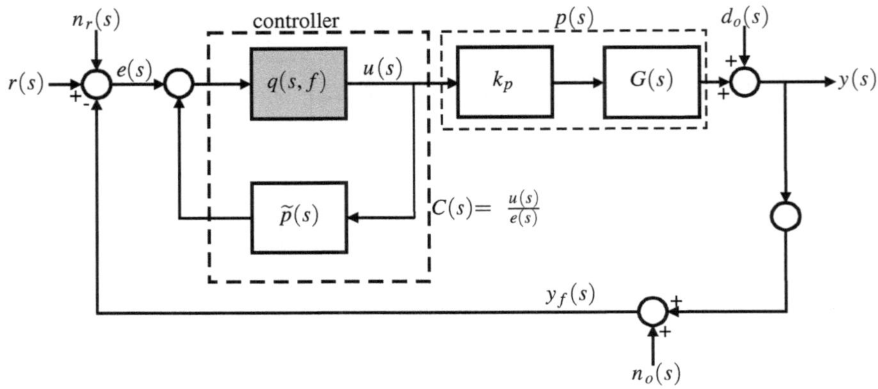

Fig. 3.19 Equivalent diagram of the internal model control principle

- **Output sensitivity or sensitivity function**

$$S_o(s) = \frac{y(s)}{d_o(s)} = \frac{1}{1 + p(s)C(s)}. \tag{3.124}$$

- **Control (command) signal sensitivity function**

$$S_u(s) = \frac{u(s)}{d_o(s)} = -\frac{C(s)}{1 + p(s)C(s)} = -S_o(s)C(s). \tag{3.125}$$

From (3.123) and (3.124) it is apparent that

$$y(s) = \frac{p(s)q(s)}{1 + (p(s) - \widetilde{p}(s))\, q(s)} r(s), \tag{3.126}$$

and

$$y(s) = \frac{1 - p(s)q(s)}{1 + (p(s) - \widetilde{p}(s))\, q(s)} d_o(s). \tag{3.127}$$

According to [5, 20], goal of the 'ideal control action' is to make the output $y(s)$ of the control loop track 'perfectly' its reference signal $r(s)$ and suppress 'perfectly' output disturbances. Those two goals can be interpreted mathematically by

$$T(s) = \frac{y(s)}{r(s)} = 1 \tag{3.128}$$

$$S_o(s) = \frac{y(s)}{d_o(s)} = 0. \tag{3.129}$$

From (3.128) it is easily seen that

$$p(s)q(s) = 1 \tag{3.130}$$

whereas from (3.131) it is found that

$$p(s) \approx \widetilde{p}(s). \tag{3.131}$$

According to Figs. 3.1 and 3.18 it is apparent that

$$p(s) \equiv k_{\mathrm{p}}G(s), \quad q(s) \equiv C(s). \tag{3.132}$$

Therefore, for a process described by

$$k_{\mathrm{p}}G(s) = k_{\mathrm{p}}\frac{1}{1+sT_{\mathrm{p}1}}\mathrm{e}^{-sT_{\mathrm{d}}} \tag{3.133}$$

the inverse $\left(k_{\mathrm{p}}G(s)\right)^{-1}$ transfer function is given by

$$\left(k_{\mathrm{p}}G(s)\right)^{-1} = \frac{1+sT_{\mathrm{p}1}}{k_{\mathrm{p}}}\mathrm{e}^{sT_{\mathrm{d}}}. \tag{3.134}$$

Of course, an implementation of the controller $q(s)$ determined by (3.134) is not feasible. For that reason, as proposed in [5, 20] the form of the implemented controller is given in this case by

$$C(s) = \frac{1+sT_{\mathrm{p}1}}{k_{\mathrm{p}}}\frac{1}{1+sf} \tag{3.135}$$

for which parameter f is chosen such that modeling errors in the approximated model are corrected. In the general case where the real and approximated plant transfer function are defined by

$$G(s) = \frac{N(s)}{D(s)}\mathrm{e}^{-sT_{\mathrm{d}}} \tag{3.136}$$

and

$$\widetilde{G}(s) = \frac{N'(s)}{D'(s)}\mathrm{e}^{-sT_{\mathrm{d}}} \tag{3.137}$$

respectively, the proposed controller according to [5, 20] depends on the characteristics of polynomial $N'(s)$. In this case, the controller's transfer function is given by

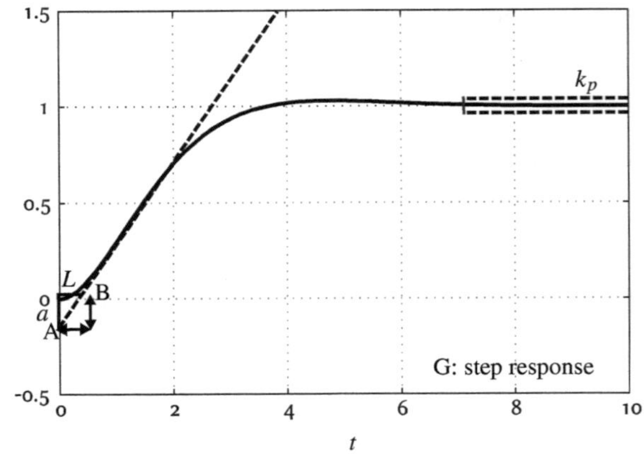

Fig. 3.20 The Ziegler–Nichols step response method

$$C_{\mathrm{IMC}}(s) = \frac{D'(s)}{N'(s)} \frac{1}{(1+fs)^r}. \tag{3.138}$$

Parameter r is named with the term 'relative order' and is equal to order of $D'(s)$ minus the order of $N'(s)$.

3.5.2 Ziegler–Nichols Step Response Method

Over the literature, two are the methods proposed by Ziegler and Nichols regarding the tuning of the PID controller. One is called the "step response method" or "process reaction curve" whereas the other one is called the "frequency response" method. A basic disadvantage of the frequency response tuning principle, which makes the method not attractive in many industry applications, is the fact that for tuning the PID controller parameters, the plant must be brought into a state where its output $y(t)$ is oscillating with constant frequency ω.

Once this frequency is measured, the PID controller parameters are determined out of expressions, which involve this frequency. On the other hand, the "step response" method introduced by Ziegler and Nichols requires three steps to determine the PID controller parameters. These steps are presented below

1. Process open loop experiment (step response of the process).
2. Calculation of the point $(t,\ y(t))$ where the maximum slope of the step response exists, see Fig. 3.20.
3. Determine values $a,\ L$ as those presented in Fig. 3.20.

From step 2, and once the point $t_1,\ y_1$ at which the maximum slope λ exists is calculated, the slope presented in Fig. 3.20 can be drawn. This slope has the form

$$y = \lambda t + b \tag{3.139}$$

where b is equal to

$$b = y_1 - \lambda t_1 \tag{3.140}$$

From (3.139), the values of $a,\ L$ can be easily calculated. These two values are used for determining the PID controller parameters according to Table 3.2.

3.5.3 Simulation Results

Three different processes investigated in Sect. 3.4 are controlled under the revised PID tuning rules via the Magnitude Optimum criterion, the IMC principle and the Ziegler–Nichols step response method. The response of the output $y(\tau)$ and the command signal $u(\tau)$ is presented. Input $d_i(s)$ and output $d_o(s)$ disturbances (step change) are also applied during the control loop's operation.

Since the IMC principle requires an approximation $\widetilde{G}(s)$ of the real process $G(s)$, and in order to have a fair comparison between the three methods, the PID controller in the case of the Magnitude Optimum principle and the Ziegler–Nichols method is tuned via an approximated model $\widetilde{G}(s)$. The control loop has been normalized with the real plant's dominant time constant, $s' = sT_{p1}$.

3.5.3.1 Plant with Five Dominant Time Constants

Consider the real process defined by

$$G(s') = \frac{1}{(1+s')^5} \tag{3.141}$$

and the approximated process defined by

$$\widetilde{G}(s') = \frac{1}{(1+s')(1+0.9s')(1+0.88s')(1+0.7s')(1+0.68s')}. \tag{3.142}$$

In this case, the PID controller regarding the revised Magnitude Optimum method and the Ziegler–Nichols step response method are tuned based on (3.142) whereas the resulting control law is applied to the real process (3.141).

Table 3.2 PID tuning formulas based on the Ziegler–Nichols step response method

Controller	k_h	K	T_i	T_d
I	1	$\frac{1}{a}$	–	–
PI	1	$\frac{0.9}{a}$	$3L$	–
PID	1	$\frac{1.2}{a}$	$2L$	$\frac{L}{2}$

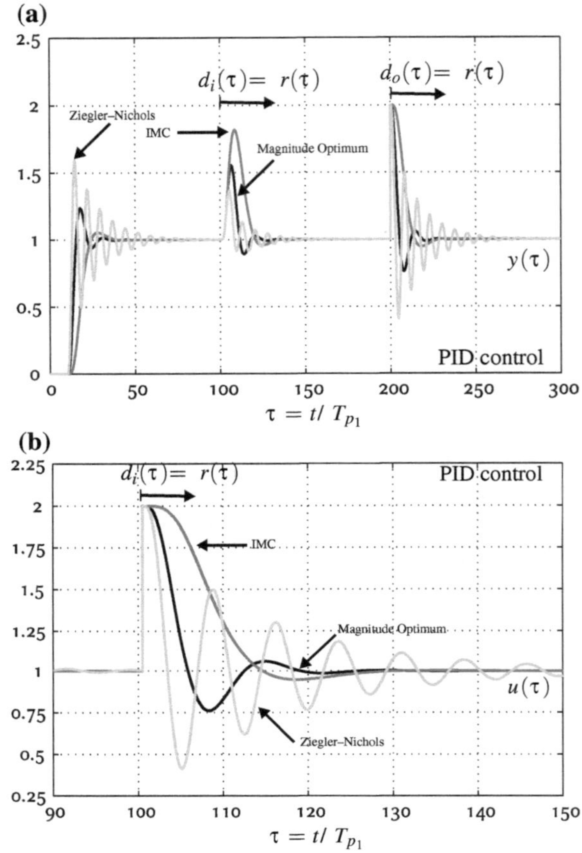

Fig. 3.21 **a** Step response of the control loop. Input and output disturbance $d_{\mathrm{i}}(s)$ and $d_{\mathrm{o}}(s)$ are applied at $t = 100\tau$ and $t = 200\tau$ respectively. **b** Response of the command signal at the presence of input disturbance $d_{\mathrm{i}}(\tau)$. Control of a process with five dominant time constants defined by (3.141). PID control action: comparison between the revised Magnitude Optimum criterion, the Internal Model Control principle (IMC, $f = 1.5$) and the Ziegler–Nichols step response method. All PID controllers are tuned based on the approximate model defined by (3.142)

From Fig. 3.21a it is apparent that the Ziegler–Nichols step response method leads to an oscillatory step response with an undesired overshoot of $\approx$60 %, which is resulted by the aggressive command signal as shown in Fig. 3.21b. On the contrary, the same method shows the minimum peak value regarding the input disturbance rejection $d_{\mathrm{i}}(s)$ and almost the same settling time compared with the other two methods.

The IMC tuning leads to a satisfactory overshoot of ($\approx$4 %) compared with the revised PID tuning method via the Magnitude Optimum criterion ($\approx$21 %) and as far as the step response is concerned. Finally, output disturbance rejection is considered acceptable only in the case of IMC, since its settling time and undershoot exhibit the minimum values with respect to the other two methods, see Fig. 3.22.

By selecting a different time constant $f = 0.5$ in the IMC tuning principle, the step response of the control loop becomes faster and so does input and output disturbance rejection. In this case the control loop in the case of internal model control outperforms the revised PID tuning method via the Magnitude Optimum criterion, see Fig. 3.22.

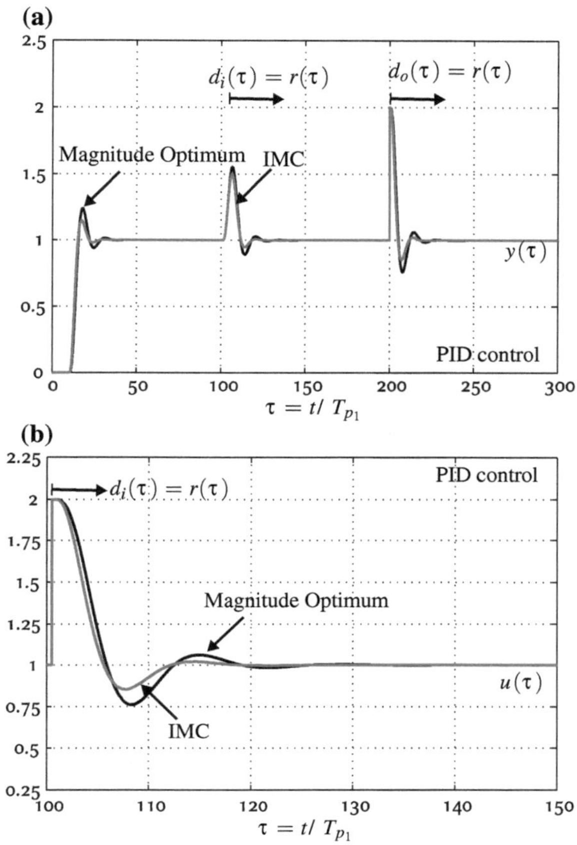

Fig. 3.22 **a** Step response of the control loop. Input and output disturbance $d_i(s)$ and $d_o(s)$ are applied at $t = 100\tau$ and $t = 200\tau$ respectively. **b** Response of the command signal at the presence of input disturbance $d_i(\tau)$. Control of a process with five dominant time constants defined by (3.141). PID control action: comparison between the revised Magnitude Optimum criterion,1 the Internal Model Control principle (IMC, $f = 0.5$) and the Ziegler–Nichols step response method. All PID controllers are tuned based on the approximate model defined by (3.142)

3.5.3.2 A Pure Time Delay Process

Consider the plant with time delay four times larger than its dominant time constant

$$G(s') = \frac{1}{(1+s')^5} e^{-4s'}, \tag{3.143}$$

and the approximated process given by

$$G(s') = \frac{1}{(1+s')(1+0.95s')(1+0.8s')(1+0.75s')(1+0.7s')} e^{-3.5s'}. \tag{3.144}$$

In this case, the Ziegler–Nichols step response method gives an unstable response, and this is why it is not depicted either in Figs. 3.23 and 3.24. In Fig. 3.23a, b the filter time constant has been chosen equal to $f = 3$ and slow disturbance rejection is

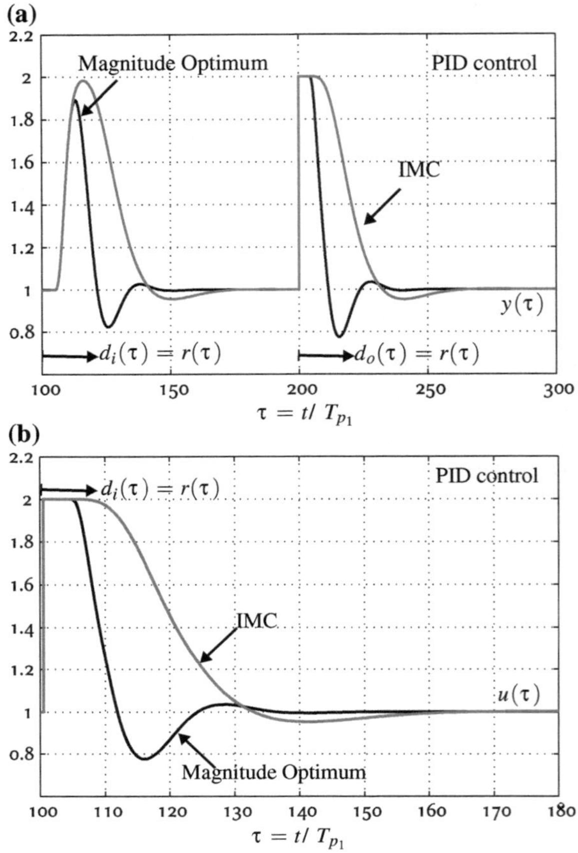

Fig. 3.23 **a** Step response of the control loop. Input and output disturbance $d_i(s)$ and $d_o(s)$ are applied at $t = 100\tau$ and $t = 200\tau$, respectively. **b** Response of the command signal at the presence of input disturbance $d_i(\tau)$. Control of a process with long time delay defined by (3.143). PID control action: comparison between the revised Magnitude Optimum criterion, the Internal Model Control principle (IMC, $f = 3$) and the Ziegler–Nichols step response method. All PID controllers are tuned based on the approximate model defined by (3.144)

observed both in the input $d_i(\tau)$ and the output $d_o(\tau)$ of the control loop compared to the Magnitude Optimum PID tuning.

If the filter time constant f is reduced from $f = 3$ to $f = 1$ the step response and disturbance rejection become faster in the case of IMC tuning, see Fig. 3.24. In this case, the IMC tuning principle outperforms the PID tuning via the revised method since it exhibits almost the same settling time regarding disturbance rejection, but with less undershoot, Fig. 3.24a.

3.5.3.3 A Nonminimum Phase Process

Consider the nonminimum phase process defined by

$$G(s') = \frac{(1 - 0.7s')(1 - 0.5s')}{(1 + s')^5} \tag{3.145}$$

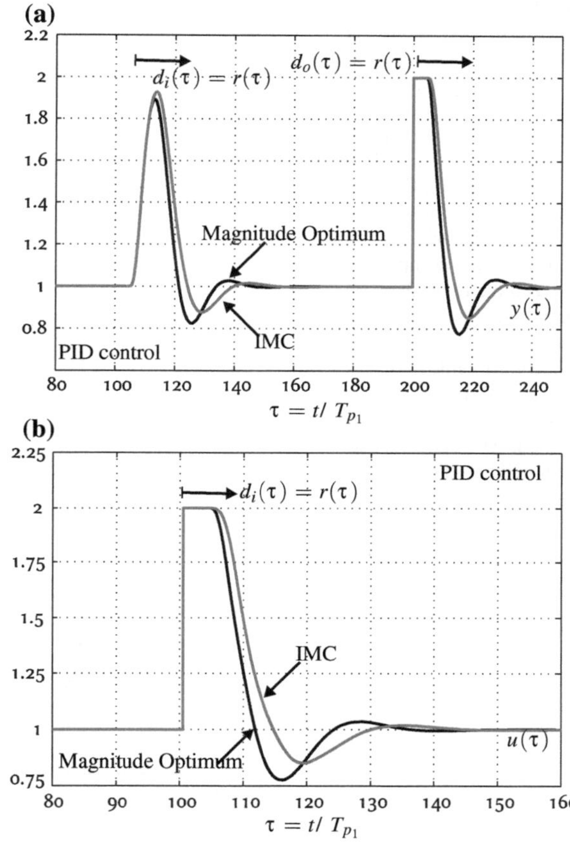

Fig. 3.24 **a** Step response of the control loop. Input and output disturbance $d_{\mathrm{i}}(s)$ and $d_{\mathrm{o}}(s)$ are applied at $t = 100\tau$ and $t = 200\tau$, respectively. **b** Response of the command signal at the presence of input disturbance $d_{\mathrm{i}}(\tau)$. Control of a process with long time delay defined by (3.143). PID control action: comparison between the revised Magnitude Optimum criterion, the Internal Model Control principle (IMC, $f = 1$) and the Ziegler–Nichols step response method. All PID controllers are tuned based on the approximate model defined by (3.144)

and the approximated process defined by

$$\widetilde{G}(s') = \frac{(1 - 0.6s')(1 - 0.4s')(1 - 0.2s')}{(1 + s')^5}. \tag{3.146}$$

In this case the PID controller via for the Ziegler–Nichols and the Magnitude Optimum criterion is tuned via the (3.146) whereas and the resulting control law is applied to (3.145).

The Ziegler–Nichols step response tuning method leads to an unstable control loop and therefore is not depicted in Figs. 3.25 and 3.26.

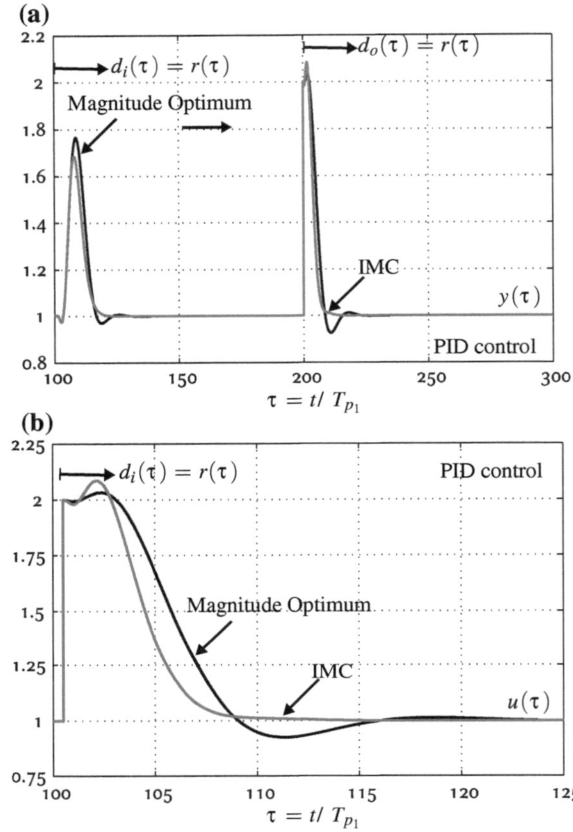

Fig. 3.25 **a** Step response of the control loop. Input and output disturbance $d_\mathrm{i}(s)$ and $d_\mathrm{o}(s)$ are applied at $t = 100\tau$ and $t = 200\tau$ respectively. **b** Response of the command signal at the presence of input disturbance $d_\mathrm{i}(\tau)$. Control of a nonminimum phase process defined by (3.146). PID control action: comparison between the revised Magnitude Optimum criterion, the Internal Model Control principle (IMC, $f = 1$) and the Ziegler–Nichols step response method. All PID controllers are tuned based on the approximate model defined by (3.146)

3.6 Explicit Tuning of PID Controllers Applied to Grid Converters

A type-I control loop within a real industry application is the typical model of an AC/DC grid connected converter. In this application, the converter connects the DC link capacitor to the grid through a grid transformer as shown in Fig. 3.27.

Its main purpose is to maintain the DC link voltage that typically supplies power for a drive, or another network. The interfaced signals involved in the control loop are described in Table 3.3 and the system parameters that should be known or estimated for controlling purpose are described in Table 3.4. The network may be modeled as a voltage source and its grid impedance that reflects its strength. The grid transformer is modeled through its magnetizing and leakage impedance.

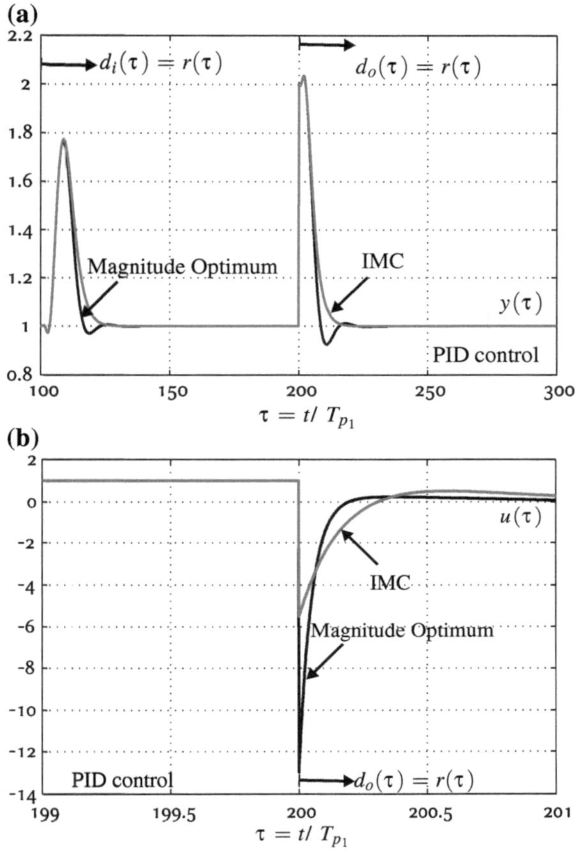

Fig. 3.26 **a** Step response of the control loop. Input and output disturbance $d_i(s)$ and $d_o(s)$ are applied at $t = 100\tau$ and $t = 200\tau$, respectively. **b** Response of the command signal at the presence of input disturbance $d_i(\tau)$. Control of a nonminimum phase process defined by (3.145). PID control action: comparison between the revised Magnitude Optimum criterion, the Internal Model Control principle (IMC, $f = 2$) and the Ziegler–Nichols step response method. All PID controllers are tuned based on the approximate model defined by (3.146)

3.6.1 Simplified Control Model and Parameters

The most classic way for controlling AC/DC grid converters is the cascaded vector control, see [14], Fig. 3.28 for which

$$C_{V_{DC}} = \frac{1 + sT_c}{sT_2}, \tag{3.147}$$

is the DC link voltage controller, and

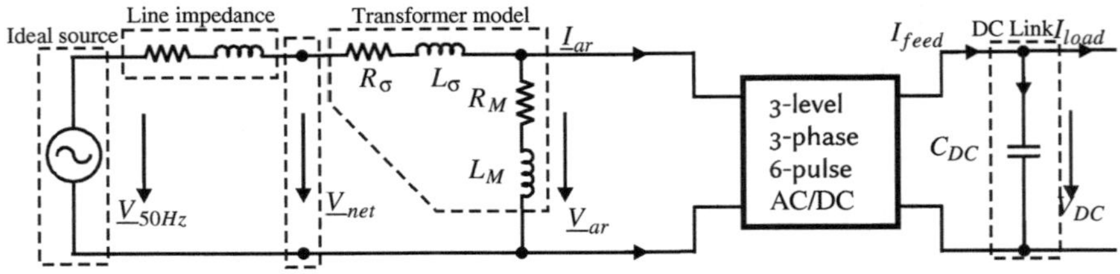

Fig. 3.27 Grid connected active rectifier on system level

Table 3.3 List of signals in the system

Signal	Unit	Description
$\underline{V}_{\text{net}}$	V	Three phase voltage measured at PCC[a]
ω_{net}	Hz	Pulsation of the network
$\underline{V}_{\text{ar}}$	V	Three phase voltage at the grid converter
$\underline{I}_{\text{ar}}$	A	Three phase current to grid converter
V_{DC}	V	DC link voltage
I_{feed}	A	Feeding current from grid converter
I_{C}	A	DC link capacitor current
I_{load}	A	Load current

[a]Point of common coupling

Table 3.4 List of parameters in the system

Parameter	Unit	Description
C_{DC}	F	DC link capacitor
L_σ	H	Leakage inductance of the transformer
R_σ	Ω	Leakage resistance of the transformer
L_{M}	H	Magnetizing inductance of the transformer
R_{M}	Ω	Magnetizing resistance of the transformer
L_{net}	H	Equivalent line inductance of the network
R_{net}	Ω	Equivalent line resistance of the network

$$C_{\text{I}} = \frac{1 + sT_{\text{LR}}}{sT_1}, \tag{3.148}$$

$$C_{\text{IF}} = \frac{1}{1 + sT_{\text{F}}}, \tag{3.149}$$

stand for the model of the current PI controller, where T_1, T_{LR} are the current controller's integrator time constant, the current controller's zero to be determined and T_{F} stands for the unmodeled controller dynamics. Let it be noted that the analysis of the control loop takes place in the $d - q$ reference frame. From the output of the

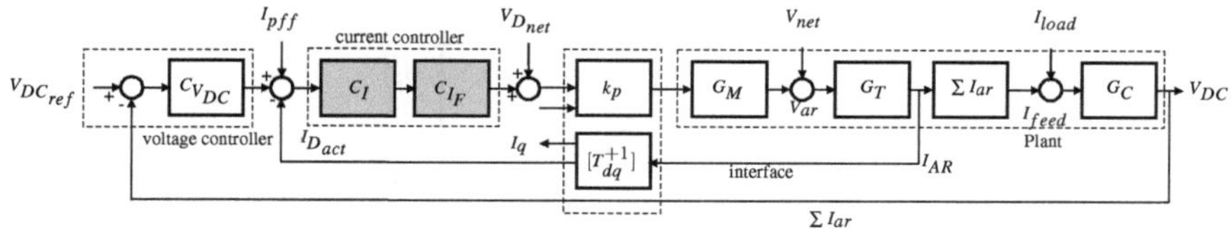

Fig. 3.28 Cascaded control loop for AC/DC grid converters

current controller, the modulation index M_{AR} and modulation angle are constructed out of the expressions

$$\sqrt{V_D^2 + V_Q^2} \quad \text{and} \quad \operatorname{atan}\left(\frac{V_D}{V_Q}\right). \tag{3.150}$$

Note that V_D, V_Q is the output of the current PI controller within the d and q path, respectively.

The modulator itself is modeled by first order process $k_{\mathrm{p}} G_{\mathrm{M}}$ model, where k_{p} stands for the modulator's gain in terms of fundamental amplitude ($k_{\mathrm{p}} = \frac{V_{\mathrm{out}_0}}{V_{\mathrm{in}_0}}$) and T_{m} is the time delay introduced from the time the controller decides the command until the final voltage is applied by the power part of the inverter.

$$G_{\mathrm{M}} = \frac{1}{1 + sT_{\mathrm{m}}}. \tag{3.151}$$

Transfer functions G_{T} and G_{C} stand for the transformer model

$$G_{\mathrm{T}} = \frac{1}{R_\sigma + sL_\sigma}, \tag{3.152}$$

and the capacitor bank path within the DC link of the inverter

$$G_{\mathrm{C}} = \frac{1}{sC_{\mathrm{DC}}} \tag{3.153}$$

respectively. The DC-link voltage controller provides the current reference to the grid current controller which itself provides a reference to the modulator through the modulation index M_{ar}. The load current I_{load} is the main perturbation of the system and a power feed-forward current I_{pff} can be provided to the current controller for enhancing its dynamics.

Although the description of the synchronization to the grid through a dedicated PLL [19] is not the scope of this section, it cannot be ignored since it provides the reference for the vector controller, Fig. 3.28. The grid voltages and currents are described in the synchronous reference frame computed by the well-known Park transformation. Only the active part of the vector control is considered in the scheme

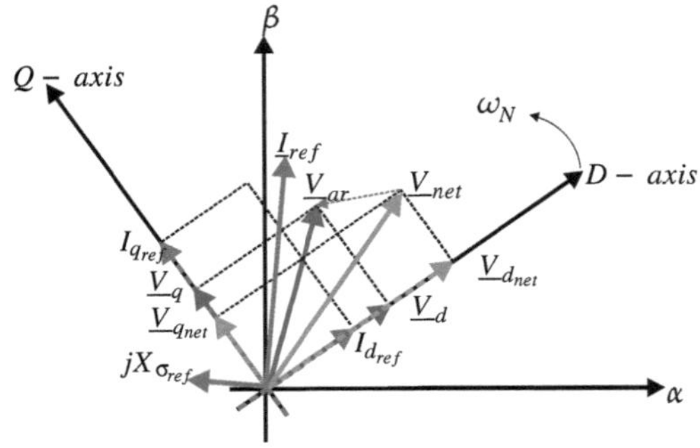

Fig. 3.29 Vector control principle

Table 3.5 List of controller signals

Signal	Unit	Description
M_{ar}	–	Modulation index of grid converter AC voltage
V_{dctrl}	V	Voltage control value (active part)
V_{qctrl}	V	Voltage control value (reactive part)
$V_{d\mathrm{net}}$	V	Grid voltage measurement (active part)
$V_{q\mathrm{net}}$	V	Grid voltage measurement (reactive part)
I_d	A	Grid current measurement (active part)
I_q	A	Grid current measurement (reactive part)
I_{pff}	A	Current feed-forward from load drive
$I_{d\mathrm{ref}}$	A	Grid current reference (active part)
$I_{q\mathrm{ref}}$	A	Grid current reference (reactive part)
$V_{d\mathrm{cref}}$	A	DC link voltage reference

depicted on Fig. 3.28, but the reactive part is also controlled by a reference set often to zero or to a nonzero value when a reactive power controller is active (Fig. 3.29).

The cross coupling due to the use of a synchronous reference frame needed in the current controller [6, 7], is also not depicted. The signals entering and generated by the controller are summarized in Table 3.5.

Considering the optimal tuning of the vector control parameters in the synchronous reference frame, the cross-coupling gain is a value corresponding to the estimation of the equivalent inductor value on the grid side of the power converter [23].

A more accurate way of decoupling this effect has been presented in [4]. Instead of cross-coupling the current components through proportional gains as in classic current controllers, one should cross-couple the error signals through integrators as depicted in Fig. 3.30. The parameters of the cross-coupling integrators are identical to the PI controllers parameters and can therefore be tuned using the optimal control action described in Sect. 3.3.

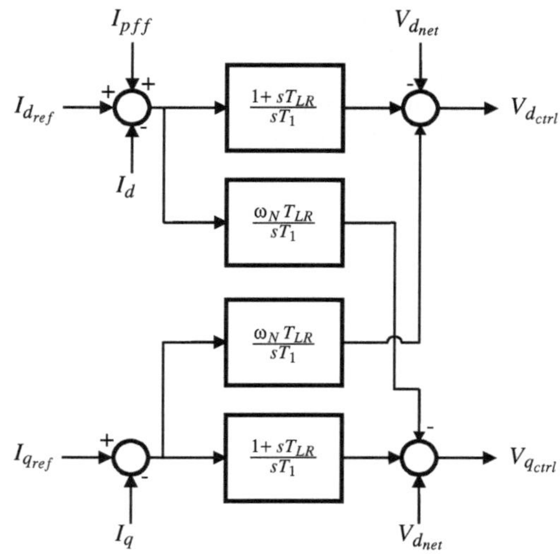

Fig. 3.30 Multivariable PI controller with voltage feed forward

The proposed method is applied to the inner control loop (current controller) since the analysis within this chapter is dedicated to type-I control loops.[3] The outer control loop is of type-II (two integrators in the open loop transfer function) and is out of the scope of this chapter. For measuring the DC gain of the process, an open loop experiment from M_{ar} to I_{ar} is carried out. In this case a good estimation of R_σ is acquired since

$$G(s) = G_{\text{M}}G_{\text{T}} = \frac{1}{(1+sT_{\text{m}})(R_\sigma + sL_\sigma)} = \frac{1}{R_\sigma\,(1+sT_{\text{m}})\,(1+s\frac{L_\sigma}{R_\sigma})} \tag{3.154}$$

for which it is apparent that

$$\begin{aligned}\lim_{t\to\infty} I_{\text{AR}}(t) &= \lim_{s\to 0} sG(s)\,M_{\text{AR}}(s) = \lim_{s\to 0} sG(s)\,\frac{1}{s} \\ &= \lim_{s\to 0}\left(\frac{1}{R_\sigma\,(1+sT_{\text{m}})\,(1+sT_k)}\right) \approx R_\sigma \end{aligned} \tag{3.155}$$

where $T_k = \frac{L_\sigma}{R_\sigma}$.

The current controller step response is evaluated alone since the optimal tuning of the DC-link controller is considered in Chap. 4. As illustrated in Fig. 3.31a, the controller is first submitted to a reference step of the current I_{ref}, then to a perturbation of the net voltage V_{net} and of the load current I_{load}. One can see that the system is completely decoupled from the voltage perturbations, this is possible only with the use of an accurate PLL that is synchronizing the system to the grid. The behavior of the feeding current I_{feed} shows the accuracy of the proposed method.

[3] Loops that track step reference signals with zero steady state error.

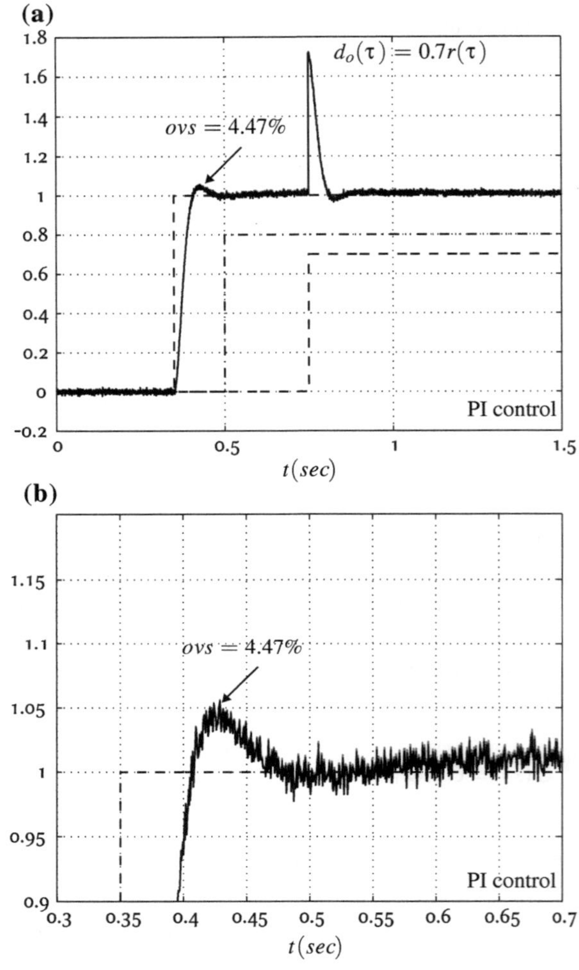

Fig. 3.31 **a** Reference and load step response of the system. **b** Detailed view on step response of degraded systems. Step response of the system

One can read on Fig. 3.31b an overshoot of some 5 % and a 7 ms response. If one modify the time constant T_M of the modulator by a factor $a = \pm 20\,\%$, the effect on the response time is obviously increasing its overshoot or its time response, showing that the proposed tuning gives a satisfactory behavior to to the current control loop.

3.7 Summary

In Sect. 3.2 the conventional PID controller tuning via the Magnitude Optimum criterion was presented. It was shown that controller parameters

1. are restricted to be tuned only with real zeros and
2. for determining controller's zeros, exact pole-zero cancellation between the process's poles and the controller's zeros has to be achieved.

Moreover, the application of the conventional Magnitude Optimum criterion has been tested only to simple process models. Based on this current state of the art, the conventional tuning method of the PID regulator has been considered poor and suboptimal. For that reason, all the aforementioned restrictions were thoroughly revised within Sect. 3.3. A new PID control law was presented that (1) determines analytically controller parameters regardless of the plant complexity (2) allows the PID zeros to become conjugate complex if needed.

An extensive performance comparison for several process models was carried out in Sect. 3.4. It was shown that the revised control action outperforms the conventional tuning rules both regarding reference tracking and disturbance rejection within the closed loop control system. It was shown that for certain processes, a 50.6 % decrease in the settling time of output disturbance rejection and reference tracking can be achieved.

Since the new tuning rules can involve all process dynamics, they can be applied directly to any linear SISO process regardless of its complexity along with the aid of system identification techniques.

Finally, in Sect. 3.6 the current control loop of a grid connected converter was presented which is of type-I.

References

1. Åström KJ (1995) Model uncertainty and robust control. Tech. rep., Department of Automatic Control, Lund University, Lund, Sweden
2. Åström KJ, Hagglund T (1995) PID controllers: theory, design and tuning, 2nd edn. Instrument Society of America
3. Åström KJ, Hagglund T (2004) Revisiting the Ziegler–Nichols step response method for PID control. J Process Control 14(6):635–650
4. Bahrani B, Kenzelmann S, Rufer A (2011) Multivariable-PI-based current control of voltage source converters with superior axis decoupling capability. IEEE Trans Ind Electron 58(7):3016–3026
5. Brosilow C, Joseph B (2002) Techniques of model-based control, 1st edn. Prentice-Hall, New Jersey
6. Bühler H (1979) Électronique de reglage et de commande. Dunod, Paris
7. Bühler HR (1997) Reglage des systemes d'electronique de puissance, vol 1, 2 and 3, Theorie, 1st edn. PPUR: Presses Polytechniques et Universitaires romandes
8. Buxbaum A, Schierau K, Straughen A (1990) Design of control systemsfor DC drives. Springer, Berlin
9. Courtiol B, Landau ID (1975) High speed adaptation system for controlled electrical drives. Automatica 11(2):119–127
10. Föllinger O (1994) Regelungstechnik. Hüthig, Heidelberg
11. Fröhr F, Orttenburger F (1982) Introduction to electronic control engineering. Siemens, Berlin
12. Goodwin GC, Graebe SF, Salgado ME (2001) Control system design. Prentice Hall, New Jersey
13. Haalman A (1965) Adjusting controllers for a dead time process. Control Engineering Practice, pp 71–73

14. Habetler TG (1993) A space vector-based rectifier regulator for AC/DC/AC converters. IEEE Trans Power Electron 8(1):30–36
15. Kessler C (1955) UG ber die Vorausberechnung optimal abgestimmter regelkreise teil III. Die optimale einstellung des reglers nach dem betragsoptimum. Regelungstechnik 3:40–49
16. Kessler C (1958) Das symmetrische optimum. Regelungstechnik, pp 395–400 and 432–426
17. Loron L (1997) Tuning of PID controllers by the non-symmetrical optimum method. Automatica 33(1):103–107
18. Lutz H, Wendt W (1998) Taschenbuch der regelungstechnik, 1st edn. Frankfurt am Main: Verlag, Harri Deutsch
19. Mohan N, Undeland TM, Robbind WF (1989) Power electronics: converters, applications and design, 1st edn. Wiley, New York
20. Morari M, Zafiriou E (1989) Robust process control, 1st edn. Prentice-Hall, New Jersey
21. Oldenbourg RC, Sartorius H (1954) A uniform approach to the optimum adjustment of control loops. Trans ASME 76:1265–1279
22. Sartorius H (1945) Die zweckmässige festlegung der frei wählbaren regelungskonstanten. Master thesis, Technische Hochscule, Stuttgart, Germany
23. Schauder C, Mehta H (1993) Vector analysis and control of advanced static VAr compensators. IEE Proc Gener, Transm Distrib 140(4):299–306
24. Umland WJ, Safiuddin M (1990) Magnitude and symmetric optimum criterion for the design of linear controlsystems: what is it and how does it compare with the others? IEEE Trans Ind Appl 26(3):489–497
25. Voda AA, Landau ID (1995) A method for the auto-calibration of PID controllers. Automatica 31(1):41–53
26. Vrančić D, Strmčnik S (1999) Practical guidelines for tuning PID controllers by using MOMI method. In: International symposium on industrial electronics, IEEE, vol 3, pp 1130–1134
27. Vrančić D, Kristiansson B, Strmčnik S (2004) Reduced MO tuning method for PID controllers. In: 5th Asian control conference, IEEE, vol 1, pp 460–465

Chapter 4
Type-II Control Loops

Abstract In this chapter, the explicit solution for tuning the PID controller parameters in the presence of integrating process is presented. The presence of one integrator coming from the plant along with one integrator coming from the PID-type control action results in a type-II control loop according to Sect. 2.5. The proposed control law is developed again in the frequency domain and lies in the principle of the symmetrical optimum criterion which, strictly speaking, is the application of the Magnitude Optimum criterion in type-II control loops. Therefore, the desired control action requires again that the magnitude of the closed-loop transfer function is equal to the unity in the widest possible frequency range. For the proof of the control law, a general transfer function process model is adopted consisting of n poles, m zeros plus unknown time delay d. The final solution determines explicitly the P, I, and D parameters as a function of all time constants involved within the control loop and irrespective of the process complexity. The potential of the proposed method is tested both (1) on benchmark process models (integrating process with dominant time constants, integrating non-minimum phase process, integrating process with long time delay). The proposed control action is tested also for the control of the actual DC link voltage in an AC/DC grid connected converter. In all cases, an extensive comparison test is presented between the conventional current state-of-the-art PID tuning and the proposed control law, justifying the potential of the proposed method.

4.1 Introduction

In the literature, the demanding problem of controlling integrating processes has driven many researchers at employing or modifying well-established control techniques [17, 22], such as the Smith predictor, see [1, 7, 14, 18, 21] and the internal model control (IMC) principle [20]. More specifically, in [1, 14], an extension of Watanabe's Smith predictor is proposed where for its tuning an accurate estimation of the input disturbance is required [1]. The proposed method involves adjustable tuning and not explicit solution for the controller's parameters, whereas in [7, 21],

K.G. Papadopoulos, *PID Controller Tuning Using the Magnitude Optimum Criterion*,
DOI 10.1007/978-3-319-07263-0_4

the proposed modified Smith predictor restricts its focus on controlling integrating processes with long dead time.

The control loops where integrating processes are involved are also called in the literature as type-II control loops [11]. The basic advantage of such control loops is the ability the output variable of the control loop exhibits, to track perfectly step and ramp reference signals with zero steady state position and velocity error respectively. On a theoretical basis and if frequency domain modeling is followed for the controller's design, type-II control loops are also characterized by the presence of two pure integrators within the open-loop transfer function.

On a practical basis, industrial examples of this case arise frequently in the area of AC/DC/AC power converters and drive systems in principle. Representative industry applications where the aforementioned AC/DC/AC configuration is met is (1) a wind energy conversion system [8], (2) a shaft generator system [6, 9], and (3) an AC/DC/AC arrangement operating in motoring mode.[1]

Taking the case of a shaft generator system as an example, the AC/DC/AC configuration has to operate often in island network mode so that the vessel's efficiency is improved.[2] Island network mode means that the grid side converter has to deliver to the grid[3] the required AC signal of certain amplitude and certain frequency, given constant DC link between the two converters. In this case, constant DC link is guaranteed by the outer DC link voltage control loop of the shaft side converter which takes the energy from the shaft generator.

Within the shaft side converter and from the control point of view, the resulting control loop of the DC link voltage is proved to be of type-II, since one integrator comes from the capacitor bank of the DC link, whereas the other integrator is coming from the PID-type controller itself. Last but not least, in the case of the motoring operation of an AC/DC/AC configuration system, the actual DC link voltage is regulated at a constant level by the grid connected converter which runs normally under a vector control scheme [2, 3]. In this case, there is again an inner current control loop and an outer voltage control loop (for regulating the actual DC link voltage) which afterwards is used by the motor side converter for driving the machine.

Motivated by such practical industrial problems, the purpose of this chapter is to provide control engineers with explicit tuning rules for the PID controller and irrespective of the complexity of the integrating process so that robust performance can be achieved by the output of a type-II control loop. To this end, development and control engineers are provided with an explicit solution which

1. allows for accurate investigation of the robustness of the controller to possible model uncertainties within the whole control loop;

[1] In this case, a grid connected converter controls the DC Link which is then used by the motor side converter which finally drives the motor.

[2] In the case of the island network, auxiliary small diesel generators are completely switched off since they are consuming expensive oil, and the energy is coming from the main diesel engine of the ship which drives the propeller.

[3] Grid of the vessel supporting the electrical load of the vessel.

2. leads to reliable results before integrating finally the whole control law on a real-time embedded system;
3. prevents on-site commissioning and service engineers from using heuristic tuning rules which most of the times lead to poor performance of the drive itself, as far as the field of power converters is concerned.

In order to develop the aforementioned control theory, the principle of the Magnitude Optimum criterion is adopted [10, 16], see Appendix A.1. Oldenbourg and Sartorius applied the Magnitude Optimum criterion in type-I systems and in succession, Kessler suggested the symmetrical optimum criterion [4, 5, 19] which in reality is the application of the Magnitude Optimum criterion to type-II control systems. In this chapter, aim of the proposed theory is to revise thoroughly the current state-of-the-art in PID tuning via the symmetrical optimum criterion by pointing out its drawbacks and improving them by

1. suggesting closed-form expressions for the PID controller's parameters and
2. achieving robust and optimal performance of the control loop both in reference tracking and disturbance rejection.

For the reasons above, and for the sake of a clear presentation of the proposed explicit solution, the sections of this chapter are organized as follows. In Sect. 4.2, a short presentation is given to the reader about the current state-of-the-art relevant to the PID tuning via the symmetrical optimum criterion. Its drawbacks are pointed out, which are basically related to (1) the simple and poor process model used till date to adopt the conventional PID tuning and (2) the pole-zero cancellation principle the current state-of-the-art method uses. Therefore, the proposed theory introduces a transfer function of integrating behavior consisting of n poles, m zeros plus unknown time delay d. Irrespective of the order of n, m, and d an explicit solution of the proposed control law is presented within the same section without using the principle of pole-zero cancellation.

The proof of the control law lies in the well-known Magnitude Optimum criterion which is presented in the Appendix B.2. Therefore, in Sect. 4.4, we apply the theoretical modeling approach on five benchmark transfer function process models. The proposed control law is also tested finally within the DC link voltage control path on an AC/DC arrangement, see Sect. 4.5. The AC/DC configuration is presented on system and closed-loop control system level, and the control loop of actual DC link is presented in the frequency domain. Controller performance is investigated in the presence of output disturbances which in this case is the load current coming from the inverter which in principle drives the electric motor.

In all examples, the proposed method is compared with the conventional state-of-the-art PID tuning via the Magnitude Optimum criterion in terms of step and ramp reference signals. Within this comparison, the output of the control loop along with the command signal of the controller (control effort) are also measured. Results and conclusions are summarized in Sect. 4.6.

4.2 Conventional PID Tuning Via the Symmetrical Optimum Criterion

In this section, the conventional PID tuning via the symmetrical optimum criterion is introduced. The same line presented in Sect. 3.2 is also followed. For that reason, the controller's design starts from I control action, proceeds with PI control action up to the PID controller tuning. In Sects. 4.2.1, and 4.2.2, it is shown that both I and PI controller design lead to an unstable control loop, whereas in Sect. 4.2.3 the proof of the end PID control action is presented.

4.2.1 I Control

Let us now consider the closed-loop system of Fig. 4.1, where $r(s)$, $e(s)$, $u(s)$, $y(s)$, $d_{\mathrm{o}}(s)$, and $d_{\mathrm{i}}(s)$ are the reference input, the control error, the input and output of the plant, the output and the input disturbances, respectively. An integrating process found in many industry applications can be defined by (4.1)

$$G(s) = \frac{1}{T_{\mathrm{m}}s(1 + T_{\mathrm{p_1}}s)(1 + T_{\Sigma_{\mathrm{p}}}s)}, \tag{4.1}$$

where T_{m} is the integrator's plant time constant, $T_{\mathrm{p_1}}$ the plant's dominant time constant and $T_{\Sigma\mathrm{p}}$ the process parasitic time constant [11]. Let it be noted that such type of modeling is frequently used in vector controlled induction motor drives. More specifically, time constant T_{m} stands for the mechanical subsystem of the motor which is the mechanism that involves the electromagnetic and load torque, the difference of which, makes the shaft rotating.

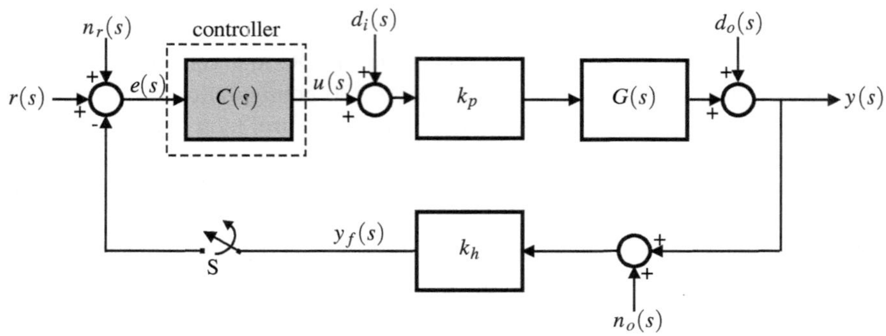

Fig. 4.1 Block diagram of the closed-loop control system. $G(s)$ is the plant transfer function, $C(s)$ is the controller transfer function, $r(s)$ is the reference signal, $d_{\mathrm{o}}(s)$ and $d_{\mathrm{i}}(s)$ are the output and input disturbance signals, respectively, and $n_{\mathrm{r}}(s)$, $n_{\mathrm{o}}(s)$ are the noise signals at the reference input and process output, respectively. k_{p} stands for the plant's dc gain, and k_{h} is the feedback path

Furthermore, time constant T_{p_1} is involved in the inner current control loop of the electrical drive and represents the stator winding time constant. Finally, $T_{\Sigma p}$ stands for the motor's unmodeled dynamics. If vector control.[4] is to be followed (control of induction motor drives), k_p stands for the pulse width modulator's gain (k_{PWM}) which is supposed to remain constant all over the whole operating range ($0 \rightarrow 1$p.u) regarding output frequency.[5] Parameter k_h is the feedback path of the output measurement and as it is proved in the sequel, k_h should satisfy condition $k_h = 1$.

Back to Fig. 4.1, for controlling (4.1), the PID controller defined by

$$C(s) = \frac{(1 + T_n s)(1 + T_v s)}{T_i s(1 + T_{\Sigma_c} s)} \tag{4.2}$$

is adopted. For its tuning, the conventional symmetrical optimum design method is employed. Time constant $T_{\Sigma c}$ stands for the controller's parasitic dynamics. If $T_n = T_v = 0$, I control cannot be applied, because the closed-loop transfer function becomes unstable. This is justified as follows. If for controlling (4.1), I control of the form

$$C(s) = \frac{1}{T_i s(1 + T_{\Sigma_c} s)} \tag{4.3}$$

is applied, then the closed-loop transfer function is given by

$$T(s) = \frac{k_p}{T_i T_m s^2 (1 + T_{p_1} s)(1 + T_\Sigma s) + k_h k_p} \tag{4.4}$$

where

$$T_{\Sigma p} T_{\Sigma c} \approx 0, \quad \text{and} \quad T_\Sigma = T_{\Sigma p} + T_{\Sigma c}. \tag{4.5}$$

From (4.4), it is evident that

$$T(s) = \frac{k_p}{T_i T_m T_{p_1} T_\Sigma s^4 + T_i T_m (T_{p_1} + T_\Sigma) s^3 + T_i T_m s^2 + k_h k_p}. \tag{4.6}$$

From (4.6), it is clear that $T(s)$ is unstable since the term of s is missing.[6]

---

[4] SFOC: stator field-oriented control, RFOC: rotor field-oriented control.

[5] In many real-world applications, k_p stands for the plant's dc gain at steady state.

[6] This kind of instability is justified by the Routh theorem. For a polynomial of the form $D(s) = a_n s^n + a_{n-1} s^{n-1} + \cdots + a_1 s + a_0$, necessary condition for $D(s)$ to be stable is $a_j > 0, j = 0, 1, 2, \ldots$ Since in (4.6) $a_1 = 0$, then according to the Routh theorem, the denominator $D(s)$ of $T(s)$ is unstable.

4.2.2 PI Control

In a similar fashion, if PI control of the form

$$C(s) = \frac{1 + T_n s}{T_i s(1 + T_{\Sigma c} s)} \tag{4.7}$$

is employed, then for determining controller parameter T_n via the conventional symmetrical optimum criterion, pole-zero cancellation must take place, $T_n = T_{p_1}$. Therefore, the dominant time constant T_{p_1} has to be evaluated and in that case, $T(s)$ becomes

$$T(s) = \frac{k_p}{T_i T_m T_\Sigma s^4 + T_i T_m T_\Sigma s^3 + T_i T_m s^2 + k_h k_p}, \tag{4.8}$$

which is unstable again for the same reason as for (4.6).

4.2.3 PID Control

Assuming again that the dominant time constant T_{p_1} is accurately measured and considering a PID controller as described by (4.2), $T_v = T_{p_1}$ is set (pole-zero cancellation, conventional symmetrical optimum design). The closed-loop transfer function becomes equal then to

$$T(s) = \frac{k_p T_n s + k_p}{T_i T_m T_\Sigma s^3 + T_i T_m s^2 + k_h k_p T_n s + k_h k_p}. \tag{4.9}$$

The magnitude of (4.9) is given as

$$|T(j\omega)| = \sqrt{\frac{k_p k_p \left[1 + (\omega T_n)^2\right]}{\left(k_p k_p - T_i T_{p_1} \omega^2\right)^2 + \omega^2 \left(k_p k_p T_n - T_i T_{p_1} T_\Sigma \omega^2\right)^2}}. \tag{4.10}$$

The denominator of (4.10) is equal to

$$\begin{aligned} D(\omega) = {} & \left(T_i T_{p_1} T_\Sigma\right)^2 \omega^6 + T_i T_{p_1} \left(T_i T_{p_1} - 2k_p k_h T_n T_\Sigma\right) \omega^4 \\ & + \left[\left(k_p k_h T_n\right)^2 - 2k_p k_h T_i T_{p_1}\right] \omega^2 + k_p^2 k_h^2. \end{aligned} \tag{4.11}$$

Thus, by setting the term of ω^4 equal to zero, see [11], results in

$$T_i T_m = 2k_p k_h T_n T_\Sigma \tag{4.12}$$

from which it is apparent that

$$T_{\mathrm{i}} = 2k_{\mathrm{p}}k_{\mathrm{h}}\frac{T_{\mathrm{n}}T_{\Sigma}}{T_{\mathrm{m}}}. \tag{4.13}$$

In similar fashion by setting the term of ω^2 equal to zero, see [11] results in

$$\left(k_{\mathrm{p}}k_{\mathrm{h}}\right)^2 T_{\mathrm{n}}^2 = 2k_{\mathrm{p}}k_{\mathrm{h}}T_{\mathrm{i}}T_{\mathrm{m}} \tag{4.14}$$

$$T_{\mathrm{i}} = \frac{1}{2}k_{\mathrm{p}}k_{\mathrm{h}}\frac{T_{\mathrm{n}}^2}{T_{\mathrm{m}}}. \tag{4.15}$$

Making equal the aforementioned equations results in

$$T_{\mathrm{n}} = 4T_{\Sigma}. \tag{4.16}$$

In that, the integrator's time constant is equal to

$$T_{\mathrm{i}} = 8k_{\mathrm{p}}k_{\mathrm{h}}\frac{T_{\Sigma}^2}{T_{\mathrm{m}}}. \tag{4.17}$$

By substituting the definitions of T_{i}, T_{n} back to (4.9), it is easily shown that for having $|T(j\omega)| \simeq 1$ then

$$k_{\mathrm{h}} = 1 \tag{4.18}$$

has to hold by. Finally, the PID control action is given by

$$\begin{bmatrix} T_{\mathrm{v}} \\ T_{\mathrm{n}} \\ T_{\mathrm{i}} \\ k_{\mathrm{h}} \end{bmatrix} = \begin{bmatrix} T_{\mathrm{p1}} \\ 4T_{\Sigma} \\ 8k_{\mathrm{p}}k_{\mathrm{h}}\frac{T_{\Sigma}^2}{T_{\mathrm{m}}} \\ 1 \end{bmatrix}. \tag{4.19}$$

Using (4.19) along with (4.9) results in

$$T(s) = \frac{1 + 4T_{\Sigma}s}{8T_{\Sigma}^3 s + 8T_{\Sigma}^2 s + 4T_{\Sigma}s + 1} \tag{4.20}$$

or finally after normalizing the frequency by substituting $s' = T_{\Sigma}s$ results in

$$T(s') = \frac{1 + 4s'}{8s'^3 + 8s'^2 + 4s' + 1}. \tag{4.21}$$

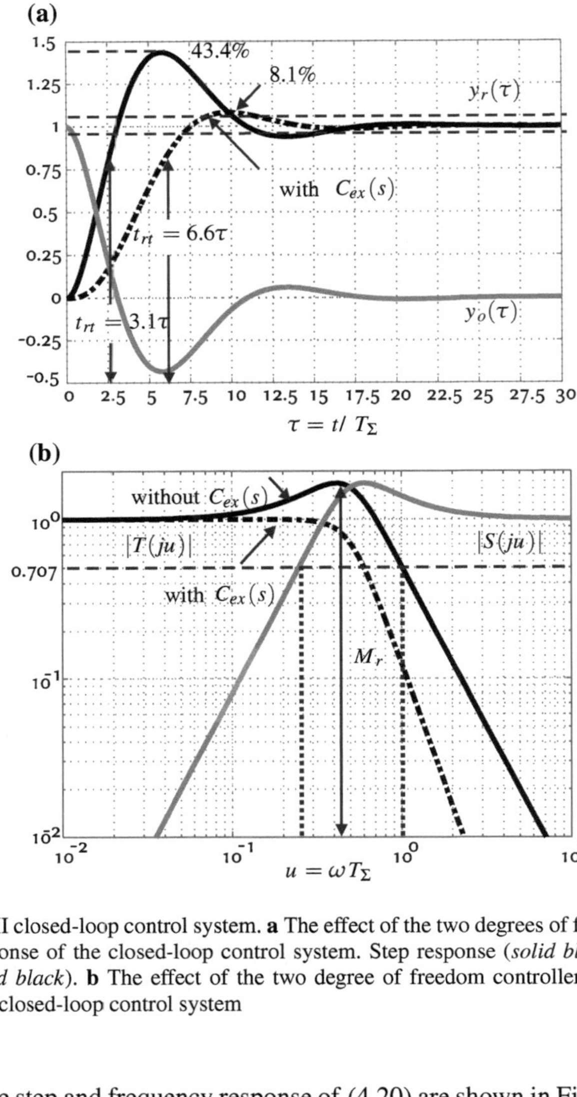

Fig. 4.2 Type-II closed-loop control system. **a** The effect of the two degrees of freedom controller to the step response of the closed-loop control system. Step response (*solid black*), filtered step response (*dotted black*). **b** The effect of the two degree of freedom controller to the frequency response of the closed-loop control system

The respective step and frequency response of (4.20) are shown in Fig. 4.2a, b. From there, it is clear that the step response of the closed-loop control system exhibits an undesired overshoot of 43.4 % in the time domain Fig. 4.2a, and a peak overshoot in the frequency domain Fig. 4.2b.

This is also justified by the open-loop frequency response Fig. 4.3 where the phase margin in the crossover frequency

$$\omega_c = \frac{1}{2T_\Sigma} \tag{4.22}$$

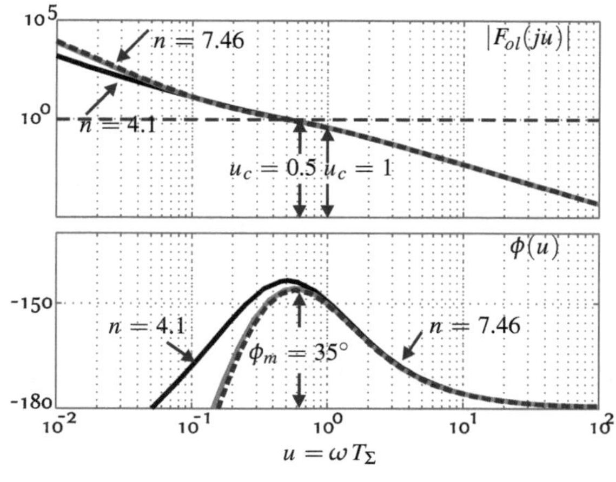

Fig. 4.3 Type-II closed-loop control system. Open-loop frequency response

is $\varphi_m \approx 35° < 45°$. Note also the symmetry of the critical frequencies $\left(\frac{1}{4T_\Sigma}, \frac{1}{T_\Sigma}\right)$ exhibited by $|F_{ol}(j\omega)|$ where its slope is equal to -1/deg around the crossover frequency $\omega_c = \frac{1}{2T_\Sigma}$, Fig. 4.3. The open-loop transfer function is given by

$$F_{ol}(s') = \frac{1 + 4s'}{8s'^2(1 + s')}. \tag{4.23}$$

In order to overcome the obstacle of 43.4 % overshoot, the reference input is filtered by adding an external controller $C_{ex}(s)$, Fig. 4.4. The great overshoot of the step response in (4.21) is owed to the zero of the transfer function, $N\left(s'\right) = 1 + 4s'$. This can be removed by including that zero as a pole in the reference filter. In that, if an

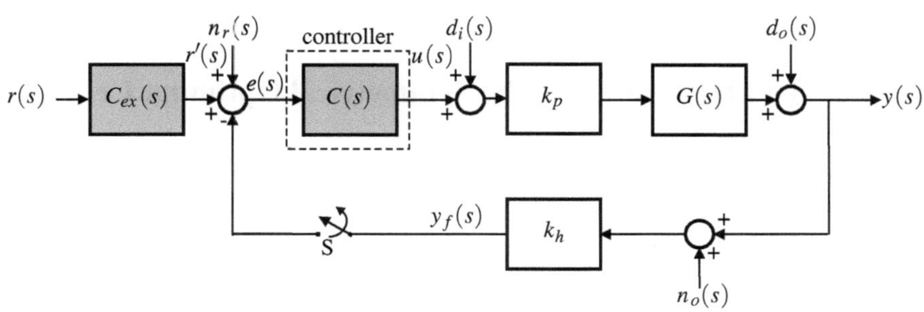

Fig. 4.4 Two degrees of freedom controller. Controller $C_{ex}(s)$ filters the reference input so that the undesired overshoot at the output $y(s)$ is diminished. Controller $C_{ex}(s)$ affects the closed-loop transfer function $T(s)$ and not the output and input disturbance transfer functions $S_o(s) = \frac{y(s)}{d_o(s)}$, $S_i(s) = \frac{y(s)}{d_i(s)}$

external filter of the form

$$C_{\text{ex}}(s') = \frac{r'(s')}{r(s')} = \frac{1}{1 + 4s'} \tag{4.24}$$

is chosen, the overshoot decreases from 43.4 to 8.1 %. Let it be noted that the rise time increases from $t_{\text{rt}} = 3.1T_{\Sigma}$ to $t_{\text{rt}} = 6.6T_{\Sigma}$. Such dynamics, can for sure be improved by adding additional dynamics in the reference filter.

4.2.4 Drawbacks of the Conventional Tuning

From the aforementioned analysis in Sect. 4.2, it becomes clear that the conventional tuning through pole-zero cancellation in the case of PI control cannot lead to a stable control loop since the final transfer function of the control loop proves to be unstable. Moreover, for tuning the PID-type controller zeros, exact pole-zero cancellation has to be achieved between the process's dominant time constant and the controller's zeros. Since this type of tuning disregards any other fundamental dynamics of the process, the resulting PID tuning is also considered suboptimal.

For these reasons, in the following section, an explicit solution for tuning the PID controller's parameters is presented. Note that the proposed control action leads also to a stable PI control action which gives the flexibility to control engineers to omit the D term depending always on the application. For the proposed control law's proof, a general transfer function of the process model is adopted, which gives the flexibility to control engineers to include in the control action all modeled process parameters and not only the dominant time constant as it happens with the conventional method. An extensive performance comparison between the conventional and the revised control law is presented within Sect. 4.4.

4.3 Revised PID Tuning Via the Symmetrical Optimum Criterion

Within common industrial control loops, the closed-loop control system of Fig. 4.1 is considered again, if modeling in the frequency domain is followed. Therefore, let the integrating process be defined by

$$G(s) = \frac{s^m\beta_m + s^{m-1}\beta_{m-1} + \cdots + s\beta_1 + 1}{s(s^{n-1}a_{n-1} + \cdots + s^3a_3 + sa_1 + 1)}\mathrm{e}^{-sT_{\mathrm{d}}} \tag{4.25}$$

where $n - 1 > m$. The proposed PID controller is given by

$$C(s) = \frac{1 + sX + s^2Y}{sT_{\mathrm{i}}^2(1 + sT_{\mathrm{p_n}})} \tag{4.26}$$

where parameter T_{p_n} stands for the parasitic controller's time constant and is considered known from the controller's implementation. Note that the flexible form of numerator $N_c(s) = 1 + sX + s^2Y$ allows parameters $X, \ Y$ to become complex conjugate if needed. Purpose of the following analysis is to determine analytically controller parameters as a function of all modeled time constants within the control loop, $X = f_1(\beta_i, a_j, T_d)$, $Y = f_2(\beta_i, a_j, T_d)$, $T_i = f_3(\beta_i, a_j, T_d)$ and in contrast to the conventional PID tuning see, [4, 5, 10, 16, 19], pole-zero cancellation does not take place. According to (4.25) and (4.26), the product $C(s)G(s)$ is defined by

$$C(s)G(s) = \frac{(1 + sX + s^2Y)\sum_{j=0}^{m}(s^j\beta_j)}{s^2T_i^2 e^{sT_d}\sum_{i=0}^{n}(s^i p_i)} \tag{4.27}$$

where

$$\sum_{i=0}^{n}(s^i p_i) = (1 + sT_{p_n})\sum_{j=0}^{n-1}(s^j a_j). \tag{4.28}$$

According to Fig. 4.1, the closed-loop transfer function is given by

$$T(s) = \frac{F_{fp}(s)}{1 + F_{ol}(s)} = \frac{k_p C(s)G(s)}{1 + k_p k_h C(s)G(s)} \tag{4.29}$$

where $F_{fp}(s)$, $F_{ol}(s)$ stand for the forward path and the open-loop transfer function respectively. Along with the aid of (4.27), $T(s)$ becomes equal to

$$T(s) = \frac{k_p(1 + sX + s^2Y)\sum_{j=0}^{m}(s^j\beta_j)}{s^2T_i^2 e^{sT_d}\sum_{i=0}^{n}(s^i p_i) + k_p k_h(1 + sX + s^2Y)\sum_{j=0}^{m}(s^j\beta_j)}. \tag{4.30}$$

In the sequel, a general-purpose time constant c_1 is considered for normalizing all time constants within the control loop. Therefore, frequency is normalized by setting $s' = sc_1$ and the following substitutions

$$x = \frac{X}{c_1}, \quad y = \frac{Y}{c_1^2}, \quad t_i = \frac{T_i}{c_1} \tag{4.31}$$

$$d = \frac{T_d}{c_1}, \quad r_i = \frac{p_i}{c_1^i}, \quad \forall i = 1, \ldots, n, \quad z_j = \frac{\beta_j}{c_1^j}, \quad \forall j = 1, \ldots, m. \tag{4.32}$$

are considered. The time delay constant is approximated by the series

$$e^{s'd} = \sum_{k=0}^{\infty}\frac{1}{k!}s^k d^k \tag{4.33}$$

see [15]. Substituting the normalized parameters along with the approximation of $e^{s'd}$ into (4.29) results in

$$T(s') = \frac{k_p(1+s'x+s'^2y)\sum_{j=0}^{m}\left(s'^j z_j\right)}{s'^2 t_i^2 e^{s'd}\sum_{i=0}^{n}\left(s'^i r_i\right) + k_p k_h\left(1+s'x+s'^2y\right)\sum_{j=0}^{m}\left(s'^j z_j\right)} \tag{4.34}$$

or in a more compact form

$$T(s') = \frac{N(s')}{D_1(s') + k_h N(s')} = \frac{N(s')}{D(s')}, \tag{4.35}$$

where

$$N(s') = k_p(1+s'x+s'^2y)\sum_{j=0}^{m}(s'^j z_j) \tag{4.36}$$

and

$$D_1(s') = s'^2 t_i^2\left(\sum_{k=0}^{7}\frac{1}{k!}(s'^k)d^k\right)\sum_{i=0}^{n}(s'^i r_i). \tag{4.37}$$

If (4.37) is expanded, results in

$$\begin{aligned} D_1(s') = {} & s'^2 t_i^2 + s'^3 t_i^2(r_1+d) + s'^4 t_i^2\left(r_2 + r_1 d + \frac{1}{2!}d^2\right) \\ & + s'^5 t_i^2\left(r_3 + r_2 d + \frac{1}{2!}d^2 r_1 + \frac{1}{3!}d^3\right) \\ & + s'^6 t_i^2\left(r_4 + r_3 d + \frac{1}{2!}d^2 r_2 + \frac{1}{3!}d^3 r_1 + \frac{1}{4!}d^4\right) + \cdots \end{aligned} \tag{4.38}$$

Substituting the constant terms of (4.38) with $q_0 = 1$, $q_1 = r_1 + d$, $q_2 = r_2 + r_1 d + \frac{1}{2!}d^2$, $q_3 = r_3 + r_2 d + \frac{1}{2!}d^2 r_1 + \frac{1}{3!}d^3$, $q_4 = r_4 + r_3 d + \frac{1}{2!}d^2 r_2 + \frac{1}{3!}d^3 r_1 + \frac{1}{4!}d^4$, results in

$$\begin{aligned} D_1(s') = {} & \cdots + s'^8 t_i^2 q_6 + s'^7 t_i^2 q_5 + s'^6 t_i^2 q_4 \\ & + s'^5 t_i^2 q_3 + s'^4 t_i^2 q_2 + s'^3 t_i^2 q_1 + s'^2 t_i^2 q_0 \end{aligned} \tag{4.39}$$

where $q_{(-2)} = q_{(-1)} = 0$. From (4.35) it becomes clear that

$$N(s') = k_p\sum_{r=0}^{p}(s'^r)(yz_{r-2} + xz_{r-1} + z_r) \tag{4.40}$$

where $z_r = 0$, if $r < 0$, and $z_0 = 1$. As a result, the final polynomial $D(s')$ of the closed-loop transfer function is defined by

$$D(s') = D_1(s') + k_h N(s') = \sum_{j=0}^{k} (t_i^2 q_j)(s')^{(j+2)} + k_h k_p \sum_{r=0}^{p} (s'^r)\left(y z_{r-2} + x z_{r-1} + z_r\right). \quad (4.41)$$

According to (4.35), (4.40) and (4.41), the resulting closed-loop transfer function is given by

$$T(s') = \frac{k_p \sum_{r=0}^{p} (s'^r)(y z_{r-2} + x z_{r-1} + z_r)}{\sum_{j=0}^{k} (t_i^2 q_j)(s')^{(j+2)} + k_p k_h \sum_{r=0}^{p} (s'^r)\left(y z_{r-2} + x z_{r-1} + z_r\right)}. \quad (4.42)$$

Since (4.42) is now written in the same form of (A.1), for determining the optimal control law the optimization conditions proved in Appendix A.1 can now be used. Eqs. (A.9)–(A.12) are used for the derivation of the optimal control law. Therefore, the problem to be solved is formulated as follows: given known the parameters of the plant, calculate explicitly the PID control action $x,\ y,\ t_i$. In Appendix B.2, the proof of the optimal control law is presented which is proved to be equal to

$$k_h = 1, \quad (4.43)$$

$$x^2 + \frac{b_1}{a_1}x + \frac{c_1}{a_1} = 0 \quad (4.44)$$

$$y = a_2 x^2 + b_2 x + c_2 \quad (4.45)$$

$$t_i^2 = \frac{1}{2}k_p k_h (x^2 - 2y) + \frac{1}{2}k_p k_h (z_1^2 - 2z_2) \quad (4.46)$$

where

$$a_1 = 2\left[q_1 (q_1 - z_1) - q_2 + z_2\right] \quad (4.47)$$

$$b_1 = -4\left(q_1^3 - 3q_1^2 z_1 + 2q_1 z_1^2 + q_1 z_2 + q_2 z_1 - q_3 + z_3 - 2z_1 z_2\right) \quad (4.48)$$

$$c_1 = \begin{bmatrix} \left(q_1^2 - 2q_1 z_1 + 2z_2\right)\left(z_1^2 + 2z_2 + 4q_2 - 4q_1 z_1\right) + \left(q_1^2 - 2q_2\right)\left(z_1^2 - 2z_2\right) \\ +4\left(q_1 z_3 + q_3 z_1 - q_4 - z_4 - q_2 z_2\right) \end{bmatrix} \quad (4.49)$$

and

$$a_2 = -\frac{1}{2} \tag{4.50}$$

$$b_2 = 2\,(q_1 - z_1) \tag{4.51}$$

$$c_2 = -\frac{1}{2}\left(z_1^2 + 2z_2 + 4q_2 - 4q_1z_1\right), \tag{4.52}$$

are process-dependent parameters. Therefore, once x is solved through (4.44), then y is calculated out of (4.45). Integrator's time constant $t_{\rm i}$ is then easily calculated out of (4.46).

4.4 Performance Comparison Between Conventional and Revised PID Tuning

In this section, a comparison performance study is presented between the conventional and the revised PID tuning rules as those proven in Sects. 4.2 and 4.3, respectively. Given the transfer function of a certain plant, the closed-loop control system is constructed and its step and frequency response is investigated. Special attention is paid also to the control effort (command signal) introduced both by the conventional and the revised design (Fig. 4.5).

4.4.1 Plant with One Dominant Time Constant

In this example, the plant is described by the transfer function

$$G(s') = \frac{1}{(1+s')(1+0.2s')(1+0.1s')(1+0.1s')(1+0.05s')}. \tag{4.53}$$

For controlling (4.53) and after applying the revised PI tuning rules the controller of

$$C_{\rm rev}(s') = \frac{1+5.73s'}{16.45s'^2(1+s't_{\rm sc})} \tag{4.54}$$

is calculated. In similar fashion, the conventional PID control action defined by

$$C_{\rm cl}(s') = \frac{(1+s't_n)(1+s't_v)}{s'^2t_{\rm i}^2(1+s't_{\rm sc})}, \tag{4.55}$$

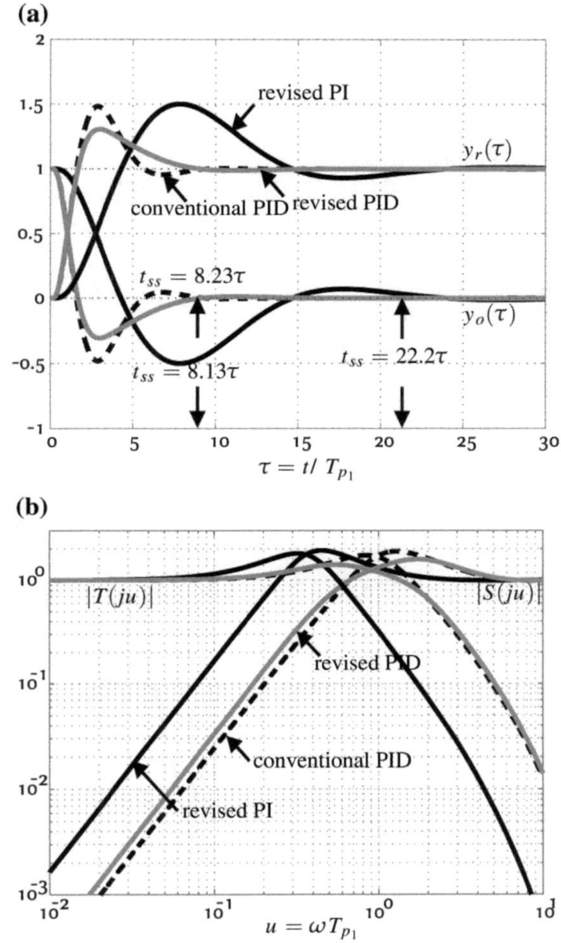

Fig. 4.5 Control of a process with one dominant time constant defined by (4.53). Comparison between the conventional and the revised PID tuning method. (*Black*) revised PI tuning, (*black dotted*) conventional PID tuning, (*gray*) revised PID tuning. **a** Step response of the control loop. **b** Frequency response of sensitivity *S* and complementary sensitivity *T*

results in

$$C_{\text{con}}(s') = \frac{(1 + 2.2s')(1 + s')}{2.42s'^2(1 + s't_{\text{sc}})}. \tag{4.56}$$

According to the revised method where the proposed controller is given by

$$C_{\text{rev}}(s') = \frac{1 + s'x + s'^2 y}{s'^2 t_{\text{i}}^2(1 + s't_{\text{sc}})} \tag{4.57}$$

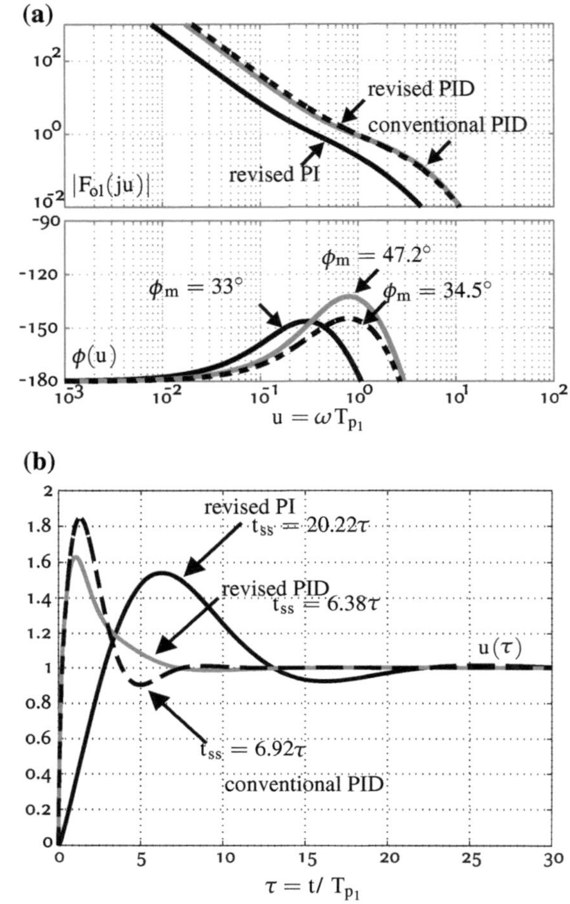

Fig. 4.6 Control of a process with one dominant time constant defined by (4.53). Comparison between the conventional and the revised PID tuning method. (*Black*) revised PI tuning, (*black dotted*) conventional PID tuning, (*gray*) revised PID tuning. **a** Frequency response, phase diagram of the open loop transfer function $F_{ol}(s)$. **b** Step response of the command signal $u(\tau)$ in the presence of a change on the reference signal $r(s)$

as the calculated parameters are defined by

$$\begin{aligned} C_{\text{rev}}(s') &= \frac{1 + 3.63s' + 3.32s^2}{3.29s'^2 (1 + s' t_{sc})} \\ &= \frac{\left[1 + s' (1.819 + 0.13i)\right]\left[1 + s' (1.819 - 0.13i)\right]}{3.29s'^2 (1 + s' t_{sc})}. \end{aligned} \tag{4.58}$$

From (4.58), it is apparent that the revised PID tuning method has led to a controller with conjugate complex zeros. In Fig. 4.6, it is shown that there is little difference both in the step and frequency response of the closed-loop control system. Specifically, after comparing the step response and output disturbance rejection of the control loop, it is clear that Fig. 4.6 settling time of disturbance rejection is 8.13τ in case of the revised tuning compared to 8.23τ in case of the conventional tuning.

From the frequency response and regarding both PID tuning methods, see Fig. 4.6 of T, S, robustness of the control loop is practically the same. Note at that point that the conventional tuning method fails to tune a PI control action, see Sect. 3.2.3, in contrast with the proposed method. From Fig. 4.6b, it is clear that the peak value of the PI control action $u(\tau)$ is significantly lower than the one provided by the PID control action. This advantage can be critical in a real-world application, since high peak command signal values might not be available by the constraints of the hardware of the actuator unit.

From the frequency response and the phase diagram, see Fig. 4.6b it is apparent that the phase margin of the $F_{\text{ol}}(s)$ is $\phi(u) = 47.2°$, whereas in the case of the revised PID tuning the phase margin is $\phi(u) = 34.5°$ in the case of PID control. The level of the phase margin is ($\phi(u) < 45°$) is justified also by the overshoot of the step response of the closed-loop control system, see Fig. 4.5 which is higher the 50 %.

4.4.2 Plant with Two Dominant Time Constants

In this example, the plant with two dominant time constants defined by

$$G(s') = \frac{1}{(1+s')(1+s')(1+0.01s')(1+0.001s')(1+0.0001s')} \tag{4.59}$$

is considered. After the application of the revised PI control law, it is found that

$$C_{\text{rev}}(s') = \frac{1+7.82s'}{30.56s'^2(1+s't_{\text{sc}})}. \tag{4.60}$$

In similar fashion, the conventional and the revised PID control action are given by (Fig. 4.7)

$$C_{\text{con}}(s') = \frac{(1+4.44s')(1+s')}{9.87s'^2(1+s't_{\text{sc}})} \tag{4.61}$$

and

$$C_{\text{rev}}(s') = \frac{1+5.1s'+6.1s^2}{6.84s'^2(1+s't_{\text{sc}})} = \frac{(1+3.15s')(1+1.88s')}{6.84s'^2(1+s't_{\text{sc}})} \tag{4.62}$$

respectively. From the step response of the closed-loop control system, see Fig. 4.8 it is found that the revised PID control action leads to faster disturbance rejection compared to the conventional tuning, since the settling time t_{ss} in the first case is $t_{\text{ss}} = 11.1\tau$ compared to $t_{\text{ss}} = 18\tau$ in the second case.

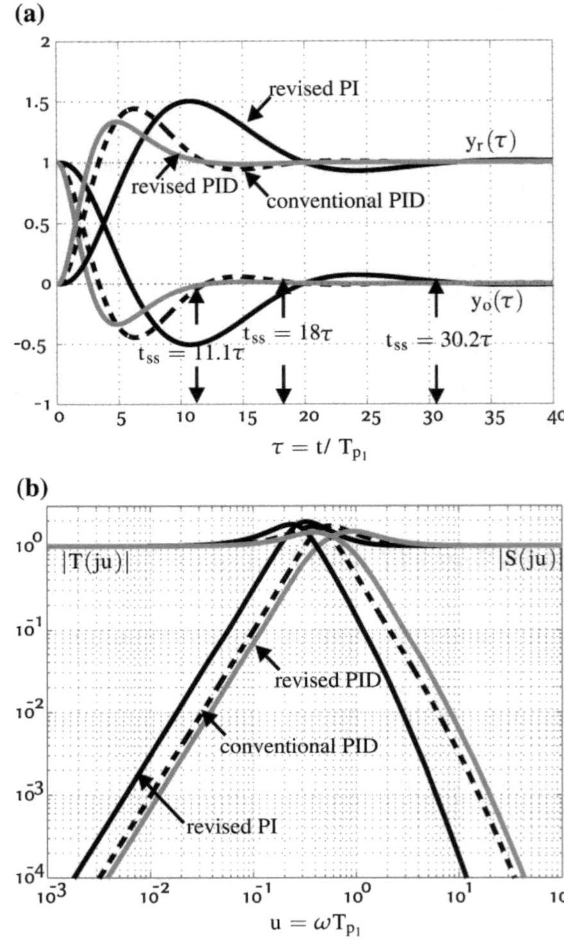

Fig. 4.7 Control of a process with two dominant time constants defined by (4.59). Comparison between the conventional and the revised PID tuning method. (*Black*) revised PI tuning, (*black dotted*) conventional PID tuning, (*gray*) revised PID tuning. **a** Step response of the control loop. **b** Frequency response of sensitivity S and complementary sensitivity T

In Fig. 4.7b, it is shown that the revised PID control action has improved the robustness of the closed-loop control system, since the magnitude of complementary sensitivity $|T(ju)|$ remains equal to one in a wider range compared to the conventional tuning. The same result holds for sensitivity S, since the amplitude of $|S(ju)|$ remains equal to zero in a wider range in the case of the revised control action.

The phase margin introduced to the closed-loop control system via the conventional tuning is equal to $\phi(u) = 36.3°$, whereas in the case of the revised tuning the phase margin is equal to $\phi(u) = 45.7°$, see Fig. 4.8a. From Fig. 4.8b, it is apparent that the revised tuning requires less effort on the command signal side since the settling time is $t_{ss} = 8.6\tau$ compared to $t_{ss} = 15.1\tau$ which is required by the conventional tuning.

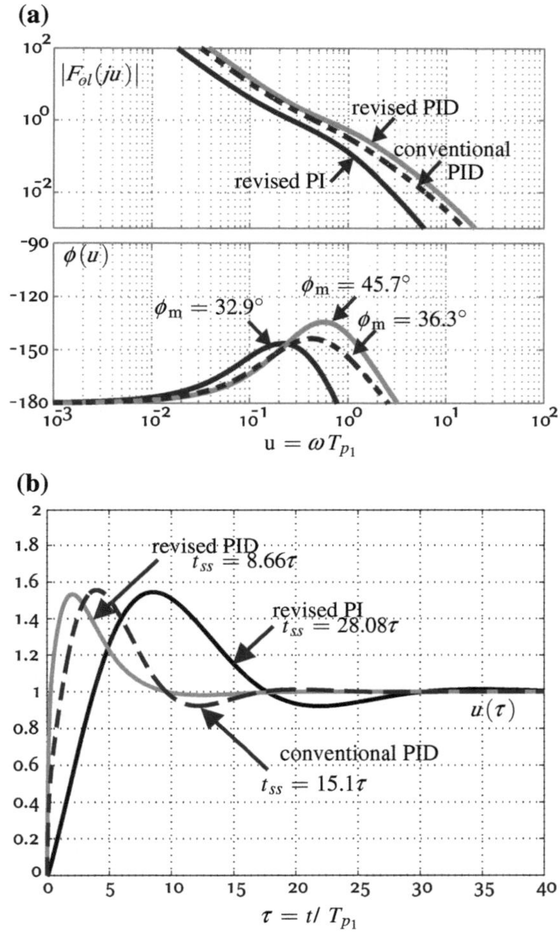

Fig. 4.8 Control of a process with two dominant time constants defined by (4.59). Comparison between the conventional and the revised PID tuning method. (*Black*) revised PI tuning, (*black dotted*) conventional PID tuning, (*gray*) revised PID tuning. **a** Frequency response, phase diagram of the open loop transfer function $F_{\mathrm{ol}}(s)$. **b** Step response of the command signal $u(\tau)$ in the presence of a change on the reference signal $r(s)$

4.4.3 A Non-minimum Phase Process

In this example, let the transfer function of the plant be defined by

$$G(s') = \frac{(1-2s')(1-1.8s')}{(1+s')^5}, \tag{4.63}$$

which introduces two zeros on the right half plane. In this case, the calculated PI control action via the revised method is given by

$$C_{\mathrm{rev}}(s') = \frac{1+29.9s'}{451.2s'^2(1+s't_{\mathrm{sc}})}. \tag{4.64}$$

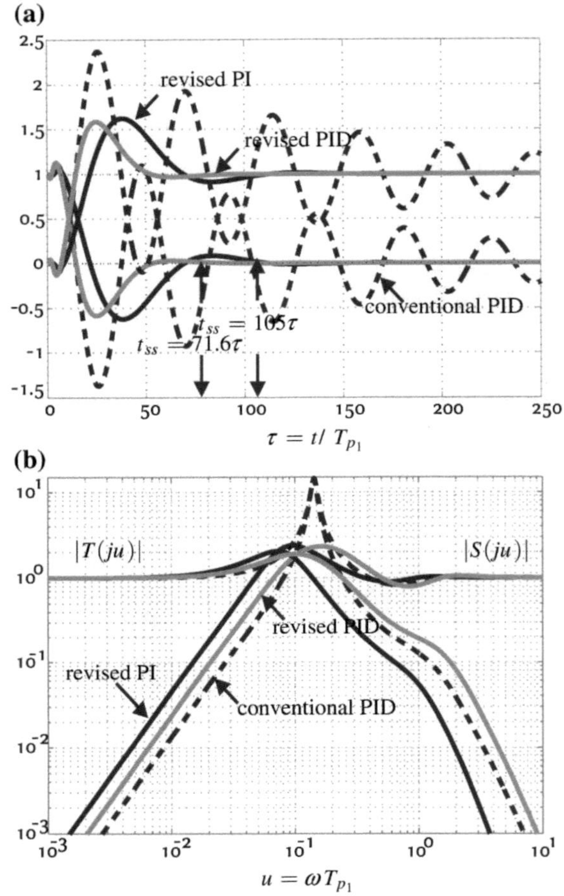

Fig. 4.9 Control of a non-minimum phase process defined by (4.63). Comparison between the conventional and the revised PID tuning method. (*Black*) revised PI tuning, (*black dotted*) conventional PID tuning, (*gray*) revised PID tuning. **a** Step response of the control loop. **b** Frequency response of sensitivity S and complementary sensitivity T

In the case of PID control, the conventional and the revised controllers are given by

$$C_{\text{con}}(s') = \frac{(1+16.4s')(1+s')}{134.48s'^2(1+s't_{\text{sc}})} \tag{4.65}$$

and

$$C_{\text{rev}}(s') = \frac{1+24.4s'+65.7s^2}{236.7s'^2(1+s't_{\text{sc}})} = \frac{(1+21.3s')(1+3.1s')}{236.7s'^2(1+s't_{\text{sc}})} \tag{4.66}$$

respectively.[7] From Fig. 4.9a, it is apparent that the conventional PID tuning method fails to tune a control loop with acceptable performance. By contrast, the revised

[7] Let it be noted that the conventional tuning has never been tested to non-minimum phase processes within the academic literature.

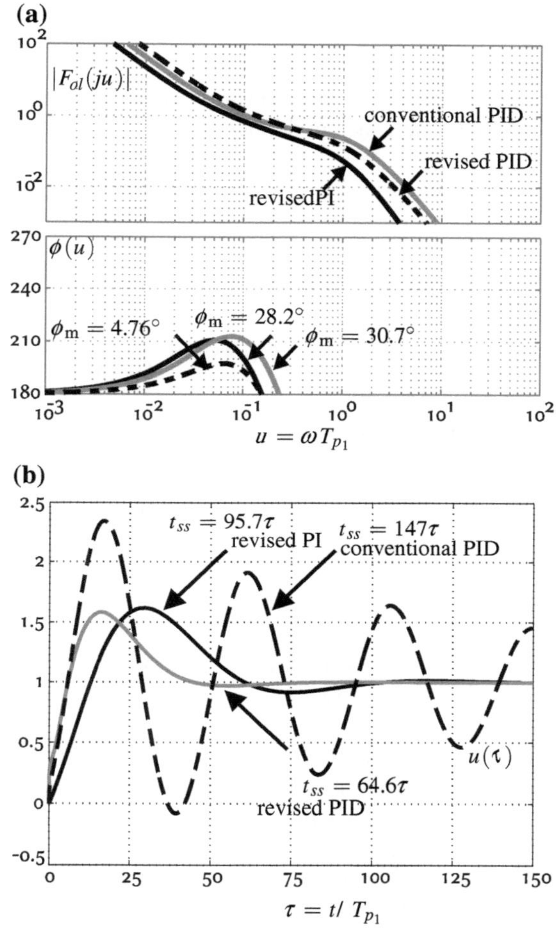

Fig. 4.10 Control of a non-minimum phase process defined by (4.63). Comparison between the conventional and the revised PID tuning method. (*Black*) revised PI tuning, (*black dotted*) conventional PID tuning, (*gray*) revised PID tuning. **a** Frequency response, phase diagram of the open loop transfer function $F_{ol}(s)$. **b** Step response of the command signal $u(\tau)$ in the presence of a change on the reference signal $r(s)$

method succeeds in tuning the PI, PID control action achieving fast disturbance suppression. The oscillatory behavior of the control loop involving the conventional tuning is also observed in the frequency domain where the magnitude of $|T(ju)|$ exhibits a high peak, ten times greater than the unity.

From Fig. 4.10b, it is clear that the oscillatory behavior in the control loop with the conventional tuning is the result of the unacceptable command signal which results from the poor tuning of the controller.

4.4.4 Plant with Long Time Delay

In this example, we consider a process with time delay five times greater than its dominant time constant defined by

$$G(s') = \frac{1}{(1+s')^5} e^{-5s'}. \tag{4.67}$$

Regarding the PI control action, the revised method results in the controller

$$C_{\text{rev}}\left(s'\right) = \frac{1+34.8s'}{606.5s'^2(1+s't_{\text{sc}})}, \tag{4.68}$$

whereas the corresponding PID controller via the conventional and the revised methods are given by

$$C_{\text{con}}(s') = \frac{(1+16.4s')(1+s')}{134.48s'^2(1+s't_{\text{sc}})} \tag{4.69}$$

and

$$C_{\text{rev}}(s') = \frac{1+27.22s'+82.3s^2}{288.22s'^2(1+s't_{\text{sc}})} = \frac{(1+23.75s')(1+3.46s')}{288.22s'^2(1+s't_{\text{sc}})} \tag{4.70}$$

respectively. Note that in this case, controller (4.69) fails to tune a stable control loop. On the contrary, the proposed method leads to a satisfactory step response and disturbance rejection, see Fig. 4.11a. Note that, the PID controller exhibits an increased robustness regarding disturbances, see Fig. 4.11b. The introduction of the D term decreases dramatically the settling time of disturbance rejection, from $t_{\text{ss}} = 123\tau$ to $t_{\text{ss}} = 81.9\tau$. This also reflected by the step response of the command signal $u(\tau)$ where the settling time of the PID control action is $t_{\text{ss}} = 71\tau$ compared to $t_{\text{ss}} = 117\tau$ in the case of PI control action, see Fig. 4.12b.

4.4.5 Plant with Large Zeros

In the last example, a process with large zeros is investigated. Its transfer function is given by

$$G(s') = \frac{(1+1.5s')}{(1+s')(1+0.9638s')(1+0.4061s')(1+0.2392s')(1+0.1751s')}. \tag{4.71}$$

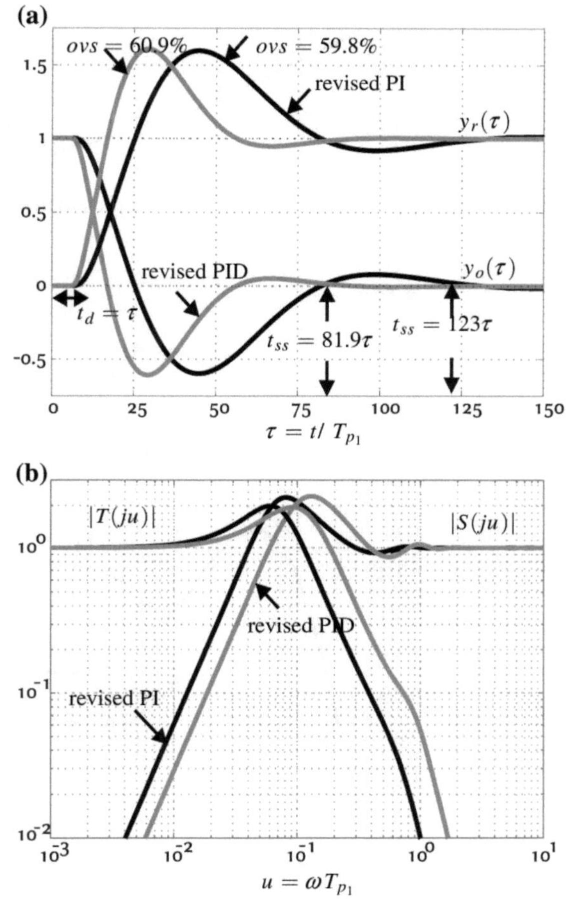

Fig. 4.11 Control of a process with long time delay defined by (4.67). Comparison between the conventional and the revised PID tuning method. (*Black*) revised PI tuning, (*black dotted*) conventional PID tuning, (*gray*) revised PID tuning. **a** Step response of the control loop. **b** Frequency response of sensitivity S and complementary sensitivity T

The aforementioned feature is reflected also in the step response of the closed-loop control system since the overshoot introduced in the case of the revised control action is almost equal to 50 % both for the PI and the PID controller, see Fig. 4.13a.

After applying PI control action, the revise controller is defined by

$$C_{\text{rev}}(s') = \frac{1 + 4.32s'}{10.49s'^2(1 + s't_{\text{sc}})}. \tag{4.72}$$

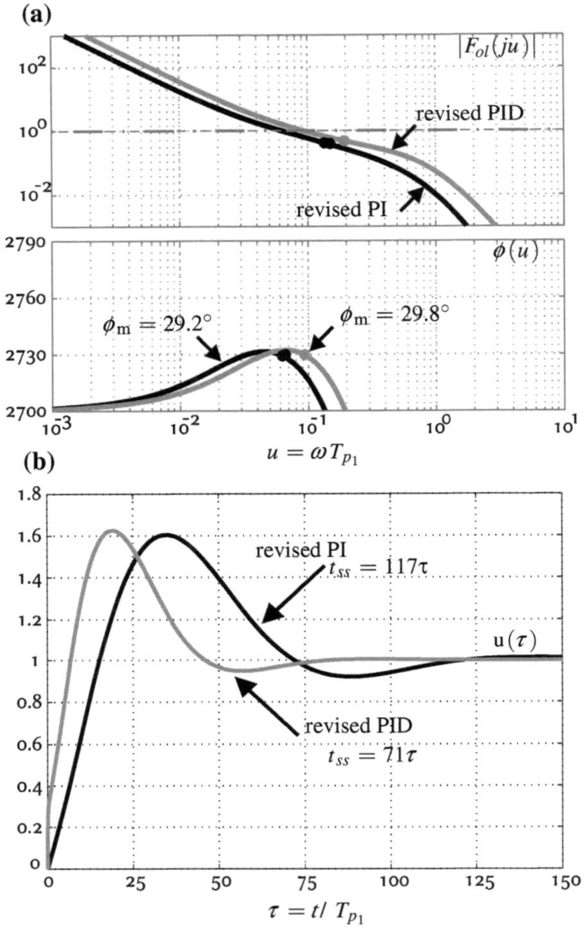

Fig. 4.12 Control of a process with long time delay defined by (4.67). Comparison between the conventional and the revised PID tuning method. (*Black*) revised PI tuning, (*black dotted*) conventional PID tuning, (*gray*) revised PID tuning. **a** Frequency response, phase diagram of the open loop transfer function $F_{ol}(s)$. **b** Step response of the command signal $u(\tau)$ in the presence of a change on the reference signal $r(s)$

whereas the corresponding PID controller for the conventional and the revised tuning are given by

$$C_{con}(s') = \frac{(1 + 7.53s')(1 + s')}{28.4s'^2(1 + s't_{sc})} \tag{4.73}$$

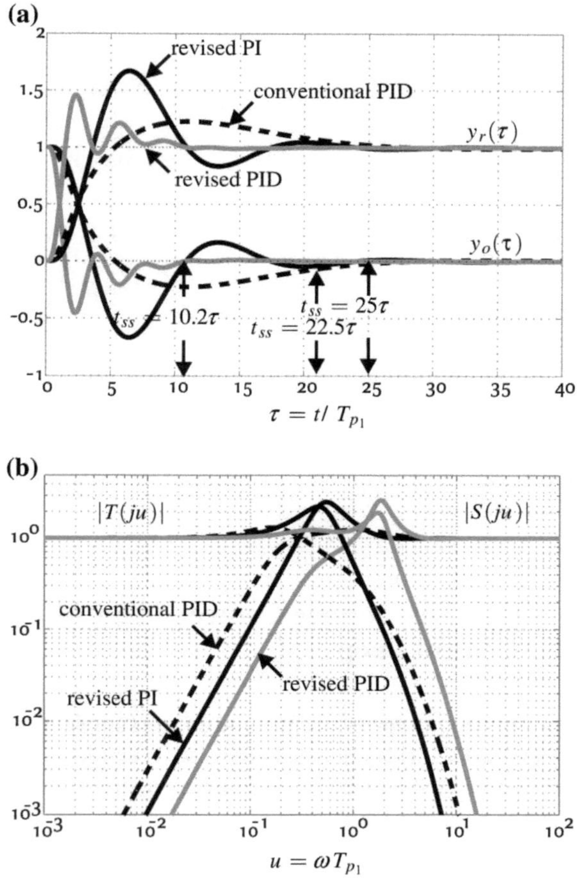

Fig. 4.13 Control of a process with large zeros defined by (4.71). Comparison between the conventional and the revised PID tuning method. (*Black*) revised PI tuning, (*black dotted*) conventional PID tuning, (*gray*) revised PID tuning. **a** Step response of the control loop. **b** Frequency response of sensitivity S and complementary sensitivity T

and

$$C_{\text{rev}}(s') = \frac{1 + 3.73s' + 4.73s^2}{3.34s'^2(1 + s't_{\text{sc}})} \tag{4.74}$$

respectively. In Fig. 4.13a, it is shown that the revised PID controller leads to a much faster output disturbance rejection, since the settling time in the case of conventional tuning is $t_{\text{ss}} = 25\tau$, whereas in the case of revised tuning is $t_{\text{ss}} = 10.2\tau$. However, the disadvantage of the revised PID control action is apparent from the phase diagram of the open-loop transfer function $F_{\text{ol}}(s)$, since the phase margin in the case of conventional tuning is $\phi(u) = 58.5 > 45°$ in contrast with the revised method which is equal to $\phi(u) = 25.4 > 45°$, see Fig. 4.14.

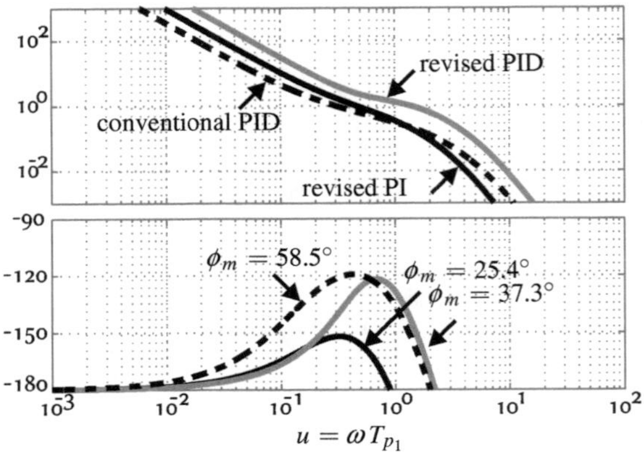

Fig. 4.14 Control of a process with large zeros defined by (4.71). Comparison between the conventional and the revised PID tuning method. (*Black*) revised PI tuning, (*black dotted*) conventional PID tuning, (*gray*) revised PID tuning. Frequency response, phase diagram of the open-loop transfer function $F_{\rm ol}(s)$

4.5 DC Link Voltage Control on an AC/DC Converter-Type-II Control Loop

For verifying the proposed method on an example from the industry, the typical model of an AC/DC grid connected converter is employed [12]. The converter connects the DC link capacitor to the grid through a grid transformer as shown in Fig. 4.15. Its main

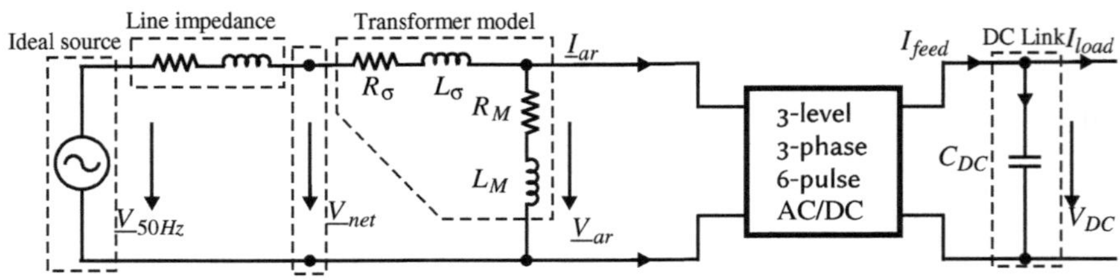

Fig. 4.15 Grid connected active rectifier on system level. The interfaced signals are: $\underline{V}_{\rm net}$ (V) and $\omega_{\rm net}$ (Hz) stand for the three-phase voltage measured at PCC; $\underline{V}_{\rm ar}$ (V) $\underline{I}_{\rm ar}$ (A) stand for the three-phase voltage and current at the grid converter respectively. $V_{\rm DC}$ (V) and $I_{\rm feed}$ (A) is the DC link voltage and the feeding current from grid converter and $I_{\rm C}$ (A), $I_{\rm load}$ (A) is the DC link capacitor current and the load current respectively. System parameters that should be known or estimated for controlling purpose are $C_{\rm DC}$ (C) from the DC link capacitor, L_σ (H) (leakage inductance of the transformer), R_σ (Ω) (leakage resistance of the transformer), $L_{\rm M}$ (H) (magnetizing inductance of the transformer), $R_{\rm M}$ (Ω) (magnetizing resistance of the transformer), $L_{\rm net}$ (H) (equivalent line inductance of the network), $R_{\rm net}$ (Ω) (equivalent line resistance of the network)

purpose is to maintain the DC link voltage that supplies power for a drive typically, or another network. The network may be modeled as a voltage source along with its grid impedance that reflects its strength. The grid transformer is modeled through its magnetizing and leakage impedance.

4.5.1 Simplified Control Model and Parameters

The most classic way for controlling AC/DC grid converters is the cascaded vector control, Fig. 4.16. The DC link voltage controller provides the current reference to the grid current controller which itself provides a reference to the modulator through the modulation index M_{ar}. The load current I_{load} is the main perturbation of the system and a power feed-forward current I_{pff} can be provided to the output of the voltage controller for enhancing its dynamics. Whereas the description of the synchronization to the grid through a dedicated PLL is not the scope of this section, it cannot be ignored since it provides the reference for the vector control, Fig. 4.16.

4.5.2 Modeling of the Control Loop in the Frequency Domain

Given the control structure in Fig. 4.16, the current control loop after neglecting the cross coupling terms is equal to

$$T_{\mathrm{I}}(s) = \frac{I_{\mathrm{d}_{\mathrm{act}}}(s)}{I_{\mathrm{d}_{\mathrm{ref}}}(s)} = \frac{1}{(1+sT_{\mathrm{p}_{\mathrm{I}}})(1+sT_{\Sigma_{\mathrm{p}}})} = \frac{1}{(1+sT_{\mathrm{p}_{\mathrm{I}}})(1+s\gamma T_{\mathrm{p}_{\mathrm{I}}})} \tag{4.75}$$

for which γ can be chosen sufficiently small compared to the current control loop's time constant, modeling any parasitic time constants of the inner current control loop itself. The transfer function regarding the DC link voltage control loop (outer control loop, see Fig. 4.16) is equal to $T_{\mathrm{v}}(s) = \frac{V_{\mathrm{DC}_{\mathrm{act}}}(s)}{V_{\mathrm{DC}_{\mathrm{ref}}}(s)}$ or

$$T_{\mathrm{v}}(s) = \frac{C_{\mathrm{v}}(s)T_{\mathrm{I}}(s)G(s)}{1+C_{\mathrm{v}}(s)T_{\mathrm{I}}(s)G(s)} = \frac{C_{\mathrm{v}}(s)T_{\mathrm{I}}(s)\dfrac{1}{sC_{\mathrm{DC}}}}{1+C_{\mathrm{v}}(s)T_{\mathrm{I}}(s)\dfrac{1}{sC_{\mathrm{DC}}}} \tag{4.76}$$

where C_{v} is the DC link voltage controller and $G(s)$ is the integrating process for the voltage control loop, which in this case is the capacitor bank path within the DC link, see Fig. 4.16. Substituting (4.75) into the voltage control loop transfer function,

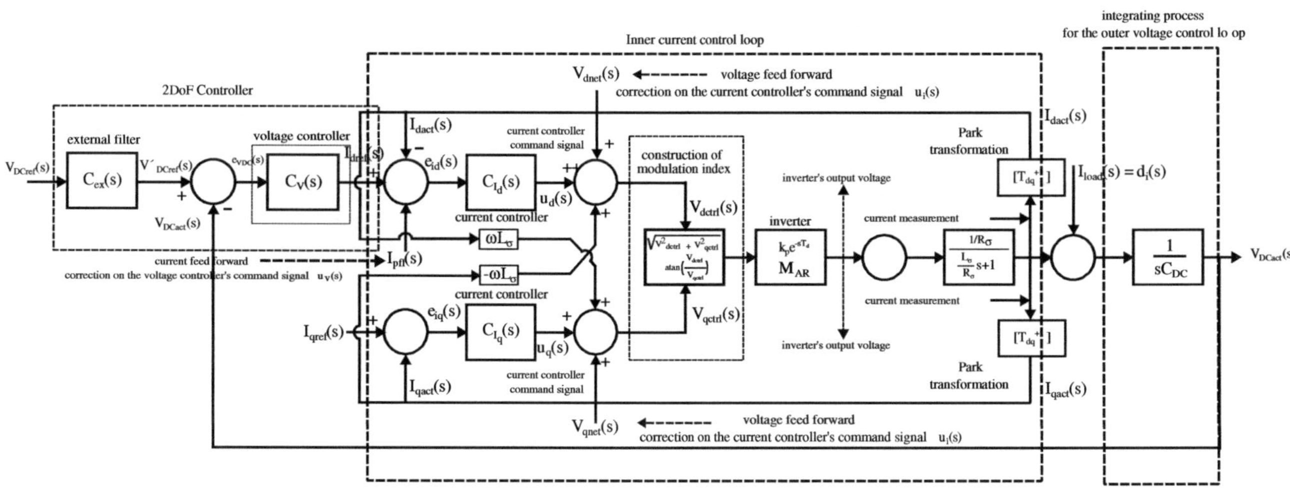

Fig. 4.16 Cascaded control loop for AC/DC grid converters. The interfaced signals are M_{ar}: modulation index of grid converter AC voltage, V_{dctrl} (V): voltage control value (active-part), V_{qctrl} (V): voltage control value (reactive-part), V_{dnet} (V): grid voltage measurement (active-part), V_{qnet} (V): grid voltage measurement (reactive-part), I_D (A): grid current measurement (active-part), I_Q (A): grid current measurement (reactive-part), I_{pff} (A): current feed-forward from load drive, I_{dref} (A): grid current reference (active-part), I_{qref} (A): grid current reference (reactive-part), V_{DCref} (V): DC link voltage reference

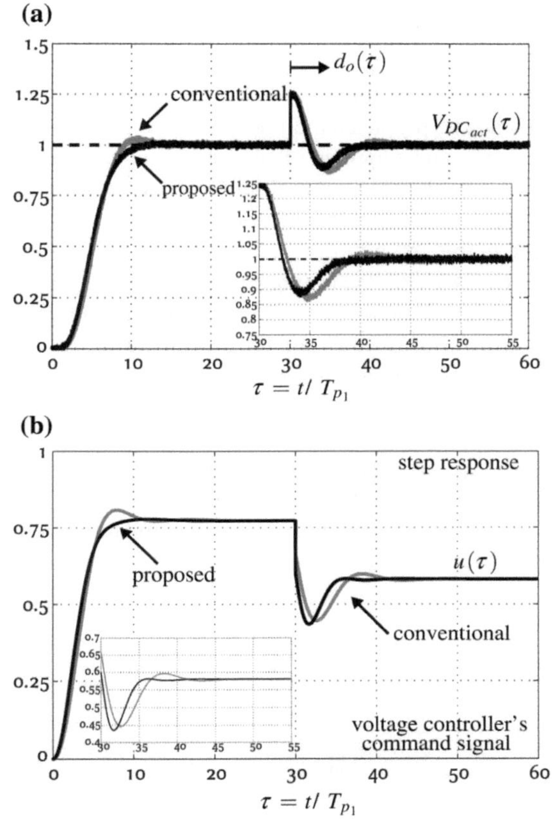

Fig. 4.17 Step response of the DC link voltage control loop. Time constants within the current and the voltage control loop have been set equal to $t_{p_1} = 1$, $t_{p_2} = 0.65$, $t_{p_1} = 0.3$, $t_{p_4} = 0.1$, $t_{p_5} = 0.05$. No feed-forward terms for correcting the command signal of the DC Link voltage PI controller is assumed. **a** Step response of the actual voltage in changes of $V_{DC_{ref}}$ in the presence of output disturbance $d_o(s)$ at $\tau = 30$. **b** Response of the command signal, output of the voltage controller (current reference signal)

results after in

$$T_v(s) = \frac{C_v(s)}{sC_{DC}(1 + sT_{p_1})(1 + s\gamma T_{p_1}) + C_v(s)} \tag{4.77}$$

where $C_v(s)$ is designed every time according to the conventional and the revised symmetrical optimum criterion presented in Sects. 4.2, 4.2.3 and 4.3.

Since the control loop is of type-II, an external filter is added on the reference signal, $V_{DC_{ref}}$ for dealing with the high overshoot in $V_{DC_{act}}$ in case of step changes in $V_{DC_{ref}}$. Note that $V_{DC_{ref}}$ is set on the electric drive, in practice it does change that often. The choice of the filter time constant of $C_{ex}(s)$ is chosen such that it cancels the zero of the calculated voltage controller as it is extensively discussed in [11, 13].

In Fig. 4.17, the step response of $V_{DC_{act}}$ is presented. Disturbance rejection applied at $\tau = 30$ shows significant improvement compared to the conventional method, decrease of settling time t_{ss} from $45.5\tau \rightarrow 32\tau$.

4.6 Summary

An explicit PID tuning solution for controlling integrating processes has been presented. The proposed method lies in the principle of the symmetrical optimum criterion and can be applied to any linear single input single output process regardless of its complexity. The control law's proof does not involve any model reduction techniques which often lead to poor tuning as it happens in the case of the conventional PID tuning procedure.

For justifying the tuning performance, the proposed control law is compared with the current state of the art relevant to the PID tuning via the symmetrical optimum criterion. This comparison focuses on the performance of the required control action, in terms of reference tracking and disturbance rejection. Since the proposed method concentrates on the PID controller which is often used in many industry applications, the control of the actual DC link voltage on an AC/DC converter arrangement was chosen as an example from the field of electric motor drives so that the feasibility in terms of the method's implementation is also justified. The presented comparison study reveals a satisfactory and promising improvement in terms of reference tracking and disturbance rejection.

References

1. Åström KJ, Hang CC, Lim BC (1994) A new Smith predictor for controlling a process with an integrator and long dead time. IEEE Trans Autom Control 39(2):343–345
2. Bahrani B, Kenzelmann S, Rufer A (2011) Multivariable-PI-based current control of voltage source converters with superior axis decoupling capability. IEEE Trans Ind Electron 58(7):3016–3026
3. Habetler TG (1993) A space vector-based rectifier regulator for AC/DC/AC converters. IEEE Trans Power Electron 8(1):30–36
4. Kessler C (1955) UG ber die Vorausberechnung optimal abgestimmter regelkreise teil III. Die optimale einstellung des reglers nach dem betragsoptimum. Regelungstechnik 3:40–49
5. Kessler C (1958) Das symmetrische optimum, Regelungstechnik pp 395–400 and 432–426
6. Li S, Pingxi Y, Lifei L, Lin C, Liuxin B, Gang C, Chao Z (2012) Research on grid-connected operation of novel variable speed constant frequency (VSCF) shaft generator system on modern ship. In: 15th international conference on electrical machines and systems. ICEMS, IEEE, Sapporo, pp 1–5
7. Mataušek MR, Micić AD (1999) On the modified Smith predictor for controlling a process with an integrator and long dead-time. IEEE Trans Autom Control 44(8):1603–1606
8. Mirecki A, Roboam X, Richardeau F (2007) Architecture complexity and energy efficiency of small wind turbines. IEEE Trans Ind Electron 54(1):660–670
9. Nishikata S, Tatsuta F (2010) A new interconnecting method for wind turbine/generators in a wind farm and basic performances of the integrated system. IEEE Trans Ind Electron 57(2):468–475
10. Oldenbourg RC, Sartorius H (1954) A uniform approach to the optimum adjustment of control loops. Trans ASME 76:1265–1279
11. Papadopoulos KG, Margaris NI (2012) Extending the symmetrical optimum criterion to the design of pid type-p control loops. J Process Control 12(1):11–25

12. Papadopoulos KG, Siemaszko D, Margaris NI (2012) Optimal automatic tuning of PID controllers applied to grid converters. In: Electrical systems for aircraft, railway and ship propulsion (ESARS). IEEE, Bologna, Italy, pp 1–6
13. Papadopoulos KG, Papastefanaki EN, Margaris NI (2013) Explicit analytical PID tuning rules for the design of type-III control loops. IEEE Trans Ind Electron 60(10):4650–4664
14. Parlmor ZJ (1996) Time delay compensation—Smith predictor and its modifications. In: Levine WS (ed) Control handbook, Boca Raton, FL: CRC Press, vol 53, pp 224–237
15. Richard JP (2003) Time-delay systems: an overview of some recent advances and open problems. Automatica 39(10):1667–1694
16. Sartorius H (1945) Die zweckmässige festlegung der frei wählbaren regelungskonstanten. Master thesis, Technische Hochscule, Stuttgart, Germany
17. Shafiei Z, Shenton AT (1994) Tuning of PID-type controllers for stable and unstable systems with time delay. Automatica 30(10):1609–1615
18. Smith OJM (1959) Closed control of loops with dead-time. Chem Eng Sci 53:217–219
19. Umland WJ, Safiuddin M (1990) Magnitude and symmetric optimum criterion for the design of linear control systems: what is it and how does it compare with the others? IEEE Trans Ind Appl 26(3):489–497
20. Wang QC, Hang CC, Yang PX (2001) Single-loop controller design via IMC principles. Automatica 37(12):2041–2048
21. Watanabe K, Ito M (1981) A process model control for linear systems with delay. IEEE Trans Autom Control 26(6):1261–1268
22. Zhang W, Xu X, Sun Y (1999) Quantitative performance design for integrating processes with time delay. Automatica 35(4):719–723

Chapter 5
Type-III Control Loops

Abstract In this chapter, the problem of designing PID type-III control loops is investigated. On a theoretical basis and if frequency domain modeling is followed, type-III control loops are characterized by the presence of three pure integrators in the open loop transfer function, see Sect. 2.1. Therefore, such a control scheme has the advantage of tracking fast reference signals since it exhibits zero steady state position, velocity and acceleration error, see Sect. 2.1. This advantage is considered critical in many industry applications, i.e. control of electrical motor drives, control of power converters, since it allows the output variable, i.e., DC-link voltage or speed, to track perfectly step, ramp and parabolic reference signals. In a similar fashion, with Chaps. 3 and 4, the proposed PID control law (1) consists of analytical expressions that involve all modeled process parameters (2) can be straightforward applied to any process regardless of its complexity since for its development a generalized transfer function process model is employed consisting of n-poles, m-zeros plus unknown time delay-d (3) allows for accurate investigation of the performance of the control action to exogenous and internal disturbances in the control loop, investigation of different operating points. For justifying the potential of the proposed control law, several examples of process models met in many industry applications are investigated.

5.1 Introduction

From a conceptual point of view, the advantage of type-III control loops compared to type-I or type-II systems is obvious, since the former are able to track a step, ramp, and parabolic reference input by achieving zero steady state position, velocity, and acceleration error, respectively. Therefore, such control loops are capable of tracking very fast reference signals.

A first attempt of designing type-III control loops for single-input single-output processes has been proposed in [3, 5, 7, 8] and is presented in Sect. 5.2.1 for the sake of completeness of the proposed theory. In this case, the design of the control loop is developed in the frequency domain, and the principle of the Symmetrical Optimum criterion is once more adopted [2, 4]. The proposed PID type-III control law is based on pole-zero cancellation as the conventional Symmetrical Optimum

K.G. Papadopoulos, *PID Controller Tuning Using the Magnitude Optimum Criterion*,
DOI 10.1007/978-3-319-07263-0_5

implies, and therefore an accurate estimation of the dominant time constants of the process is required. Further to this constraint, the PID controller zeros are restricted to be tuned only with real values, and not with conjugate complex if needed. Moreover, the process model used to develop the aforementioned control law is simple (second order process model), and therefore other dynamics of the process are neglected.

All aforementioned constraints, regarding the conventional PID type-III control law proposed in [5] can be summarized as follows:

1. the PID controller parameters are tuned as a function of the process's dominant time constants. Unmodeled dynamics of the plant are approximated by a first order lag time constant,
2. the principle of pole-zero cancellation is followed,
3. the PID controller zeros are allowed to be tuned only with real values,
4. a simple second order model is employed for the development of the proposed PID control law regarding type-III control loops.

From the above, it is apparent that when the complexity of the process increases, the conventional PID type-III control law presented in Sect. 5.2.1, and according to [5, 8] fails sometimes to tune a stable control loop as it is shown in the sequel. One way to improve the control law presented in Sect. 5.2.1 is to introduce a more complex process model, and explicitly tune the PID parameters without following model reduction techniques, see Sect. 5.2.2 and Appendix B.3.

To cope with this model reduction approximation issue, a first attempt of designing type-III control loops without using a simple process model has been reported in [8]. In this work, for modelling the process, a transfer function consisting of n poles and unknown time delay d has been employed whereas any zeros of the process are not taken into account. The potential of this PID control law is tested on a nonminimum phase process and a process with dominant time constants achieving promising results.

For that reason and motivated by the promising results in [8], scope of this chapter is to tune analytically a PID type controller, regardless of the process complexity (n poles, m zeros plus time delay d), so that the final closed loop control system exhibits zero steady state, position velocity and acceleration error. At this point, the assumptions presented in [5, Sect. 5.2.1] are disregarded and for developing the proposed theory

1. the PID controller parameters are tuned explicitly as a function of all n poles, m zeros plus time delay d,
2. the principle of pole-zero cancellation is not followed,
3. a more flexible form is introduced and the PID controller zeros are allowed to be tuned both with real values and conjugate complex values if needed, see [9],
4. no model reduction techniques are going to be followed, see Sect. 5.2.2.

In this case, zeros of the PID controller are not forced to be compensated by the plant's dominant time constants since a more flexible form of the PID controller is introduced. This form tunes the zeros of the controller as a function of all modeled process parameters allowing its values to become conjugate complex if needed. To this end, control engineers are able to design PID type-III control loops regardless of the process complexity and analyze accurately the control loop's performance before proceeding on a real time implementation.

For developing the proposed control law, once more the concept of Symmetrical Optimum criterion is employed [2, 4, 9]. Thus, for extracting the explicit solution for the proposed PID control law, the principle of the Magnitude Optimum criterion presented in Appendix A.1 is utilized once more.

For justifying the proposed control law, several process models are employed for testing the control loop's response to step, ramp, and parabolic reference signals. The proposed control law is compared with the conventional PID tuning via the Symmetrical Optimum criterion of Sect. 5.2.1. The proposed method achieves satisfactory performance in terms of reference tracking (zero steady state position, velocity, and acceleration error) compared to the conventional tuning, where its resulting response is oscillatory and most of the times unstable, Sect. 5.2.3. The robustness of the proposed control law to model uncertainties is also discussed, see Sect. 5.2.3.4. As a result, control engineers are given the ability apart from designing a type-III control loop, to test on a simulation basis the performance of the proposed control law before integrating it on a real time application.

5.2 PID Tuning Rules for Type-III Control Loops

In Sect. 5.2.1 a first attempt of designing type-III control loops is presented. The control action is of PID-lead-lag and the principle of pole-zero cancellation is adopted. For determining controller's zeros, only the dominant time constants of the plant are considered. Therefore, controller's zeros are forced to be tuned only with real values. In Sect. 5.2.2, and in a similar fashion with Sect. 4.2 the explicit solution for type-III control loops is presented.

5.2.1 Pole-Zero Cancellation Design

According to the design of type-II closed loop control systems, see Sect. 4.2.3, a similar methodology for the design of type-I, type-II control loops is proposed. For the following analysis, again the integrating process of the form (Fig. 5.1)

$$G(s) = \frac{1}{sT_{\mathrm{m}}(1 + sT_{\mathrm{p}_1})(1 + sT_{\Sigma_{\mathrm{p}}})} \tag{5.1}$$

is adopted, where T_{p_1} stands for the dominant time constant of the process and T_{m}, $T_{\Sigma\mathrm{p}}$ stand for the integrator's time constant and the unmodeled plant dynamics, respectively [1]. Supposing that the dominant time constant T_{p_1} is evaluated, the proposed I-PID controller is defined by

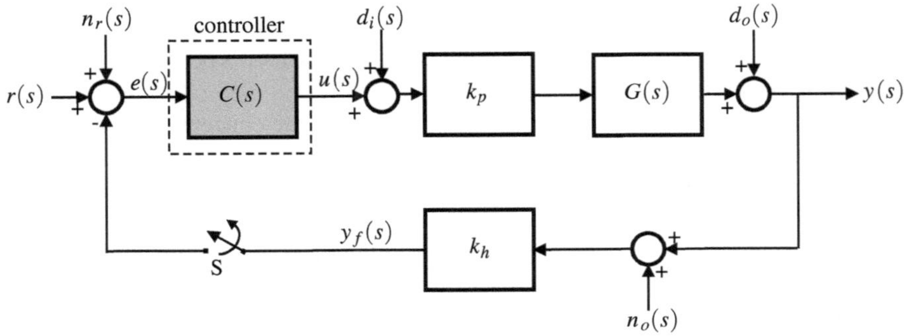

Fig. 5.1 Block diagram of the closed-loop control system. $G(s)$ is the plant transfer function, $C(s)$ is the controller transfer function, $r(s)$ is the reference signal, $d_{\mathrm{o}}(s)$ and $d_{\mathrm{i}}(s)$ are the output and input disturbance signals, respectively, and $n_{\mathrm{r}}(s)$, $n_{\mathrm{o}}(s)$ are the noise signals at the reference input and process output, respectively. k_{p} stands for the plant's dc gain and k_{h} is the feedback path

$$C(s) = \frac{(1+sT_{\mathrm{n}})(1+sT_{\mathrm{v}})(1+sT_x)}{s^2 T_{\mathrm{i}}(1+sT_{\Sigma_{\mathrm{c}_1}})(1+sT_{\Sigma_{\mathrm{c}_2}})} \tag{5.2}$$

where $T_{\Sigma_{\mathrm{c}_1}}$, $T_{\Sigma_{\mathrm{c}_2}}$ are known and sufficiently small time constants compared to T_{p_1}, arising from the controller's implementation. By setting $T_x = T_{\mathrm{p}_1}$ (pole-zero cancellation) and assuming that

$$T_{\Sigma_{\mathrm{c}}} = T_{\Sigma_{\mathrm{c}_1}} + T_{\Sigma_{\mathrm{c}_2}}, \quad \text{and} \quad T_{\Sigma_{\mathrm{c}_1}} T_{\Sigma_{\mathrm{c}_2}} \approx 0 \tag{5.3}$$

the transfer function of the closed loop control system is equal to

$$T(s) \approx \frac{s^2 k_{\mathrm{p}} T_{\mathrm{n}} T_{\mathrm{v}} + s k_{\mathrm{p}}(T_{\mathrm{n}} + T_{\mathrm{v}}) + k_{\mathrm{p}}}{\left[T_{\mathrm{i}} T_{\mathrm{m}} T_{\Sigma} s^4 + T_{\mathrm{i}} T_{\mathrm{m}} s^3 + s^2 k_{\mathrm{p}} k_{\mathrm{h}} T_{\mathrm{n}} T_{\mathrm{v}} + s k_{\mathrm{p}} k_{\mathrm{h}} (T_{\mathrm{n}} + T_{\mathrm{v}}) + k_{\mathrm{p}} k_{\mathrm{h}}\right]} \tag{5.4}$$

where $T_{\Sigma} = T_{\Sigma_{\mathrm{c}}} + T_{\Sigma_{\mathrm{p}}}$. The magnitude of (5.4) is given by

$$|T(j\omega)| = \sqrt{\frac{k_{\mathrm{p}}^2(1 - T_{\mathrm{n}} T_{\mathrm{v}} \omega^2)^2 + k_{\mathrm{p}}^2 (T_{\mathrm{n}} + T_{\mathrm{v}})^2 \omega^2}{\begin{bmatrix} T_{\mathrm{i}} T_{\mathrm{m}} T_{\Sigma} \omega^4 \\ + k_{\mathrm{p}} k_{\mathrm{h}} (1 - T_{\mathrm{n}} T_{\mathrm{v}} \omega^2) \end{bmatrix}^2 + \omega^2 \begin{bmatrix} (k_{\mathrm{p}} k_{\mathrm{h}} (T_{\mathrm{n}} + T_{\mathrm{v}}) \\ - T_{\mathrm{i}} T_{\mathrm{m}} \omega^2) \end{bmatrix}^2}}. \tag{5.5}$$

The denominator of (5.4) is defined by

$$D(\omega) = \sqrt{\begin{array}{l} (T_{\mathrm{i}} T_{\mathrm{m}} T_{\Sigma})^2 \omega^8 + T_{\mathrm{i}} T_{\mathrm{m}} \left(T_{\mathrm{i}} T_{\mathrm{m}} - 2 k_{\mathrm{p}} k_{\mathrm{h}} T_{\mathrm{n}} T_{\mathrm{v}} T_{\Sigma}\right) \omega^6 \\ + k_{\mathrm{p}} k_{\mathrm{h}} \left[2 T_{\mathrm{i}} T_{\mathrm{m}} T_{\Sigma} - 2 (T_{\mathrm{n}} + T_{\mathrm{v}}) T_{\mathrm{i}} T_{\mathrm{m}} + k_{\mathrm{p}} k_{\mathrm{h}} T_{\mathrm{i}}^2 T_{\mathrm{m}}^2\right] \omega^4 \\ + (k_{\mathrm{p}} k_{\mathrm{h}})^2 \left(T_{\mathrm{n}}^2 + T_{\mathrm{v}}^2\right) \omega^2 + (k_{\mathrm{p}} k_{\mathrm{h}})^2 \end{array}} \tag{5.6}$$

One way to optimize the magnitude of (5.5) is to set the terms of ω^j, $j = 2, 4, 6, \ldots$, in (5.6), equal to zero, starting again from the lower frequency range [6–8]. Setting $k_\mathrm{h} = 1$ and the term of ω^6 equal to zero leads to

$$T_\mathrm{i} = \frac{2k_\mathrm{p}k_\mathrm{h}T_\mathrm{n}T_\mathrm{v}T_\Sigma}{T_\mathrm{m}}. \tag{5.7}$$

In a similar fashion, setting the term of ω^4 equal to zero and along with the aid of (5.7) results in

$$4T_\Sigma^2 - 4\big(T_\mathrm{n} + T_\mathrm{v}\big)T_\Sigma + T_\mathrm{n}T_\mathrm{v} = 0. \tag{5.8}$$

If $T_\mathrm{v} = nT_\Sigma$ is chosen, then (5.8) becomes

$$T_\mathrm{n} = \frac{4(n-1)}{n-4}T_\Sigma. \tag{5.9}$$

Summarizing the relations (5.7) and (5.9), the aforementioned PID control law defined in (5.2) results in

$$\begin{bmatrix} T_x \\ T_\mathrm{v} \\ T_\mathrm{n} \\ T_\mathrm{i} \\ k_\mathrm{h} \end{bmatrix} = \begin{bmatrix} T_{\mathrm{p}1} \\ nT_\Sigma \\ \dfrac{4(n-1)}{n-4}T_\Sigma \\ 8k_\mathrm{p}k_\mathrm{h}\dfrac{n(n-1)T_\Sigma^3}{(n-4)T_\mathrm{m}} \\ 1 \end{bmatrix}. \tag{5.10}$$

Proper selection of parameter n, ($n > 4$ must hold by) leads to a feasible I-PID control law. Substituting Eq. (5.10) into the closed loop transfer function results in

$$T(s) = \frac{4n(n-1)T_\Sigma^2 s^2 + (n^2-4)T_\Sigma s + n - 4}{\begin{bmatrix} 8n\,(n-1)\,T_\Sigma^4 s^4 + 8n(n-1)T_\Sigma^3 s^3 + 4n(n-1)T_\Sigma^2 s^2 \\ +(n^2-4)T_\Sigma s + (n-4) \end{bmatrix}}. \tag{5.11}$$

Normalizing again the time by setting $s' = sT_\Sigma$, (5.11) becomes equal to

$$T(s') = \frac{4n(n-1)s'^2 + (n^2-4)s' + (n-4)}{\begin{bmatrix} 8n(n-1)s'^4 + 8n(n-1)s'^3 + 4n(n-1)s'^2 \\ +(n^2-4)s' + (n-4) \end{bmatrix}}. \tag{5.12}$$

Note that the control loop defined in (5.12) is of type-III, since the terms of s'^j, $j = 0, 1, 2$, are equal, $a_0 = b_0$, $a_1 = b_1$, $a_2 = b_2$, see Sect. 2.5. The respective step and frequency responses of (5.12) for two different values of parameter n, are presented

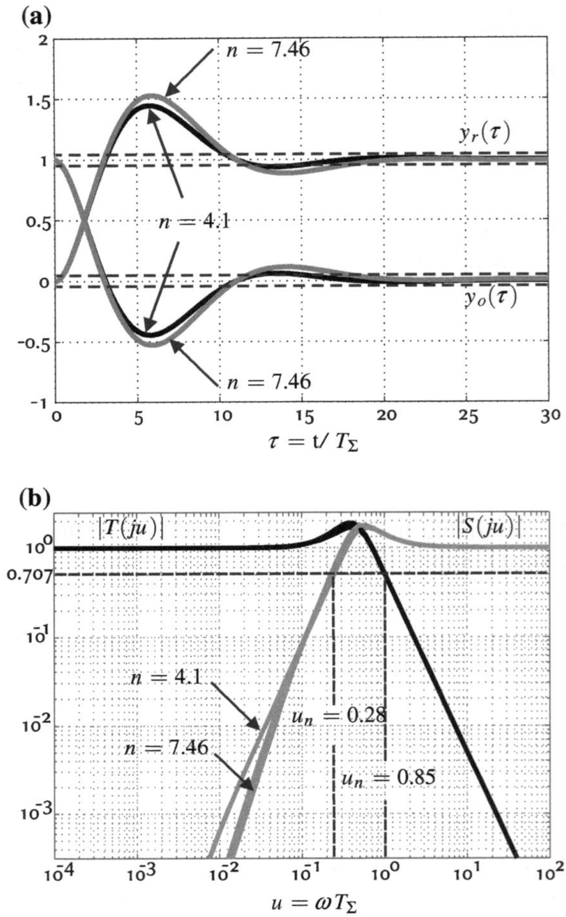

Fig. 5.2 Step and frequency response of a type-III closed loop control system. **a** Step response and output disturbance rejection of type-III closed loop control system. **b** Frequency response of type-III closed loop control system

in Fig. 5.2. In addition, in Fig. 5.3 the open loop frequency response is shown. Its transfer function is given by

$$F_{\mathrm{ol}}(s) = \frac{4n(n-1)T_\Sigma^2 s^2 + (n^2-4)T_\Sigma s + n - 4}{8n(n-1)T_\Sigma^3 s^3(1+sT_\Sigma)}. \tag{5.13}$$

From Fig. 5.2b it is concluded that the magnitude of the complementary sensitivity $|T(ju)|$ is practically independent of the parameter n. Sensitivity $|S(ju)|$ becomes maximum if $n = 4.1$ and minimum, if $n = 7.46$. If $n = 7.46$ then $T_{\mathrm{n}} = T_{\mathrm{v}}$ holds by. For every other value of parameter n, the shape of the open loop frequency response is preserved exactly as presented in Fig. 5.2b.

The same conclusion holds also for the overshoot of the step response of the type-III control loop which remains almost equal to 50 % regardless of the parameter n.

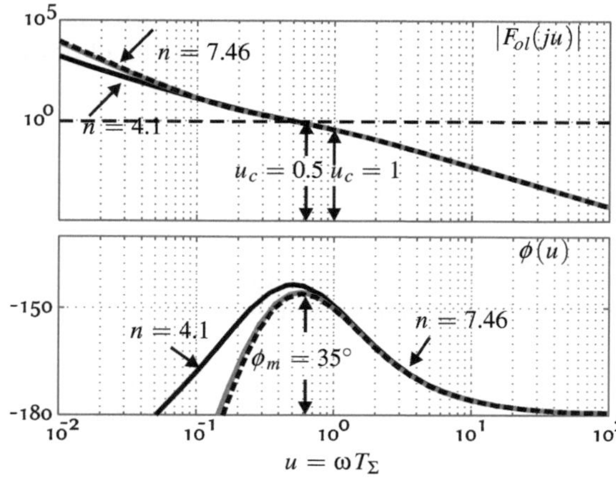

Fig. 5.3 Open loop frequency response of type-III closed loop control system

Since the phase margin is $\varphi_{\rm m} = 35° < 45°$, we expect an undesired overshoot in the step response of the closed loop system, Fig. 5.2a, which can be decreased along with the aid of an external filter $C_{\rm ex}(s)$ as mentioned in Sect. 4.2.3.

5.2.2 Revised PID Tuning Rules

For the derivation of the optimal control law a general type-0 stable process model defined by

$$G(s) = \frac{s^m\beta_m + s^{m-1}\beta_{m-1} + \cdots + s^2\beta_2 + s\beta_1 + 1}{s^{n-1}a_{n-1} + \cdots + s^3a_3 + s^2a_2 + sa_1 + 1}\mathrm{e}^{-sT_{\rm d}} \tag{5.14}$$

is adopted where $n - 1 > m$. Since the target of the design is a type-III control loop, according to the analysis presented in Sect. 2.5, three integrators in $F_{\rm ol}(s) = k_{\rm h}k_{\rm p}G(s)C(s)$ must exist. Therefore, the proposed I-I-PID controller is given by

$$C(s) = \frac{1 + sX + s^2Y}{s^3T_{\rm i}^3(1 + sT_{\rm p_n})}. \tag{5.15}$$

Parameter $T_{\rm p_n}$ stands for the parasitic controller time constant as mentioned in Sect. 5.2.1. In contrast with Sect. 5.2.1, the flexible form of the numerator $N_{\rm c}(s) = 1 + sX + s^2Y$ defined in (5.15) allows its parameters $X,\ Y$ to become complex conjugates if possible, see [9].

Purpose of this section is to determine explicitly controller's parameters, as a function of all plant parameters, without following the principle of pole-zero cancellation and ignoring other possible fundamental dynamics of the process. In that case,

X, Y, T_i are determined at the end of this section as functions $X = f_1(a_j, b_j, T_\mathrm{d})$, $Y = f_2(a_j, b_j, T_\mathrm{d})$, $T_\mathrm{i} = f_3(a_j, b_j, T_\mathrm{d})$ of all process parameters. To this end, the product $k_\mathrm{p}C(s)G(s)$ is defined by

$$k_\mathrm{p}C(s)G(s) = k_\mathrm{p}\frac{\sum_{j=0}^{m}(s^j\beta_j)(1 + sX + s^2Y)}{s^3T_\mathrm{i}^3\sum_{i=0}^{n}(s^ip_i)}\mathrm{e}^{-sT_\mathrm{d}} \tag{5.16}$$

where $\sum_{i=0}^{n}(s^ip_i) = (1 + sT_{\mathrm{p}_n})\sum_{j=0}^{n-1}(s^ja_j)$. According to Fig. 5.1, the closed loop transfer function is given by

$$T(s) = \frac{k_\mathrm{p}C(s)G(s)}{1 + k_\mathrm{p}k_\mathrm{h}C(s)G(s)} \tag{5.17}$$

and along with the aid of (5.16) results in

$$T(s) = \frac{k_\mathrm{p}(1 + sX + s^2Y)\sum_{j=0}^{m}(s^j\beta_j)}{s^3T_\mathrm{i}^3\sum_{j=1}^{n}(s^jp_j)\mathrm{e}^{sT_\mathrm{d}} + k_\mathrm{p}k_\mathrm{h}(1 + sX + s^2Y)\sum_{j=0}^{m}(s^j\beta_j)} \tag{5.18}$$

where $p_0 = 1$, $\beta_0 = 1$. For the need of the analysis, a general purpose time constant c_1 is considered. Therefore all time constants involved within the control loop are normalized by setting $s' = sc_1$. This results in the following substitutions

$$x = \frac{X}{c_1},\quad y = \frac{Y}{c_1^2},\quad t_\mathrm{i} = \frac{T_\mathrm{i}}{c_1}, \tag{5.19}$$

$$d = \frac{T_\mathrm{d}}{c_1},\quad r_j = \frac{p_j}{c_1^i},\quad \forall j = 1, \ldots, n,\quad z_i = \frac{\beta_i}{c_1^i} \tag{5.20}$$

$\forall i = 1, \ldots, m$. Time delay constant $\mathrm{e}^{s'd}$ is substituted with the "all pole" series approximation

$$\mathrm{e}^{s'd} = \sum_{k=0}^{7}\left(\frac{1}{k!}s'^kd^k\right). \tag{5.21}$$

Substituting (5.19)–(5.21) into (5.18) results in

$$T(s') = \frac{N(s')}{D(s')} = \frac{N(s')}{D_1(s') + k_\mathrm{h}N(s')}$$
$$= \frac{k_\mathrm{p}(1 + s'x + s'^2y)\sum_{j=0}^{m}(s'^jz_j)}{\left[s'^3t_\mathrm{i}^3\sum_{i=0}^{n}(s'^jr_j)\sum_{k=0}^{7}(\frac{1}{k!}s'^kd^k) + k_\mathrm{h}k_\mathrm{p}(1 + s'x + s'^2y)\sum_{j=0}^{m}(s'^jz_j)\right]} \tag{5.22}$$

where $r_0 = z_0 = 1$. Polynomials $D_1(s')$ and $N(s')$ are proved to be equal to

$$\begin{aligned} D_1(s') &= s'^3 t_i^3 + s'^4 t_i^3 (r_1 + d) + s'^5 t_i^3 \left(r_2 + r_1 d + \frac{1}{2!} d^2 \right) \\ &+ s'^6 t_i^3 \left(r_3 + d r_2 + \frac{1}{2!} d^2 r_1 + \frac{1}{3!} d^3 \right) \\ &+ s'^7 t_i^3 \left(r_4 + d r_3 + \frac{1}{2!} d^2 r_2 + \frac{1}{3!} d^3 r_1 + \frac{1}{4!} d^4 \right) \\ &+ s'^8 t_i^3 \left(r_5 + d r_4 + \frac{1}{2!} d^2 r_3 + \frac{1}{3!} d^3 r_2 + \frac{1}{4!} d^4 r_1 + \frac{1}{5!} d^5 \right) \\ &+ \cdots \end{aligned} \tag{5.23}$$

Substituting the constant terms of (5.23) with

$$\begin{bmatrix} q_1 \\ q_2 \\ q_3 \\ q_4 \\ q_5 \end{bmatrix} = \begin{bmatrix} r_1 + d \\ r_2 + r_1 d + \frac{1}{2!} d^2 \\ r_3 + r_2 d + \frac{1}{2!} d^2 r_1 + \frac{1}{3!} d^3 \\ r_4 + r_3 d + \frac{1}{2!} d^2 r_2 + \frac{1}{3!} d^3 r_1 + \frac{1}{4!} d^4 \\ r_5 + r_4 d + \frac{1}{2!} d^2 r_3 + \frac{1}{3!} d^3 r_2 + \frac{1}{4!} d^4 r_1 + \frac{1}{5!} d^5 \end{bmatrix} \tag{5.24}$$

results in the polynomial $D_1(s')$ to be rewritten in the form of

$$\begin{aligned} D_1(s') = \cdots &+ s'^8 t_i^3 q_5 + s'^7 t_i^3 q_4 + s'^6 t_i^3 q_3 + s'^5 t_i^3 q_2 \\ &+ s'^4 t_i^3 q_1 + s'^3 t_i^3, \end{aligned} \tag{5.25}$$

or finally

$$D_1(s') = \sum_{j=0}^{k} \left(t_i^3 q_j \right) s'^{(j+3)} \tag{5.26}$$

where

$$q_k = \sum_{i=0}^{k} r_{(k-i)} \left(\frac{1}{i!} d^i \right) \tag{5.27}$$

and $k = 0, 1, 2, \ldots, n$ with $r_0 = 1$, $z_r = 0$ if $r < 0$ and $z_0 = 1$. Polynomial $k_h N(s')$ is equal to

$$k_{\mathrm{h}}N(s') = k_{\mathrm{h}}k_{\mathrm{p}}\sum_{r=0}^{p} s'^{(r)}\left(yz_{(r-2)} + xz_{(r-1)} + z_{(r)}\right). \tag{5.28}$$

Finally,

$$\begin{aligned} D(s') &= D_1(s') + k_{\mathrm{h}}N\left(s'\right) \\ &= \sum_{j=0}^{k}(t_{\mathrm{i}}^3 q_j)s'^{(j+3)} + k_{\mathrm{h}}k_{\mathrm{p}}\sum_{r=0}^{p} s'^{(r)}\left(yz_{(r-2)} + xz_{(r-1)} + z_{(r)}\right). \end{aligned} \tag{5.29}$$

As a result the final closed loop transfer function is given by

$$T(s') = \frac{k_{\mathrm{p}}\sum_{r=0}^{p} s'^{(r)}(yz_{(r-2)} + xz_{(r-1)} + z_{(r)})}{\sum_{j=0}^{k}(t_{\mathrm{i}}^3 q_j)s'^{(j+3)} + k_{\mathrm{h}}k_{\mathrm{p}}\sum_{r=0}^{p} s'^{(r)}\left(yz_{(r-2)} + xz_{(r-1)} + z_{(r)}\right)}. \tag{5.30}$$

For determining explicitly the parameters $x,\ y,\ t_{\mathrm{i}},\ k_{\mathrm{h}}$ of the proposed PID controller, the Magnitude Optimum criterion presented in Appendix A.1 is adopted.

There, it is shown that for maintaining $|T(j\omega)| \simeq 1$ in the wider possible frequency range, certain optimization conditions have to hold by. These optimization conditions are proved in Appendix A.1 and are applied in (5.30). The proof of the optimal control law are also given in Appendix B.3. From there, it is shown that in a similar fashion with Sects. 4.2.3 and 5.2.1 the optimal control law $(x,\ y,\ t_{\mathrm{i}},\ k_{\mathrm{h}})$ is finally given by

$$k_{\mathrm{h}} = 1 \tag{5.31}$$

$$\sum_{j=0}^{8} C_j x^j = 0 \tag{5.32}$$

$$y = \frac{1}{2}x^2 + \frac{1}{2}(z_1^2 - 2z_2) \tag{5.33}$$

$$t_{\mathrm{i}}^3 = \frac{k_{\mathrm{h}}k_{\mathrm{p}}}{2}\frac{\left[y^2 - (z_1^2 - 2z_2)^2 + z_2^2 - 2z_1z_3 + 2z_4\right]}{x + z_1 - q_1}. \tag{5.34}$$

Let it be noted that parameters $C_j,\ z_1,\ z_2,\ z_3,\ z_4,\ q_1, k_{\mathrm{p}}$ are coming from the model of the process $G(s)$ and assumed measurable. Therefore, x is calculated out of (5.32) and is substituted into (5.33) for calculating y. Finally, integrator's time constant t_{i} is calculated out of (5.34).

5.2.3 Simulation Results

In this section a comparison between the proposed (Sect. 5.2.2) and the conventional PID tuning (Sect. 5.2.1) takes place when within the control loop the same integrating process is involved. In each example, two sets of comparative responses are presented.

1. The step response of the conventional tuning presented in Sect. 4.2.3 is compared with the revised control law presented in 5.2.2. Response of the output $y(\tau)$ and the control effort $u(\tau)$ is investigated in the presence of reference tracking $r(\tau)$, input $d_i(\tau)$ and output $d_o(\tau)$ disturbance rejection.
2. The ramp ($r(\tau) = \tau$) and parabolic ($r(\tau) = \tau^2$) response $y(\tau)$ of both the conventional tuning, see Sect. 4.2.3 and the revised control law 5.2.2 is also investigated.

In the sequel, three benchmark integrating processes are considered: (1) a process with dominant time constants (2) a process with time delay equal with the plant's dominant time constant and (3) a nonminimum phase process.

Note that for deriving a type-III control loop, the process is assumed to have an integrating behavior and therefore one more integrator is added within the PID controller so that it becomes I-PID.

5.2.3.1 Process with Dominant Time Constants

For testing the potential of the proposed method the process defined by

$$G(s') = \frac{0.254}{s'(1+s')^5} \tag{5.35}$$

is considered, consisting of five equal dominant time constants. The normalizing constant has been chosen equal to $s' = sT_{p_1}$. In Fig. 5.4a the step response of the type-III control loop is presented, where $r(s') = \frac{1}{s'}$ and $d_o(s') = \frac{0.25}{s'}$, $d_i(s') = r(s')$. Input and output disturbance $d_i(s')$, $d_o(s')$ act at $\tau = 400$ and $\tau = 800$, respectively.

From Fig. 5.4a it is apparent that the conventional PID tuning leads almost to an unstable control loop. Figure 5.4b presents the control effort $u(s)$ in the presence of the aforementioned input and output disturbances. For filtering the reference signal so that great overshoot at the output of the process is avoided, see Sect. 4.2.3, an external filter $C_{ex}(s')$ of the form

$$C_{ex_1}(s') = \frac{1}{1 + a_2(t_n + t_v)s' + a_1 t_n t_v s'} \tag{5.36}$$

$$C_{ex_2}(s') = \frac{1}{1 + a_2 x s' + a_1 y^2 s'} \tag{5.37}$$

is selected.

Note that x, y and t_n, t_v are the solutions coming from the conventional and the revised control law, respectively, so that the comparison between the two tuning

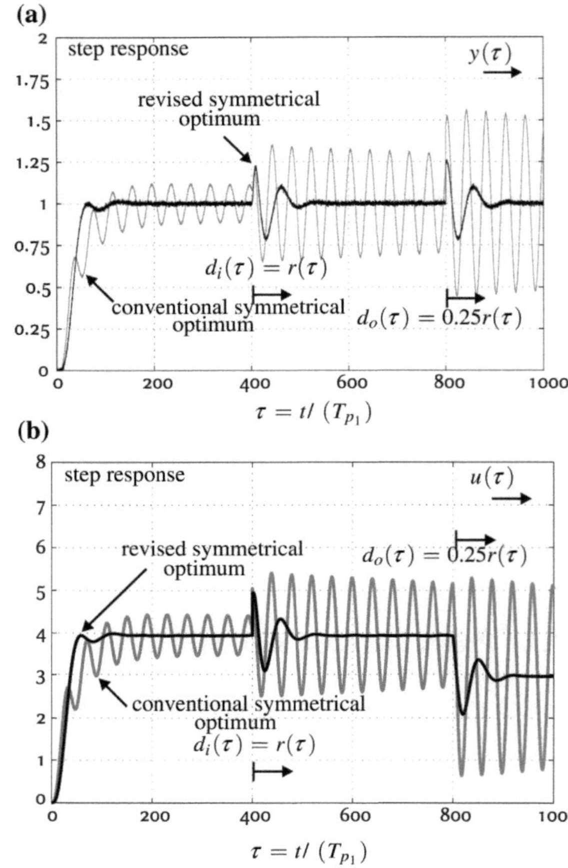

Fig. 5.4 Step response of a PID type-III closed loop control system. Plant with five dominant time constants defined by (5.35). A step input $d_{\mathrm{i}}(\tau) = r(\tau)$ and output $d_{\mathrm{o}}(\tau) = 0.25r(\tau)$ disturbance is applied at $t = 400\tau$ and $t = 800\tau$, (*black*) revised tuning, (*gray*) conventional tuning. **a** Response of the output $y(\tau)$ in the presence of input and output disturbance. **b** Response of the command signal $u(\tau)$ in the presence of input and output disturbance

methods remains one to one. Parameters $a_1,\ a_2$ have been chosen equal to $a_1 = 1.25,\ a_2 = 1.2$. Parameter n regarding the conventional control law, see Sect. 5.2.1 has been chosen equal to $n = 7.46$. The corresponding PID controllers regarding the conventional and the revised tuning are given by

$$
\begin{aligned}
C_{\text{PID-SO}}(s') &= \frac{(1+s't_{\text{n}})(1+s't_{\text{v}})(1+s't_x)}{s'^2 t_{\text{i}}(1+s't_{\text{sc}_1})(1+s't_{\text{sc}_2})} \\
&= \frac{(1+7.46s')(1+30.58s')(1+s')}{s'^2 475.75(1+0.1s')(1+0.1s')}, \qquad (5.38)
\end{aligned}
$$

$$
\begin{aligned}
C_{\text{PID}}(s') &= \frac{(1+s'x+s'^2 y)}{s'^2 t_{\text{i}}(1+s't_{\text{sc}_1})} \\
&= \frac{(1+29.03s'+421.5s'^2)}{s'^2 942.8(1+0.1s')}. \qquad (5.39)
\end{aligned}
$$

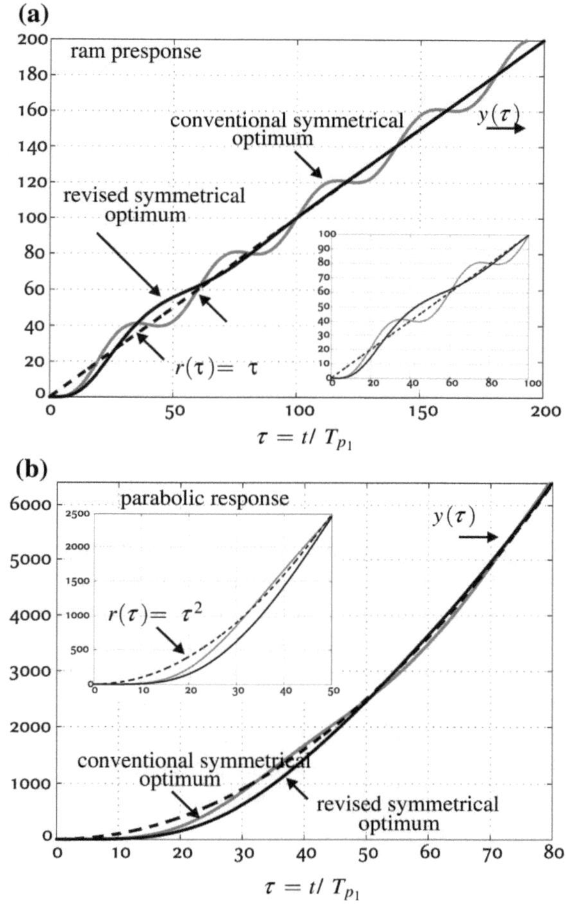

Fig. 5.5 Comparison between the conventional and the revised PID control law. Control of a process with five dominant time constants. **a** Ramp response of the output $y(\tau)$ of the control loop. **b** Parabolic response of the output $y(\tau)$ of the control loop

Fig. 5.5a, b present the ramp and parabolic response of the final closed loop control system when the PID controller is tuned via (5.38) and (5.39), respectively. From Fig. 5.5a it is apparent that the revised tuning reaches steady state at $\tau = 64$ in contrast with the conventional tuning where its response remains practically unstable.

5.2.3.2 Process with Time Delay Equal to Its Dominant Time Constant

In this example the process to be controlled is defined by

$$G(s') = \frac{0.254}{s'(1+s')(1+0.5s')(1+0.2s')^2(1+0.1s')} e^{-s'} \tag{5.40}$$

which exhibits a time delay constant T_d equal with the dominant time constant T_{p1}, $d = \frac{T_d}{T_{p1}} = 1$. The resulting PID controller parameters according to Sects. 4.2.3 and 5.2.2 are equal to

$$
\begin{aligned}
C_{\text{PID-SO}}(s') &= \frac{(1+s't_n)(1+s't_v)(1+s't_x)}{s'^2 t_i(1+s't_{sc_1})(1+s't_{sc_2})} \\
&= \frac{(1+7.46s')(1+2.98s')(1+s')}{s'^2 4.52(1+0.1s')(1+0.1s')}
\end{aligned}
\tag{5.41}
$$

and

$$
\begin{aligned}
C_{\text{PID}}(s') &= \frac{(1+s'x+s'^2y)}{s'^2 t_i(1+s't_{sc_1})} \\
&= \frac{(1+13.63s'+92.9s'^2)}{s'^2 97.6(1+0.1s')}
\end{aligned}
\tag{5.42}
$$

respectively. In Fig. 5.6a, b the response of the control loop for $y(\tau)$, $u(\tau)$ to a step reference input $r(s') = \frac{1}{s'}$, a step output and input disturbance $d_o(s') = \frac{0.25}{s'}$, $d_i(s') = r(s')$ is presented both for the conventional and the revised control law, respectively.

The PID controller via the conventional PID tuning (Symmetrical Optimum criterion) leads to unacceptable response in terms of overshoot, input and output disturbance rejection, see Fig. 5.6a, b. Let it be noted that in both cases, conventional and revised PID tuning, the reference $r(s')$ is filtered by an external controller $C_{ex}(s')$ defined by (5.36) and (5.37), respectively. Ramp response of the revised tuning, settles faster than the conventional tuning, Fig. 5.7a.

5.2.3.3 Non Minimum Phase Process

In this example a nonminimum phase process is considered defined by

$$
G(s') = \frac{1.58(1-0.7s')(1-0.3s')}{(1+s')(1+0.9s')(1+0.8s')(1+0.1s')(1+0.05s')}.
\tag{5.43}
$$

The resulting PID control law according to the conventional and the revised tuning are given by

$$
\begin{aligned}
C_{\text{PID-SO}}(s') &= \frac{(1+s't_n)(1+s't_v)(1+s't_x)}{s'^2 t_i(1+s't_{sc_1})(1+s't_{sc_2})} \\
&= \frac{(1+14.55s')(1+7.47s')(1+s')}{s'^2 669.4(1+0.1s')(1+0.1s')}
\end{aligned}
\tag{5.44}
$$

$$
C_{\text{PID}}(s') = \frac{(1+s'x+s'^2y)}{s'^2 t_i(1+s't_{sc_1})}
$$

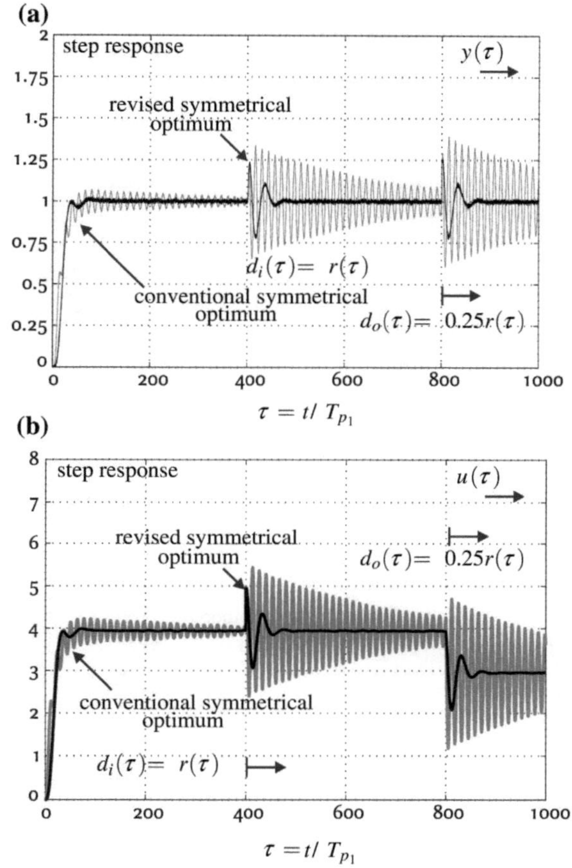

Fig. 5.6 Step response of a PID type-III closed loop control system. Plant with time delay equal with its dominant time constant. A step input $d_i(\tau) = r(\tau)$ and output $d_o(\tau) = 0.25r(\tau)$ disturbance is applied at $\tau = 400$ and $\tau = 800$, (*black*) revised tuning, (*gray*) conventional tuning.
a Response of the output $y(\tau)$ in the presence of input and output disturbance.
b Response of the command signal $u(\tau)$ in the presence of input and output disturbance

$$= \frac{\left(1 + 22.02s' + 242.8s'^2\right)}{s'^2 2577.41(1 + 0.1s')}. \tag{5.45}$$

Once more the conventional PID tuning fails to tune a stable type-III control loop, Fig. 5.8a, b. Ramp response of the conventional tuning reaches the steady state operation much faster than the conventional control loop, Fig. 5.9.

5.2.3.4 Robustness Analysis

The robustness of the proposed control law is investigated in this section. The model of the process to be controlled is given by

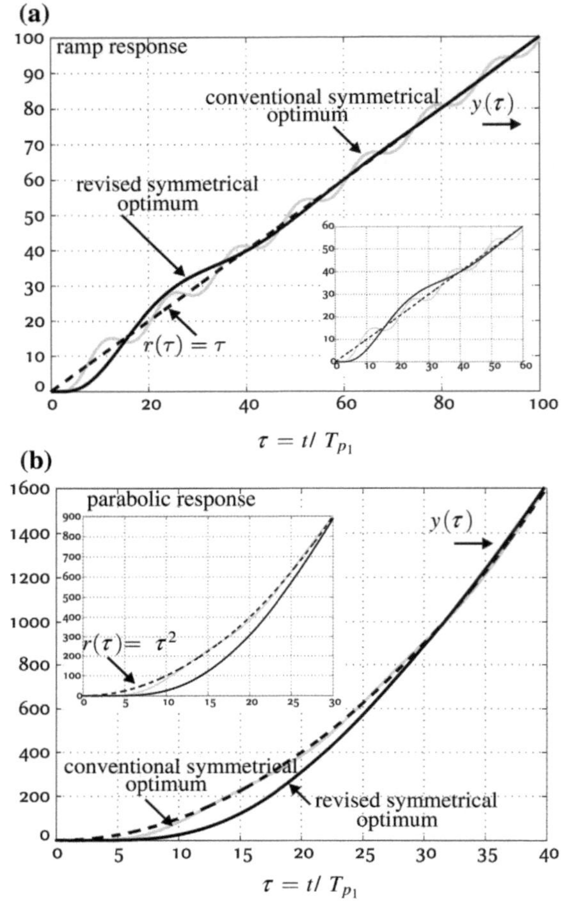

Fig. 5.7 Comparison between the conventional and the revised PID control law. Control of a process with time delay equal to its dominant time constant. **a** Ramp response of the output $y(\tau)$ of the control loop. **b** Parabolic response of the output $y(\tau)$ of the control loop

$$G(s') = \frac{1}{(1+s')^5}\mathrm{e}^{-4s'}. \tag{5.46}$$

For controlling (5.46) an approximation of $G(s')$ defined by

$$\widetilde{G}(s') = \frac{1+a}{(1+s')^5}\mathrm{e}^{-4(1+b)s'} \tag{5.47}$$

If $a = b = 0$ then $\widetilde{G}(s') = G(s')$. In Fig. 5.10a the proposed controller is tuned based on (5.47) when $a = 0$ and $b = -0.25$. Settling time of both input and output disturbance rejection remains practically unaltered. In both cases external controller $C_{\mathrm{ex}}(s)$ for filtering the reference $r(s)$ has been chosen equal to

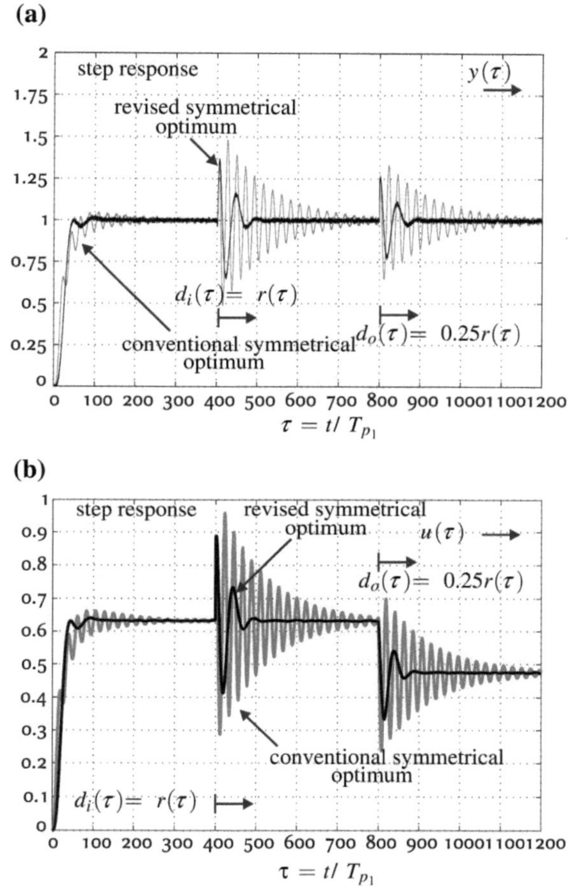

Fig. 5.8 Step response of a PID type-III closed loop control system involving a nonminimum phase process. A step input $d_i(\tau) = r(\tau)$ and output $d_o(\tau) = 0.25r(\tau)$ disturbance is applied at $\tau = 400$ and $\tau = 800$, (*black*) revised tuning, (*gray*) conventional tuning. **a** Response of the output $y(\tau)$ in the presence of input and output disturbance. **b** Response of the command signal $u(\tau)$ in the presence of input and output disturbance

$$C_{ex}(s') = \frac{1}{1 + xs' + y^2 s'}. \tag{5.48}$$

In Fig. 5.10b a variation of 25 % is forced on the dc gain of the process. From Fig. 5.10b it is apparent that the response of the control loop stays within the range of the optimal response in terms of reference tracking and disturbance rejection. The peak (undershoot) of output and input disturbance rejection has increased by 10 %.

5.3 Explicit PID Tuning Rules for Type-p Control Loops

An extension of the Symmetrical Optimum criterion for the design of PID type-p closed loop control systems is proposed. Type-p control loops are characterized by the presence of p integrators in the open-loop transfer function. For designing a PID

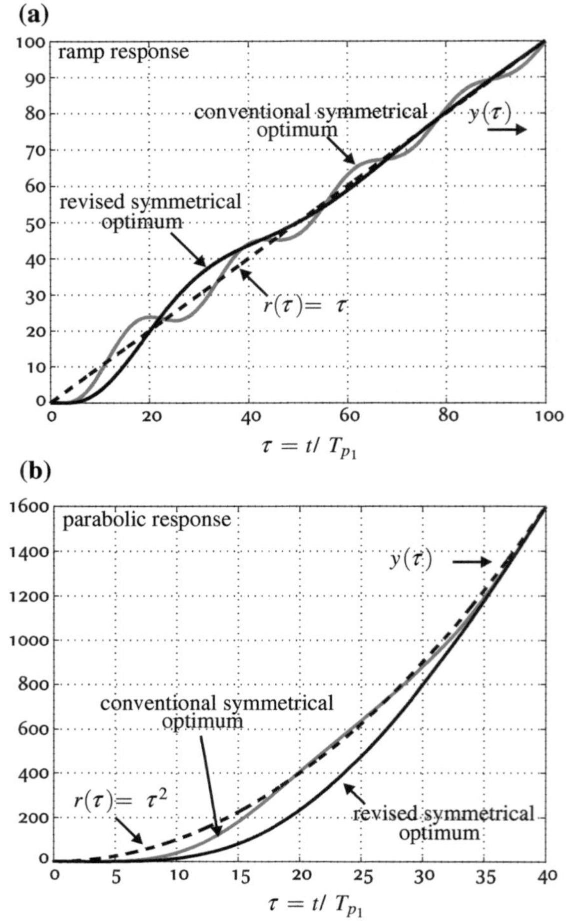

Fig. 5.9 Comparison between the conventional and the revised PID control law. Control of a nonminimum phase process. **a** Ramp response of the output $y(\tau)$ of the control loop. **b** Parabolic response of the output $y(\tau)$ of the control loop

type-p control loop there should exist an $\mathrm{PI}^p D$, or $\mathrm{PI}^{(p-1)}\mathrm{D}$, or PID and so on, if the process is of type-0 or type-I or type-$(p-1)$, respectively. A type-II control loop achieves zero steady state position and velocity error, a type-III control loop achieves zero steady state position, velocity and acceleration error, and therefore a type-p control loop is expected to track both faster reference signals and eliminate higher order errors at steady state.

For deriving the proposed control law, a transfer function containing dominant time constants and the plant's unmodeled dynamics has been considered in the frequency domain. The final control law consists of analytical expressions that involve both dominant dynamics and model uncertainty of the plant. For justifying the potential of the proposed theory, simulation results for representative processes met in many real world applications are presented.

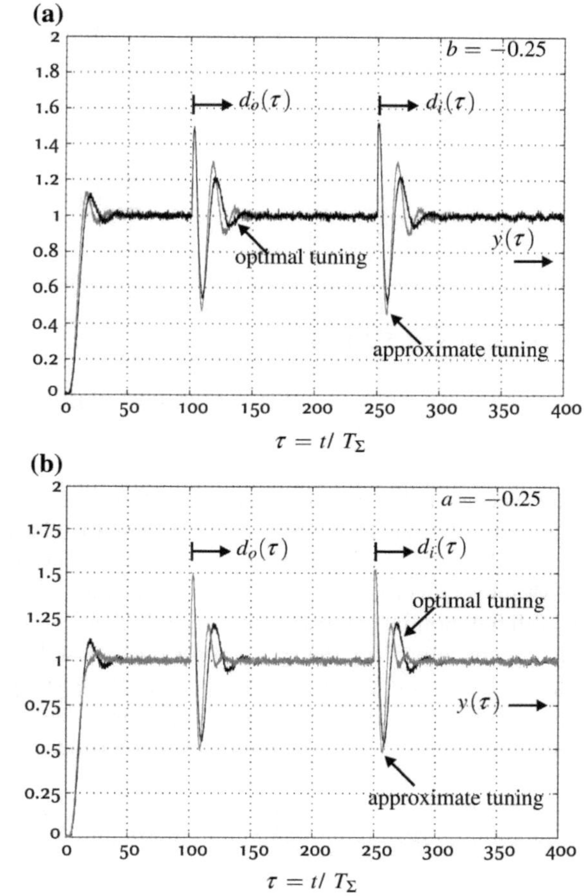

Fig. 5.10 Response of the control loop in the presence of plant's parameters variations. In both cases the proposed PID control law is tuned with the wrong values. **a** Time delay constant d is underestimated, $b = -0.25$. **b** A variation in the plant's dc gain is forced, $a = -0.25$

5.3.1 Extending the Design to Type-p Control Loops

According to the analysis presented in Sect. 5.2.1 a similar analysis for tuning the PID type controller's parameters is presented, regarding the design of type-p control loops. Note that parameter p stands for the free integrators of the open-loop transfer function. Therefore, let the process be defined by

$$G(s) = \frac{1}{T_m s^q \prod_{j=1}^{n_m}(1 + T_{m_j}s)\prod_{k=1}^{n_s}(1 + T_{s_k}s)}, \tag{5.49}$$

consisting of q integrators and T_m one of the integrator's time constant. Assuming that the plant's dominant time constants are defined by T_{m_j}, $(j = 1, 2, \ldots, n_m)$ and the process unmodeled dynamics by T_{s_k}, $(k = 1, 2, \ldots, n_s)$ we can substitute in (5.49), without loss of generality with the approximation

$$\prod_{k=1}^{n_s}(1+T_{s_k}s) = 1 + T_{\Sigma_s}s \tag{5.50}$$

where

$$T_{\Sigma_s} = \sum_{k=1}^{n_s} T_{s_k} \tag{5.51}$$

stands for the process' small unmodeled time constants. Since the target of the design is the final closed-loop control system to be of type-p, according to the analysis presented in Sect. 5.2.1, the proposed PID type controller is given by

$$C(s) = \frac{\prod_{j=1}^{n_m}\left(1+T_{m_j}s\right)\prod_{r=1}^{p-1}\left(1+T_{n_r}s\right)}{T_i s^{p-q}\prod_{z=1}^{n_c}\left(1+T_{c_z}s\right)}. \tag{5.52}$$

Thus, according the design of type-II (4.2.3), type-III control loops (5.2.1), the PID type controller has to contain n_m zeros equal to the T_{m_j} dominant time constants ($j = 1, 2, \ldots, n_m$) so that exact pole-zero cancellation is achieved.

Moreover, in order the denominator of the final closed-loop transfer function $T(s)$ is a full polynomial in terms of the s^j coefficients, it is easily proved after manipulating algebraically $T(s)$, that $p-1$ zeros must exist. Furthermore, the controller must introduce $p-q$ integrators, so that the final closed-loop is of type-p. Finally, in order the controller transfer function is strictly causal, denominator's order must be greater or equal to $p-1+n_m$. The unmodeled controller's dynamics are represented by

$$\prod_{z=1}^{n_c}\left(1+T_{c_z}s\right) = 1 + T_{\Sigma_c}s \tag{5.53}$$

where

$$T_{\Sigma_c} = \sum_{z=1}^{n_c} T_{c_z}. \tag{5.54}$$

In that case, the open-loop transfer function becomes

$$F_{ol}(s) = k_p k_h G(s) C(s) = k_p k_h \frac{\prod_{r=1}^{p-1}\left(1+T_{n_r}s\right)}{T_i T_m s^p \prod_{k=1}^{n_s}(1+T_{s_k}s)\prod_{z=1}^{n_c}(1+T_{c_z}s)} \tag{5.55}$$

or by substituting (5.50) and (5.52)–(5.54) results in

$$F_{ol}(s) = k_p k_h \frac{\prod_{r=1}^{p-1}(1+T_{n_r}s)}{T_i T_m s^p (1+T_\Sigma s)} \tag{5.56}$$

where $T_\Sigma = T_{\Sigma_s} + T_{\Sigma_c}$ and $T_{\Sigma_s} T_{\Sigma_c} \approx 0$.

Finally, the closed-loop transfer function is equal to

$$T(s) = \frac{k_p \prod_{r=1}^{p-1}(1+T_{n_r}s)}{T_i T_m T_\Sigma s^{p+1} + T_i T_m s^p + k_p k_h \prod_{r=1}^{p-1}(1+T_{n_r}s)}. \tag{5.57}$$

In that, Eq. (5.57) yields

$$T(s) = \frac{b_{p-1}s^{p-1} + b_{p-2}s^{p-2} + \cdots + b_3 s^3 + b_2 s^2 + b_1 s + b_0}{a_{p+1}s^{p+1} + a_p s^p + a_{p-1}s^{p-1} + \cdots + a_3 s^3 + a_2 s^2 + a_1 s + a_0} \tag{5.58}$$

where

$$b_{p-1} = \prod_{j=1}^{p-1} T_{p_j} = T_{p_1}T_{p_2}\cdots T_{p_{p-1}}, \quad b_3 = k_p \sum_{i\neq j\neq k=1}^{p-1} T_{n_i}T_{n_j}T_{n_k}, \tag{5.59}$$

$$b_2 = k_p \sum_{i\neq j=1}^{p-1} T_{n_i}T_{n_j}, \quad b_1 = k_p \sum_{i=1}^{p-1} T_{n_i}, \quad b_0 = k_p, \tag{5.60}$$

and

$$a_{p+1} = T_i T_m T_\Sigma, \quad a_p = T_i T_m, \tag{5.61}$$

$$a_3 = k_p k_h \sum_{i\neq j\neq k=1}^{p-1} T_{n_i}T_{n_j}T_{n_k}, \quad a_2 = k_p k_h \sum_{i\neq j=1}^{p-1} T_{n_i}T_{n_j}, \tag{5.62}$$

$$a_1 = k_p k_h \sum_{i=1}^{p-1} T_{n_i}, \quad a_0 = k_p k_h. \tag{5.63}$$

According to (A.9), if $a_0 = b_0$ then

$$k_h = 1. \tag{5.64}$$

Since the goal is to determine parameters T_i, T_{n_r}, $(r = 1, \ldots, p-1)$ the magnitude of (5.58) is optimized according to the Appendix A.1. For every order p, the optimal integral gain is given by

$$T_i = 2k_p k_h \frac{T_\Sigma}{T_m} \prod_{r=1}^{p-1} T_{n_r}. \tag{5.65}$$

This can be proved as follows. For a process of one dominant time constant defined by (4.1) where $(q = 1)$, then in order the final control loop is of type-II $p = 2$, the PID type controller (according to the Symmetrical Optimum criterion) is given by

$$C(s) = \frac{(1 + T_{n_1}s)(1 + T_{n_2}s)}{T_i s(1 + T_{\Sigma_c}s)}, \tag{5.66}$$

for which

$$T_{n_2} = T_{p_1} \tag{5.67}$$

and $(1 + sT_{\Sigma_p})(1 + sT_{\Sigma_c}) \approx 1 + sT_\Sigma$ have been set. In that, the open-loop transfer function is given by

$$F_{ol}(s) = k_p k_h \frac{(1 + T_{n_1}s)}{T_i T_m s^2 (1 + T_\Sigma s)} \tag{5.68}$$

and the closed-loop transfer function is then given by

$$T(s) = \frac{k_p(1 + T_{n_1}s)}{T_i T_m s^2(1 + T_\Sigma s) + k_p k_h (1 + T_{n_1}s)}. \tag{5.69}$$

According to the analysis presented in Sect. 5.2.1, the integrator's time constant is calculated if

$$a_2^2 = 2a_1a_3 \tag{5.70}$$

is set, as another means of optimizing the magnitude of (5.69) [11, 12]. The resulting integrator's time constant proves to be equal to

$$T_i = 2k_p k_h \frac{T_\Sigma}{T_m} T_{n_1} \tag{5.71}$$

and if $T_{n_1} = 4T_\Sigma$ is chosen, then $T_i = 8k_p k_h \frac{T_\Sigma^2}{T_m}$, see Sect. 5.2.1.

According to Sect. 5.2.1, for a process of one dominant time constant defined again by (5.1) where ($q = 1$) then in order the final control loop is of type-III $p = 3$, the PID type controller is given by

$$C(s) = \frac{(1 + T_{n_1}s)(1 + T_{n_2}s)(1 + T_{n_3}s)}{T_i s^2 (1 + T_{\Sigma_c}s)}. \tag{5.72}$$

Assuming again pole-zero cancellation

$$T_{n_3} = T_{p_1} \tag{5.73}$$

and $(1 + sT_{\Sigma_p})(1 + sT_{\Sigma_c}) \approx 1 + sT_\Sigma$ the open-loop transfer function $F_{ol}(s)$ becomes

$$F_{ol}(s) = k_p k_h \frac{(1 + T_{n_1}s)(1 + T_{n_2}s)}{T_i T_m s^3 (1 + T_\Sigma s)}. \tag{5.74}$$

Therefore the closed-loop transfer function is equal to

$$T(s) = \frac{k_{\mathrm{p}}\left(1 + T_{\mathrm{n}_1} s\right)\left(1 + T_{\mathrm{n}_2} s\right)}{T_{\mathrm{i}} T_{\mathrm{m}} s^3 \left(1 + T_{\Sigma} s\right) + k_{\mathrm{p}} k_{\mathrm{h}} \left(1 + T_{\mathrm{n}_1} s\right)\left(1 + s T_{\mathrm{n}_2}\right)}. \tag{5.75}$$

According to (5.70) and since $n = 2$, the integrator's time constant is calculated via

$$a_3^2 = 2a_2a_4. \tag{5.76}$$

Finally, after some algebraic manipulation it was shown that the integrator's time constant is equal to

$$T_{\mathrm{i}} = 2k_{\mathrm{p}}k_{\mathrm{h}}\frac{T_{\Sigma}}{T_{\mathrm{m}}}T_{\mathrm{n}_1}T_{\mathrm{n}_2}. \tag{5.77}$$

In a similar fashion, for a process of one dominant time constant defined by (5.1) and if $n = k - 1$, in order the final control loop is of type-p, the PID type controller is given by

$$C(s) = \frac{(1 + T_{\mathrm{n}_1}s)(1 + T_{\mathrm{n}_2}s)\cdots(1 + T_{\mathrm{n}_k}s)}{T_{\mathrm{i}}s^{k-1}(1 + T_{\Sigma_{\mathrm{c}}}s)}. \tag{5.78}$$

According to the analysis presented previously, it can be claimed regarding the integrator's time constant $T_{i_{k-1}}$, that

$$T_{i_{k-1}} = 2k_{\mathrm{p}}k_{\mathrm{h}}\frac{T_{\Sigma}}{T_{\mathrm{m}}}\prod_{j=1}^{k-1}T_{\mathrm{n}_j}. \tag{5.79}$$

Therefore, for $n = k$, it has to be proved that

$$\begin{aligned} T_{\mathrm{i}} &= 2k_{\mathrm{p}}k_{\mathrm{h}}\frac{T_{\Sigma}}{T_{\mathrm{m}}}\prod_{j=1}^{k}T_{\mathrm{n}_j} \\ &= 2k_{\mathrm{p}}k_{\mathrm{h}}\frac{T_{\Sigma}}{T_{\mathrm{m}}}\left(\prod_{j=1}^{k-1}T_{\mathrm{n}_j}\right)T_{\mathrm{n}_k} = T_{i_k}T_{\mathrm{n}_k}. \end{aligned} \tag{5.80}$$

According to the design of type-p control loops, the PID type controller is given by

$$C(s) = \frac{(1 + T_{\mathrm{n}_1}s)(1 + T_{\mathrm{n}_2}s)\cdots(1 + T_{\mathrm{n}_k}s)(1 + T_{\mathrm{n}_{k+1}}s)}{T_{\mathrm{i}}s^{k}(1 + T_{\Sigma_{\mathrm{c}}}s)} \tag{5.81}$$

for which $T_{\mathrm{n}_{k+1}} = T_{\mathrm{p}1}$ is set, assuming design via pole-zero cancellation. Since again $(1 + sT_{\Sigma_{\mathrm{p}}})(1 + sT_{\Sigma_{\mathrm{c}}}) \approx 1 + sT_{\Sigma}$, the open and closed-loop transfer functions are given by

$$F_{\mathrm{ol}}(s) = k_{\mathrm{p}}k_{\mathrm{h}} \frac{\prod_{j=1}^{k}(1 + T_{\mathrm{n}_j}s)}{T_{\mathrm{i}}T_{\mathrm{m}}s^{k+1}(1 + T_{\Sigma}s)}, \tag{5.82}$$

$$T(s) = \frac{k_{\mathrm{p}}\prod_{j=1}^{k}(1 + T_{\mathrm{n}_j}s)}{T_{\mathrm{i}}T_{\mathrm{m}}s^{k+1}(1 + T_{\Sigma}s) + k_{\mathrm{p}}k_{\mathrm{h}}\prod_{j=1}^{k}(1 + T_{\mathrm{n}_j}s)}, \tag{5.83}$$

or

$$T(s) = \frac{k_{\mathrm{p}}\left(r_k s^k + r_{k-1}s^{k-1} + \cdots + r_2 s^2 + r_1 s + 1\right)}{T_{\mathrm{i}}T_{\mathrm{m}}T_{\Sigma}s^{k+2} + T_{\mathrm{i}}T_{\mathrm{m}}s^{k+1} + k_{\mathrm{p}}k_{\mathrm{h}}\begin{pmatrix} r_k s^k + r_{k-1}s^{k-1} + \cdots \\ +r_2 s^2 + r_1 s + 1 \end{pmatrix}} \tag{5.84}$$

respectively. Then, according to (5.84), T_{i} is calculated by

$$a_{k+1}^2 = 2a_{k+2}a_k \tag{5.85}$$

or

$$(T_{\mathrm{i}}T_{\mathrm{m}})^2 = 2k_{\mathrm{p}}k_{\mathrm{h}}T_{\mathrm{i}}T_{\mathrm{m}}T_{\Sigma}r_k \tag{5.86}$$

or

$$T_{\mathrm{i}}T_{\mathrm{m}} = 2k_{\mathrm{p}}k_{\mathrm{h}}T_{\Sigma}r_k. \tag{5.87}$$

Finally, along with the aid of (5.85), it is obtained

$$\begin{aligned} T_{\mathrm{i}} &= 2k_{\mathrm{p}}k_{\mathrm{h}}\frac{T_{\Sigma}}{T_{\mathrm{m}}}r_k = 2k_{\mathrm{p}}k_{\mathrm{h}}\frac{T_{\Sigma}}{T_{\mathrm{m}}}\prod_{j=1}^{k}T_{\mathrm{n}_j} \\ &= 2k_{\mathrm{p}}k_{\mathrm{h}}\frac{T_{\Sigma}}{T_{\mathrm{m}}}\left(\prod_{j=1}^{k-1}T_{\mathrm{n}_j}\right)T_{\mathrm{n}_k} = T_{i_k}T_{\mathrm{n}_k} \end{aligned} \tag{5.88}$$

which is equal to (5.80). In that case, if (5.88) is substituted into (5.83), results in

$$T(s) = \frac{k_{\mathrm{p}}\prod_{r=1}^{p-1}(1 + T_{\mathrm{n}_r}s)}{2k_{\mathrm{p}}k_{\mathrm{h}}T_{\Sigma}^2\prod_{r=1}^{p-1}T_{\mathrm{n}_r}s^{p+1} + 2k_{\mathrm{p}}k_{\mathrm{h}}T_{\Sigma}\prod_{r=1}^{p-1}T_{\mathrm{n}_r}s^{p} + k_{\mathrm{p}}k_{\mathrm{h}}\prod_{r=1}^{p-1}(1 + T_{\mathrm{n}_r}s)}. \tag{5.89}$$

For determining now parameters T_{n_r}, it is shown that in order the magnitude of (5.89) satisfies condition $|T(j\omega) \simeq 1|$, controller time constants T_{n_r} must satisfy condition

$$4T_{\Sigma}^{2}\sum\prod_{i=1}^{p-3}T_{\mathrm{n}_i}-4T_{\Sigma}\sum\prod_{i=1}^{p-2}T_{\mathrm{n}_i}+\prod_{i=1}^{p-1}T_{\mathrm{n}_i}=0. \tag{5.90}$$

This is justified as follows. In type-II control loops for determining parameter T_{n_1} we make use of $a_1^2-2a_2a_0=0$ [see (A.11)]. This results in

$$k_{\mathrm{p}}^{2}T_{\mathrm{n}_1}^{2}=2k_{\mathrm{p}}(2k_{\mathrm{p}}T_{\mathrm{n}_1}T_{\Sigma}) \tag{5.91}$$

or finally

$$T_{\mathrm{n}_1}-4T_{\Sigma}=0. \tag{5.92}$$

In a similar fashion, in type-III control loops for determining parameters T_{n_1}, T_{n_2} we make use of $a_2^2-2a_3a_1+2a_4a_0=0$, see (A.11). This results in

$$4T_{\Sigma}^{2}T_{\mathrm{n}_1}T_{\mathrm{n}_2}-4T_{\Sigma}T_{\mathrm{n}_1}T_{\mathrm{n}_2}(T_{\mathrm{n}_1}+T_{\mathrm{n}_2})+T_{\mathrm{n}_1}^{2}T_{\mathrm{n}_2}^{2}=0 \tag{5.93}$$

or finally,

$$T_{\mathrm{n}_1}T_{\mathrm{n}_2}-4T_{\Sigma}(T_{\mathrm{n}_1}+T_{\mathrm{n}_2})+4T_{\Sigma}^{2}=0. \tag{5.94}$$

According to the above, and based on (5.79) if the closed-loop control system is of type-p, then for determining parameters T_{i} and T_{n_j}, $(j=1,2,\ldots,k)$, the following optimization conditions are claimed to be,

$$a_{k-1}^{2}=2a_{k-2}a_{k}-2a_{k-3}a_{k+1}, \tag{5.95}$$
$$a_{k}^{2}=2a_{k+1}a_{k-1} \tag{5.96}$$

the ones that satisfy condition $|T(j\omega)\simeq 1|$ in a wide range of frequencies.

Therefore, if $n=k-1$ then controller $C(s)$ is defined by (5.78), and the closed-loop transfer function is given by

$$T(s)=\frac{k_{\mathrm{p}}(r_{k-1}s^{k-1}+r_{k-2}s^{k-2}+\cdots+r_2s^2+r_1s+1)}{T_{\mathrm{i}}T_{\mathrm{m}}T_{\Sigma}s^{k+1}+T_{\mathrm{i}}T_{\mathrm{m}}s^{k}+k_{\mathrm{p}}k_{\mathrm{h}}\begin{pmatrix}r_{k-1}s^{k-1}+r_{k-2}s^{k-2}\\+\cdots+r_2s^2+r_1s+1\end{pmatrix}}. \tag{5.97}$$

In (5.87) it was shown that $T_{\mathrm{i}}T_{\mathrm{m}}=2k_{\mathrm{p}}k_{\mathrm{h}}T_{\Sigma}r_k$. By applying (5.95)–(5.97) we obtain

$$k_{\mathrm{p}}^{2}r_{k-1}^{2}-2r_{k-2}k_{\mathrm{p}}\left(2k_{\mathrm{p}}T_{\Sigma}r_{k-1}\right)+2k_{\mathrm{p}}r_{k-3}\left(2k_{\mathrm{p}}T_{\Sigma}r_{k-1}\right)T_{\Sigma}=0, \tag{5.98}$$

which yields

$$r_{k-1}-4T_{\Sigma}r_{k-2}+4T_{\Sigma}^{2}r_{k-3}=0. \tag{5.99}$$

If $n = k$, then we are going to show that

$$r_k - 4T_\Sigma r_{k-1} + 4T_\Sigma^2 r_{k-2} = 0. \tag{5.100}$$

If $n = k$ then the closed-loop transfer function is given by (5.84). Since $T_\mathrm{i}T_\mathrm{m} = 2k_\mathrm{p}k_\mathrm{h}T_\Sigma r_k$ then by applying $a_k^2 = 2a_{k-1}a_{k+1} - 2a_{k-2}a_{k+2}$ to (5.84) we obtain

$$k_\mathrm{p}^2 r_k^2 - 2r_{k-1}k_\mathrm{p}(2k_\mathrm{p}T_\Sigma r_k) + 2k_\mathrm{p}r_{k-2}(2k_\mathrm{p}T_\Sigma r_k)T_\Sigma = 0 \tag{5.101}$$

which yields

$$4T_\Sigma^2 r_{k-2} - 4T_\Sigma r_{k-1} + r_k = 0 \tag{5.102}$$

or finally

$$4T_\Sigma^2 \sum \prod_{i=1}^{p-2} T_{\mathrm{n}_i} - 4T_\Sigma \sum \prod_{i=1}^{p-1} T_{\mathrm{n}_i} + \prod_{i=1}^{p} T_{\mathrm{n}_i} = 0. \tag{5.103}$$

The above equation is rewritten in the form of

$$4T_\Sigma^2 T_{\mathrm{n}_p} \sum \prod_{i=1}^{p-3} T_{\mathrm{n}_i} - 4T_\Sigma T_{\mathrm{n}_p} \sum \prod_{i=1}^{p-2} T_{\mathrm{n}_i} + T_{\mathrm{n}_p} \prod_{i=1}^{p-1} T_{\mathrm{n}_i} = 0 \tag{5.104}$$

or finally

$$\left(4T_\Sigma^2 \sum \prod_{i=1}^{p-3} T_{\mathrm{n}_i} - 4T_\Sigma \sum \prod_{i=1}^{p-2} T_{\mathrm{n}_i} + \prod_{i=1}^{p-1} T_{\mathrm{n}_i}\right) T_{\mathrm{n}_p} = 0 \tag{5.105}$$

which is true, since (5.98) holds by.

Obviously, the number of combinations of the T_{n_i} optimal parameters that satisfy (5.105) is infinite. More specifically, by applying condition (5.105) for the design of up to type-V control loops results in

Type-V control loops:

$$\begin{aligned} &4\begin{pmatrix} T_{\mathrm{n}_1}T_{\mathrm{n}_2} + T_{\mathrm{n}_1}T_{\mathrm{n}_3} + T_{\mathrm{n}_1}T_{\mathrm{n}_4} \\ +T_{\mathrm{n}_2}T_{\mathrm{n}_3} + T_{\mathrm{n}_2}T_{\mathrm{n}_4} + T_{\mathrm{n}_3}T_{\mathrm{n}_4} \end{pmatrix} T_\Sigma^2 \\ &\quad - 4\begin{pmatrix} T_{\mathrm{n}_1}T_{\mathrm{n}_2}T_{\mathrm{n}_3} + T_{\mathrm{n}_1}T_{\mathrm{n}_2}T_{\mathrm{n}_4} \\ +T_{\mathrm{n}_2}T_{\mathrm{n}_3}T_{\mathrm{n}_4} + T_{\mathrm{n}_1}T_{\mathrm{n}_3}T_{\mathrm{n}_4} \end{pmatrix} T_\Sigma + T_{\mathrm{n}_1}T_{\mathrm{n}_2}T_{\mathrm{n}_3}T_{\mathrm{n}_4} = 0 \end{aligned} \tag{5.106}$$

Type-IV control loops:

$$4\left(T_{n_1}+T_{n_2}+T_{n_3}\right)T_\Sigma^2-4\left(T_{n_1}T_{n_2}+T_{n_2}T_{n_3}+T_{n_1}T_{n_3}\right)T_\Sigma+T_{n_1}T_{n_2}T_{n_3}=0. \tag{5.107}$$

Type-III control loops:

$$4T_\Sigma^2-4\left(T_{n_1}+T_{n_2}\right)T_\Sigma+T_{n_1}T_{n_2}=0. \tag{5.108}$$

Type-II control loops:

$$T_{n_1}=4T_\Sigma. \tag{5.109}$$

Note that (5.108) and (5.109) are equal to (4.16), respectively. In similar fashion with type-III control loops and for the sake of simplicity of the analysis, if we choose

$$T_{n_1}=T_{n_2}=\cdots=T_{n_{p-1}}=nT_\Sigma \tag{5.110}$$

the respective open $F_{ol}(s)$ and closed-loop $T(s)$ transfer functions are given by

$$F_{ol}(s)\approx\frac{(1+nT_\Sigma s)^{p-1}}{2n^{p-1}T_\Sigma^p s^p(1+T_\Sigma s)} \tag{5.111}$$

and

$$T(s)\approx\frac{(1+nT_\Sigma s)^{p-1}}{2n^{p-1}T_\Sigma^{p+1}s^{p+1}+2n^{p-1}T_\Sigma^p s^p+(1+nT_\Sigma s)^{p-1}}. \tag{5.112}$$

The optimal value of parameter n depends on the type of the control loop we want to design. If we substitute (5.110) into (5.107)–(5.109), we have consequently,

Type-V control loops:

$$n^2\left(n^2-16n+24\right)T_\Sigma^4=0 \quad\Rightarrow\quad n_{opt}=14.32. \tag{5.113}$$

Type-IV control loops:

$$n\left(n^2-12n+12\right)T_\Sigma^3=0 \quad\Rightarrow\quad n_{opt}=10.89. \tag{5.114}$$

Type-III control loops:

$$\left(n^2-8n+4\right)T_\Sigma^2=0 \quad\Rightarrow\quad n_{opt}=7.46. \tag{5.115}$$

With respect to the above, for the design of a type-IV control loop, a PID type controller of three zeros in its transfer function is required. Therefore, if we chose

$$T_{n_1} = T_{n_2} = nT_\Sigma \tag{5.116}$$

according to (5.110), we obtain from (5.107) that

$$T_{n_3} = \frac{4n(n-2)}{n^2 - 8n + 4} T_\Sigma = \frac{4n\,(n-2)}{(n-0.536)(n-7.464)} T_\Sigma \tag{5.117}$$

Based on the above, the corresponding $F_{ol}(s)$ and $T(s)$ transfer functions are given by

$$F_{ol}(s) = \frac{\begin{bmatrix} 4n^3(n-2)T_\Sigma^3 s^3 + n^2(n^2-12)T_\Sigma^2 s^2 \\ +2n^2(n-6)T_\Sigma s + (n-0.536)(n-7.464) \end{bmatrix}}{8n^3(n-2)T_\Sigma^5 s^5 + 8n^3(n-2)T_\Sigma^4 s^4}, \tag{5.118}$$

$$T(s) = \frac{b_3 s^3 + b_2 s^2 + b_1 s + b_0}{a_5 s^5 + a_4 s^4 + a_3 s^3 + a_2 s^2 + a_1 s + a_0} \tag{5.119}$$

where

$$b_3 = 4n^3\,(n-2)\,T_\Sigma^3, \quad b_2 = n^2 \frac{(n^2-12)}{n-2} T_\Sigma^2 \tag{5.120}$$

$$b_1 = 2n^2 \frac{(n-6)}{n-2} T_\Sigma, \quad b_0 = \frac{(n-0.536)\,(n-7.464)}{n-2} \tag{5.121}$$

and

$$a_5 = 8n^3(n-2)T_\Sigma^5, \quad a_4 = 8n^3 T_\Sigma^4 \tag{5.122}$$

$$a_3 = 4n^3 T_\Sigma^3, \quad a_2 = n^2 \frac{(n^2-12)}{n-2} T_\Sigma^2 \tag{5.123}$$

$$a_1 = 2n^2 \frac{(n-6)}{n-2} T_\Sigma, \quad a_0 = \frac{(n-0.536)\,(n-7.464)}{n-2}. \tag{5.124}$$

According to (5.119), the closed-loop control system is of type-IV since, $a_j = b_j$, $j = 0, 1, 2, 3$, see Sect. 2.5. If $n < 7.464$ the closed-loop control system is unstable. As a result, for having a feasible PID type control law, $n > 7.464$ has to hold by, see (5.120). In Fig. 5.11b the frequency response of sensitivity S and complementary sensitivity T of the type-IV closed-loop is presented, for several variations of parameter n, $n \in [7.5, \infty)$. From there, it is obvious that variations of parameter n do not lead to critical variations of both functions $T(s)$, $S(s)$ in the frequency domain. Sensitivity S is affected only in the lower frequency region.

Note that, in a similar fashion with type-III control loops, sensitivity $S(s)$ becomes minimum when all controller zeros are equal, $T_{n_1} = T_{n_2} = T_{n_3}, n = 10.89$, Fig. 5.14.

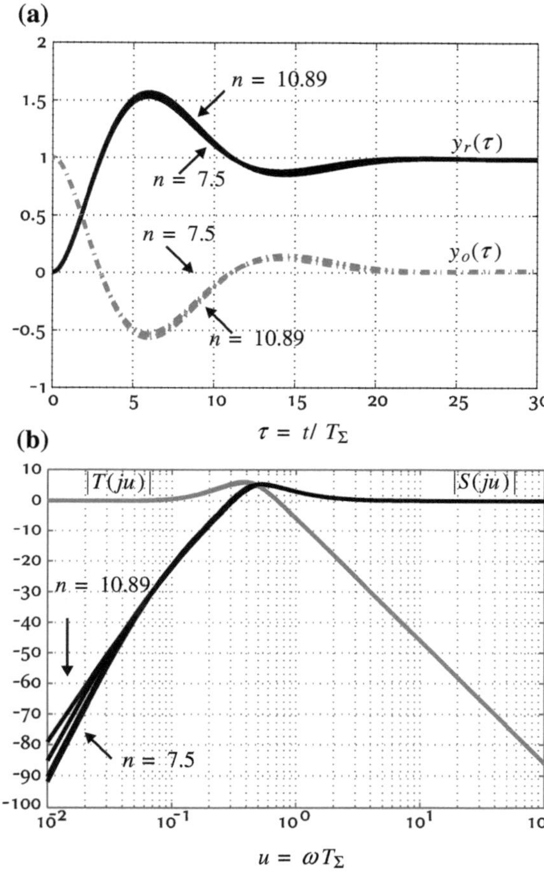

Fig. 5.11 Type-IV control loop. Step and frequency response of the final closed-loop control system for various values of parameter n. **a** Step response and output disturbance rejection of the final closed-loop control system for various values of parameter n. **b** Frequency response of the final closed-loop control system for various values of parameter n

There, it is shown how the controller's zeros are affected in case of variations in design parameter n. Similar results are also observed in the time domain, Fig. 5.11a.

The step response of the type-IV closed-loop control system exhibits an overshoot of 50 %, which is justified by the phase margin ($\phi = 32° < 45°$) of the open-loop $F_{ol}(s)$ frequency response, Fig. 5.12. For decreasing the overshoot of the final closed-loop control system, the two degrees of freedom controller structure is again be exploited. If $n = 10.89$, then the closed-loop transfer function in terms of time constants form is given by

$$T(s) = \frac{N_1(s)}{D_1(s)D_2(s)D_3(s)}, \tag{5.125}$$

where

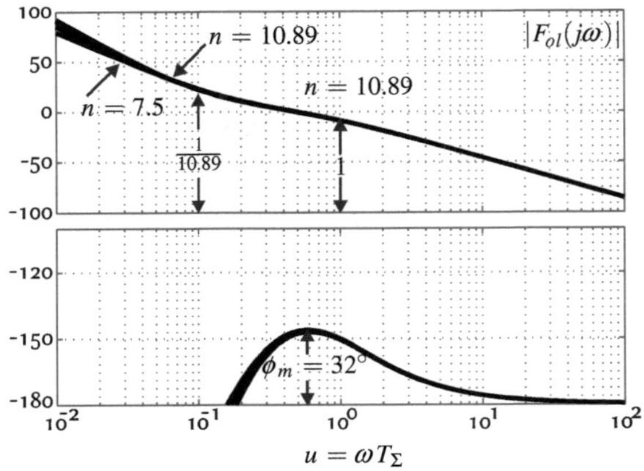

Fig. 5.12 Open-loop frequency response of a type-IV control loop for various values of parameter n

$$N_1(s) = (1 + 10.89T_\Sigma s)^3 \quad D_1(s) = (1 + 2.3T_\Sigma s), \tag{5.126}$$

$$D_2(s) = (2.274)^2 T_\Sigma^2 s^2 + 0.99(2.274)T_\Sigma s + 1, \tag{5.127}$$

$$D_3(s) = (14.75)^2 T_\Sigma^2 s^2 + 1.9(14.75)T_\Sigma s + 1. \tag{5.128}$$

Thus, by choosing an external controller of the form (Fig. 5.14)

$$C_{\text{ex}}(s) = \frac{(1 + 2.3T_\Sigma s)\left[(14.75)^2 T_\Sigma^2 s^2 + 1.9(14.75)T_\Sigma s + 1\right]}{(1 + 10.89T_\Sigma s)^3(1 + T_\Sigma s)} \tag{5.129}$$

overshoot is reduced to 14.75 %, Fig. 5.13.

5.3.2 Simulation Results

For justifying the control's law potential simulation examples of type-II, type-III, type-IV, type-V control loops are presented. According to the control law presented in Sect. 5.3.1 the I-I-PID type controller for controlling a type-0 process is given by

$$C(s) = \frac{(1 + T_{\text{n}_1}s)(1 + T_{\text{n}_2}s)(1 + T_{\text{n}_3}s)}{T_{\text{i}}s^3(1 + T_{\Sigma_{\text{c}_1}}s)(1 + T_{\Sigma_{\text{c}_2}}s)}. \tag{5.130}$$

In all three examples, it is assumed that the sum T_Σ of all time constants of the process is accurately measured. Time constant

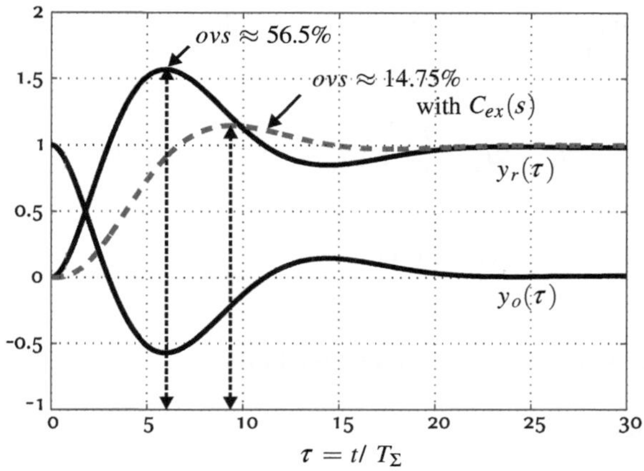

Fig. 5.13 The effect of the two degrees of freedom controller structure to the step response of the type-IV closed-loop control system

Fig. 5.14 Variations of parameters T_{n_1}, T_{n_2}, T_{n_3} according to variations of parameter n

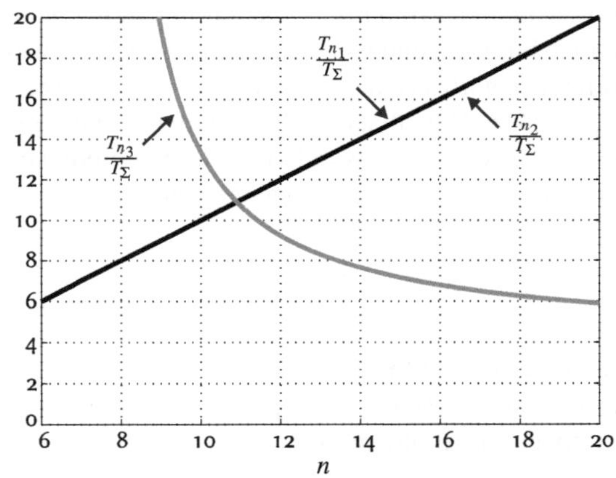

$$T_\Sigma = \sum_{j=1}^{k} T_{p_j} + T_{\Sigma c} \tag{5.131a}$$

$$T_{\Sigma c} = T_{\Sigma_{c1}} + T_{\Sigma_{c2}} \tag{5.131b}$$

includes both plant's and controller's unmodeled dynamics. Since type-III control loops are designed

$$T_{n_1} = T_{p_1}, \tag{5.132a}$$

$$T_{n_2} = \frac{4(n-1)}{n-4} T_\Sigma, \tag{5.132b}$$

$$T_{n_3} = nT_\Sigma. \tag{5.132c}$$

Parameter n has been chosen equal to $n = 7.46$. The integrator's time constant is calculated through

$$T_{\mathrm{i}} = 2k_{\mathrm{p}}k_{\mathrm{h}}\frac{T_{\Sigma}}{T_{\mathrm{m}}}\prod_{r=1}^{p-1} T_{\mathrm{n}_r} = 2k_{\mathrm{p}}k_{\mathrm{h}}T_{\mathrm{n}_2}T_{\mathrm{n}_3}T_{\Sigma}. \tag{5.133}$$

In all three cases $T_{\mathrm{m}} = 1$ has been set.

5.3.2.1 Process with Dominant Time Constants

The process described by

$$G(s') = \frac{2}{(1+s')\,(1+0.84s')(1+0.78s')(1+0.57s')(1+0.28s')} \tag{5.134}$$

is considered. From Fig. 5.15a it is apparent that the type-III closed-loop control system exhibits an undesired overshoot of 87.4 % which is decreased by the filtering the reference with an external controller $C_{\mathrm{ex}_1}(s)$. Settling time remains almost unaltered, $t_{\mathrm{ss}} = 143\tau$. Note that disturbance rejection has remained the same since the external controller $C_{\mathrm{ex}_1}(s)$ acts only only at the reference signal outside of the control loop.

For manipulating the overshoot of the output, if

$$C_{\mathrm{ex}_2}(s) = \frac{1}{(t_{\mathrm{n}_2}t_{\mathrm{n}_3})s^2 + (t_{\mathrm{n}_2} + t_{\mathrm{n}_3})s + 1} \tag{5.135}$$

reference filter is to be used, then the overshoot is decreased to 6.2 %. Since the closed-loop control system is of type-III, the output of the process can track perfectly both ramp and parabolic reference signals, Fig. 5.16. External filter of the form

$$C_{\mathrm{ex}}(s) = \frac{1}{(0.45t_{\mathrm{n}_2}t_{\mathrm{n}_3})s^2 + (t_{\mathrm{n}_2} + 0.45t_{\mathrm{n}_3})s + 1} \tag{5.136}$$

is used for decreasing the overshoot of the output.

5.3.2.2 Process with Time Delay

A process with time delay of the form

$$G(s') = \frac{2}{(1+s')(1+0.99s')(1+0.57s')(1+0.28s')(1+0.1s')}\mathrm{e}^{-s'} \tag{5.137}$$

is assumed in this example. Note that the proposed control law does not take into account the effect of the time delay and therefore in this example the robustness

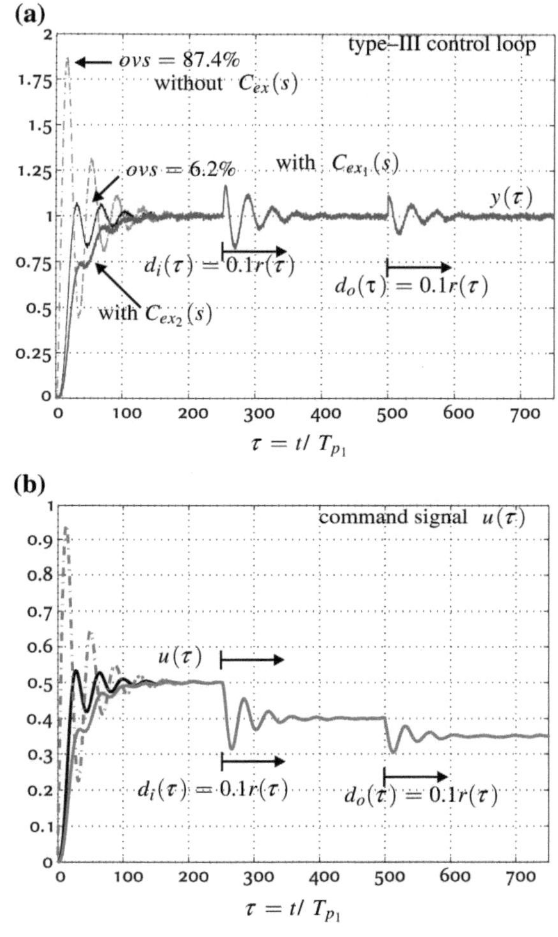

Fig. 5.15 Type-III closed-loop control system, $G(s)$ defined by (5.134). Output disturbance rejection is applied at $\tau = 250$. Input disturbance $d_{\mathrm{i}}(s) = 0.1r(s)$ is applied at $\tau = 250$ and output disturbance $d_{\mathrm{i}}(s) = 0.1r(s)$ is applied at $\tau = 500$. **a** Response of the output $y(\tau)$ of the control system in the presence of input and output disturbances. **b** Response of the command signal $u(\tau)$ in the presence of input and output disturbances

of the method to model uncertainties is also tested. If no external filter is used for reference tracking, the control loop exhibits an overshoot of 100.4 %, Fig. 5.17a.

The use of both $C_{\mathrm{ex}_1}(s)$, $C_{\mathrm{ex}_2}(s)$ eliminates the overshoot to 9.4 and 0 %, respectively, Fig. 5.17a. Disturbance rejection remains unaltered. $C_{\mathrm{ex}_2}(s)$ is of the same form as in the previous example. Note that control signal $u(\tau)$ is improved in case the reference signal is filtered, Figs. 5.15b and 5.17b. External filter of the form

$$C_{\mathrm{ex}_1}(s) = \frac{1}{\left(0.45t_{\mathrm{n}_2}t_{\mathrm{n}_3}\right)s^2 + \left(t_{\mathrm{n}_2} + 0.45t_{\mathrm{n}_3}\right)s + 1} \tag{5.138}$$

is used for decreasing the overshoot of the output.

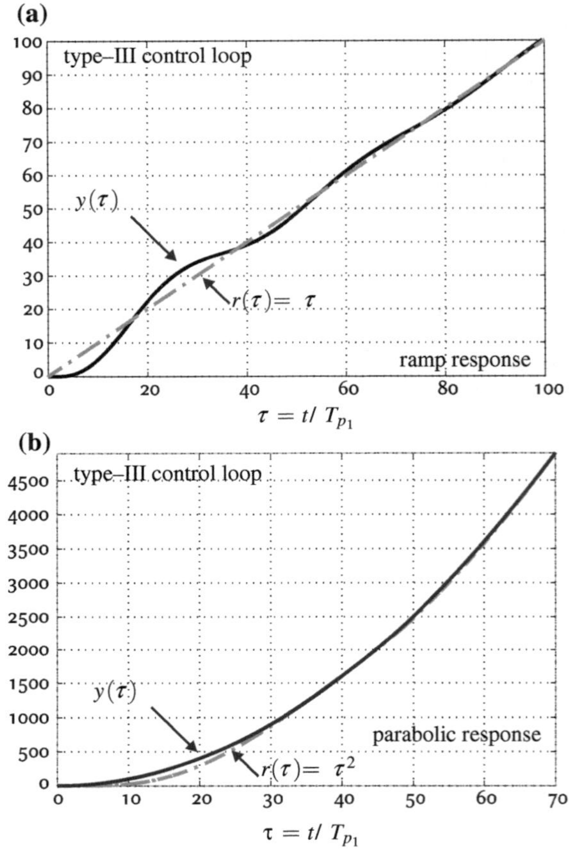

Fig. 5.16 Type-III closed-loop control system, $G(s)$ defined by (5.134). **a** Ramp response of the closed loop control system. **b** Parabolic response of the closed-loop control system

5.3.2.3 A Nonminimum Phase Process

Although the proposed theory does not take into account the existence of zeros in the process model, a non minimum phase process of the form

$$G(s') = \frac{1.34(1 - 0.771s')}{(1+s')(1+0.33s')(1+0.12s')(1+0.056s')(1+0.038s')} \tag{5.139}$$

is adopted for testing the robustness of the proposed control law. The step response of (5.139) is presented in Fig. 5.18. In addition, in Fig. 5.19a, b the step response of the output $y(\tau)$ and the control signal $u(\tau)$ are presented, respectively. If no external filter is used, the overshoot of the step response is 59.9 %. Since this is undesirable, if $r(s)$ is filtered by $C_{ex_1}(s)$, $C_{ex_2}(s)$ then the overshoot is reduced to 0 % in both cases. Output and input disturbance rejection remain unaltered since the external filter does not participate into

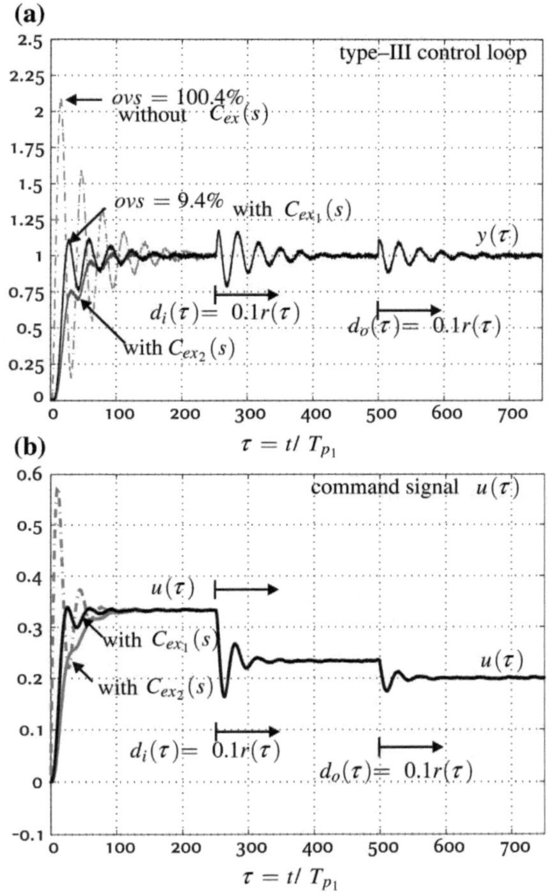

Fig. 5.17 Type-III closed-loop control system, $G(s)$ defined by (5.137). Output disturbance rejection is applied at $\tau = 250$. Input disturbance $d_{\mathrm{i}}(s) = 0.1r(s)$ is applied at $\tau = 250$ and output disturbance $d_{\mathrm{i}}(s) = 0.1r(s)$ is applied at $\tau = 500$. **a** Response of the output $y(\tau)$ of the control system in the presence of input and output disturbances. **b** Response of the command signal $u(\tau)$ of the control system in the presence of input and output disturbances

$$S_{\mathrm{i}}(s) = \frac{y(s)}{d_{\mathrm{i}}(s)} \tag{5.140}$$

and

$$S_{\mathrm{o}}(s) = \frac{y(s)}{d_{\mathrm{i}}(s)} \tag{5.141}$$

respectively. External filter of the form

$$C_{\mathrm{ex}_1}(s) = \frac{1}{\left(0.45 t_{\mathrm{n}_2} t_{\mathrm{n}_3}\right) s^2 + \left(t_{\mathrm{n}_2} + 0.45 t_{\mathrm{n}_3}\right) s + 1} \tag{5.142}$$

is used for decreasing the overshoot of the output.

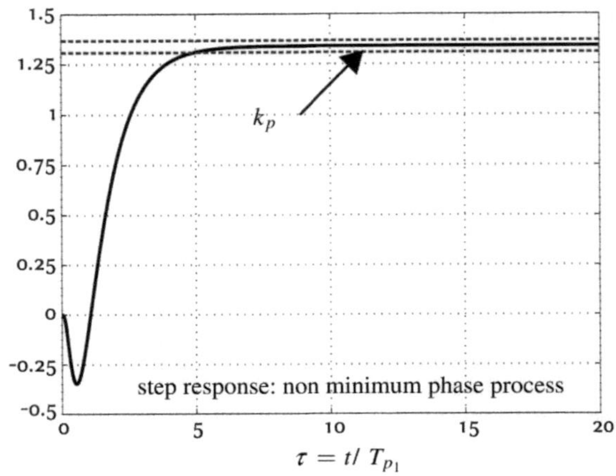

Fig. 5.18 Step response of the nonminimum phase process defined by (5.139)

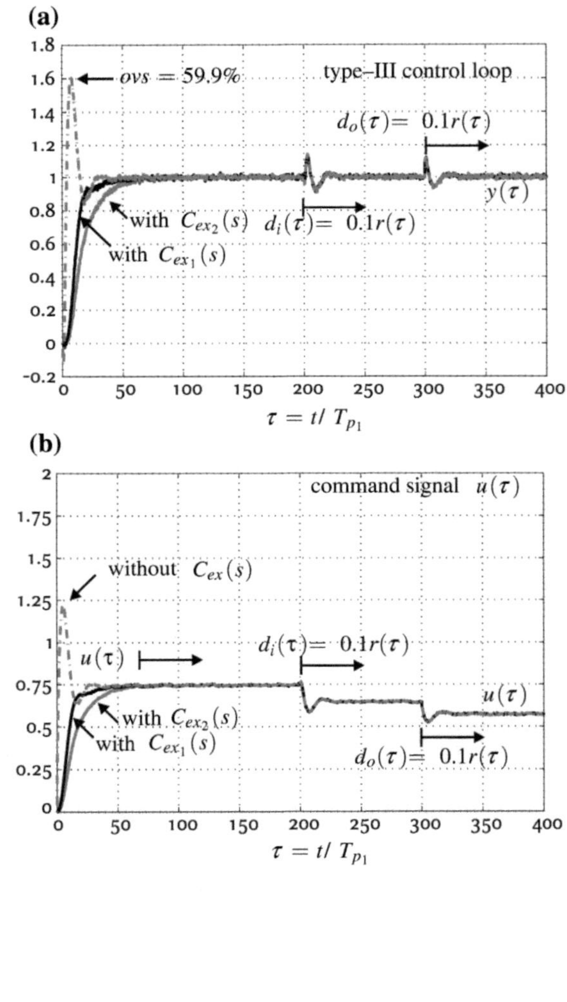

Fig. 5.19 Type-III closed-loop control system for a nonminimum phase process, $G(s)$ defined by (5.139). Output disturbance rejection is applied at $\tau = 200$. Input disturbance $d_i(s) = 0.1r(s)$ is applied at $\tau = 250$ and output disturbance $d_i(s) = 0.1r(s)$ is applied at $\tau = 300$. **a** Response of the output $y(\tau)$ of the control system in the presence of input and output disturbances. **b** Response of the command signal $u(\tau)$ of the control system in the presence of input and output disturbances

5.3.2.4 A Type-IV and Type-V Control Loop

From the Laplace transformation it is known that if $r(t) = t^n$ then

$$\mathcal{L}\{y(t)\} = \frac{n!}{s^{n+1}}. \tag{5.143}$$

For example, if $n = 1$ then $L\{r(t)\} = \frac{1}{s^2}$ and the system is of type-II, or if $n = 2$ then $L\{r(t)\} = \frac{2}{s^3}$ and the system is of type-III. For a type-IV and type-V control loop the Laplace transformation of the reference signal is given if $n = 3$ and $n = 4$ for which we have

$$L\{r(t)\} = \frac{3!}{s^{3+1}} \tag{5.144a}$$

$$L\{r(t)\} = \frac{4!}{s^{4+1}} \tag{5.144b}$$

respectively. According to the proposed theory for a type-IV, type-V control loop the proposed PID type controllers are given by

$$C(s) = \frac{\left(1 + T_{n_1}s\right)\left(1 + T_{n_2}s\right)\left(1 + T_{n_3}s\right)\left(1 + T_{n_4}s\right)}{T_i s^4 (1 + T_{\Sigma_{c_1}}s)(1 + T_{\Sigma_{c_2}}s)}, \tag{5.145}$$

$$C(s) = \frac{(1 + T_{n_1}s)(1 + T_{n_2}s)(1 + T_{n_3}s)(1 + T_{n_4}s)(1 + T_{n_5}s)}{T_i s^5 (1 + T_{\Sigma_{c_1}}s)(1 + T_{\Sigma_{c_2}}s)}. \tag{5.146}$$

respectively. For determining parameters T_{n_1}, T_{n_2}, T_{n_3}, T_{n_4}, T_i in (5.145) according to the proposed theory, we set $T_{n_4} = T_{p_1}$ and $T_{n_1} = T_{n_2} = nT_\Sigma$ according to (5.110). For that reason, (5.107) becomes

$$4(2nT_\Sigma + T_{n_3}) - 4(n^2T_\Sigma + 2nT_{n_3}) + n^2T_{n_3} = 0 \tag{5.147}$$

or finally

$$T_{n_3} = \frac{4n(n-2)}{(n^2 - 8n + 4)}T_\Sigma. \tag{5.148}$$

Integrator's time constant for the type-IV control loop is equal to

$$T_i = 2k_p k_h T_{n_1} T_{n_2} T_{n_3} T_\Sigma. \tag{5.149}$$

In a similar fashion, for the (5.146) PID type controller and since the control loop is of type-V, we set $T_{n_5} = T_{p_1}$ and $T_{n_1} = T_{n_2} = T_{n_3} = nT_\Sigma$. Accordingly, (5.106) becomes

$$4(3n^2T_\Sigma^2 + 3nT_\Sigma T_{n4})T_\Sigma - 4(n^3T_\Sigma^3 + 3n^2T_\Sigma^2 T_{n4}) + n^3T_\Sigma^2 T_{n4} = 0 \quad (5.150)$$

which after some algebraic manipulation yields

$$T_{n4} = \frac{4n^2\,(n-3)}{n\left(n^2 - 12n + 12\right)} T_\Sigma = \frac{4n\,(n-3)}{n^2 - 12n + 12} T_\Sigma. \quad (5.151)$$

Integrator's time constant for the type-V control loop is equal to

$$T_i = 2k_p k_h T_{n_1} T_{n_2} T_{n_3} T_{n_4} T_\Sigma. \quad (5.152)$$

The process in this example is defined by (5.154). The respective response to $r(t) = t^3$ and $r(t) = t^4$ reference signals for the type-IV and the type-V control loop are presented in Fig. 5.20b.

5.3.3 Robustness Performance

In this section the robustness of the proposed design tuning procedure is tested.

5.3.3.1 Controller Tuning Without Pole Zero Cancellation

For testing the robustness of the proposed control law to parameter uncertainties, a type-III closed loop control system is designed where the PID controller does not achieve pole-zero cancellation. Therefore, parameter T_{n_1} is determined by $T_{n_1} = (1 + a)T_{p_1}$ where a is the error when measuring T_{p_1}. The process is given by

$$G(s') = \frac{1.23}{(1 + s')(1 + 0.872s')(1 + 0.367s')(1 + 0.287s')(1 + 0.11s')}. \quad (5.153)$$

From Fig. 5.21a, b it is apparent that if an error of 30 % when measuring T_{p_1} occurs, a small change is observed in the overshoot of the closed loop control system. In addition, both input and output disturbance rejection remain almost unaltered.

5.3.3.2 Comparison Between a Type-I and a Type-III Control Loop

For showing the advantages of designing a higher order faster control loop, the following process

$$G(s') = \frac{1.23}{(1 + s')(1 + 0.992s')(1 + 0.692s')(1 + 0.139s')(1 + 0.107s')} \quad (5.154)$$

is adopted. For this process, a type-I, type-III closed control loop is designed. For designing the PID type-I control loop the conventional Magnitude Optimum criterion

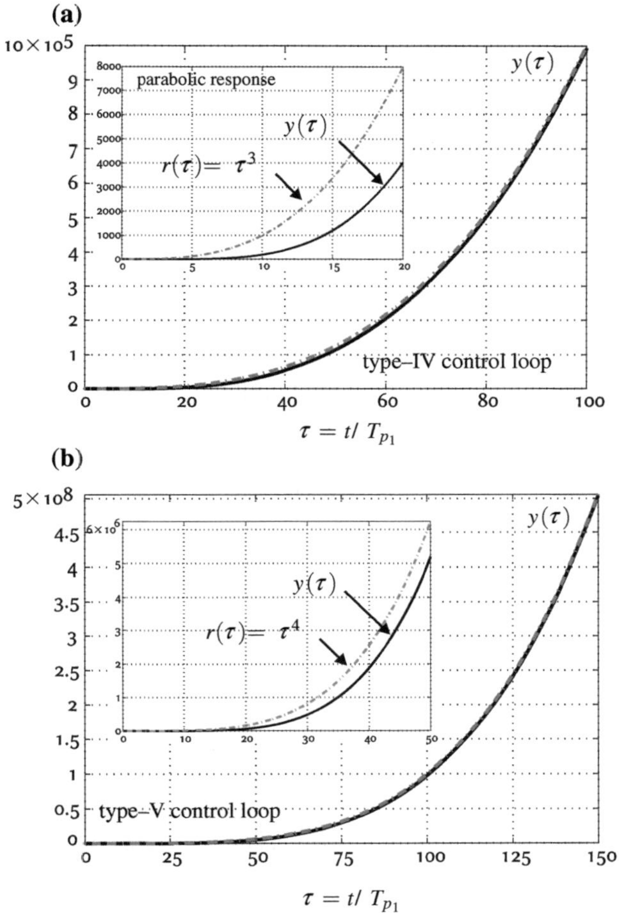

Fig. 5.20 Response of a type-IV and a type-V control loop. Parameter n has been chosen equal to $n = 14.32$ according to (5.114). **a** Response of the type-IV control loop to reference signal $r(\tau) = \tau^3$, parameter n has been chosen equal to $n = 10.89$ according to (5.114). **b** Response of the type-V control loop to reference signal $r(\tau) = \tau^4$

(see 3.2.4 and 5.2.1) is employed. Note that for determining controller's zeros, exact pole zero cancellation has to take place (see 3.2.4 and 5.2.1), [10]. From Fig. 5.22 it is apparent that the type-I control loop fails to track both the ramp and the parabolic reference signal exhibiting nonzero steady state velocity and acceleration error.

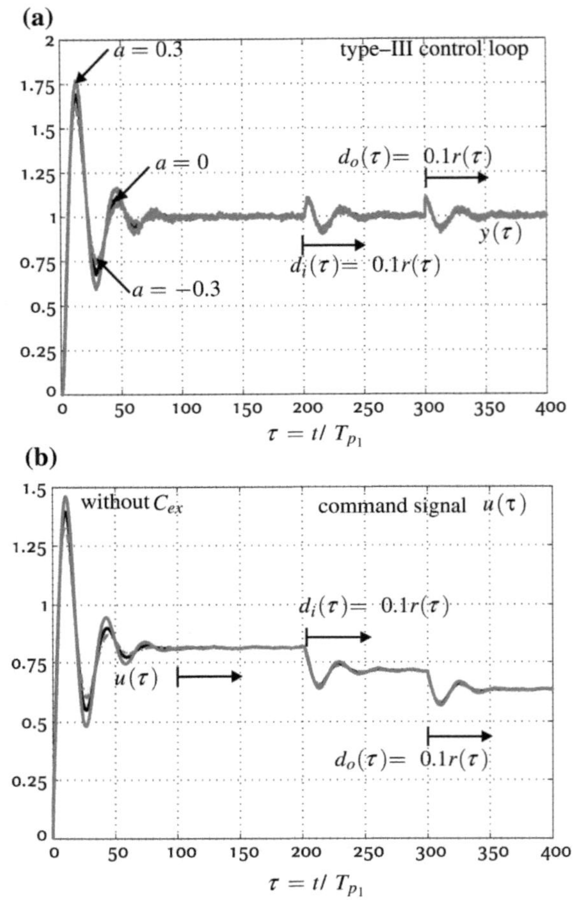

Fig. 5.21 Type-III closed-loop control system. The PID controller is tuned without pole zero cancellation $a = 0.3,\ a = -0.3$. The PID controller is tuned via exact pole-zero cancellation $a = 0$. **a** Response of the output $y(\tau)$ in the presence of input and output disturbance. **b** Response of the command signal $u(\tau)$ in the presence of input and output disturbance

5.3.3.3 Effect of the Process Unmodeled Dynamics to the Control Performance

The effect of the process unmodeled dynamics is discussed in this example. The process defined by

$$G(s) = \frac{1}{(1+s')(1+as')(1+a^2s')(1+a^3s')(1+a^4s')} \tag{5.155}$$

is adopted. As proved in Sects. 4.2.3 and 5.2.1 the proposed control law depends on pole-zero cancellation and time constant T_Σ which models the process' unmodeled dynamics (poles of the process far from the origin), see (4.5) and (5.3) where $T_\Sigma = T_{\Sigma_c} + T_{\Sigma_p}$ and T_{Σ_p} is the process parasitic time constant and $T_{\Sigma_c} \ll T_{\Sigma_p}$. In Fig. 5.23 the process is modeled by $a = 0.15$ containing a relatively large dominant time

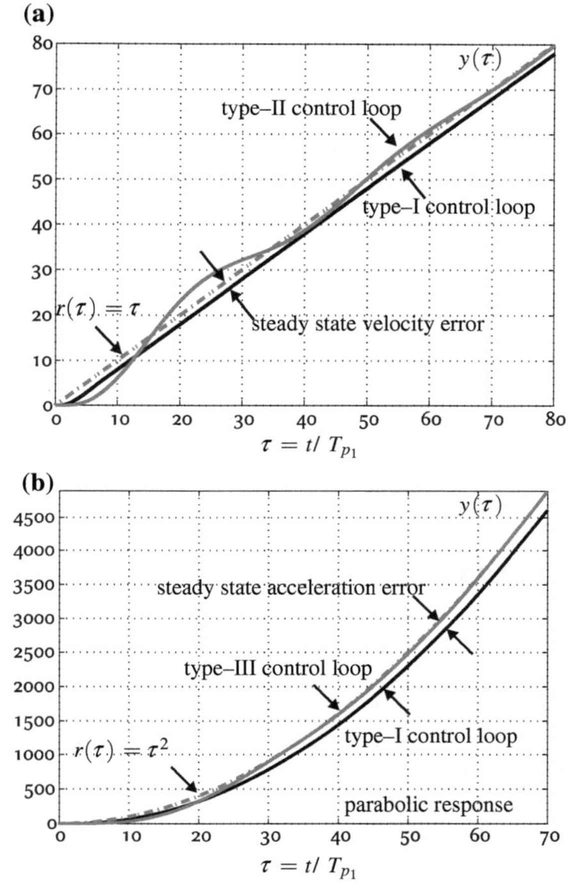

Fig. 5.22 Comparison between a type-I, type-III PID control loop. **a** The type-I control loop fails to track the ramp $r(\tau) = \tau$ reference signal since constant steady state velocity error is observed. **b** The type-I control loop fails to track the parabolic $r(\tau) = \tau^2$ reference signal since constant steady state velocity and acceleration error is observed

constant and in the next case $a = 0.6$ the parasitic time constant of the process is comparable to its dominant time constant.

Since

$$\frac{T_{\Sigma \mathrm{p}}}{T_{\mathrm{p}_1}} = \sum_{j=1}^{4} a^j, \tag{5.156}$$

it is apparent that when $a = 0.15$ then $T_{\Sigma \mathrm{p}} = T_{\mathrm{p}_1} \sum_{j=1}^{4} a^j = 0.1764 T_{\mathrm{p}_1}$ and when $a = 0.6$ then $T_{\Sigma \mathrm{p}} = T_{\mathrm{p}_1} \sum_{j=1}^{4} a^j = 1.3056 T_{\mathrm{p}_1}$. The conclusion according to Fig. 5.23 is that the less accurate the model of the process in terms of zeros, time delay, poles compared to the dominant time constant ($T_\Sigma \approx T_{\mathrm{p}_j}$), the poorer the performance becomes, (see settling time of the output and input disturbance rejection Fig. 5.23).

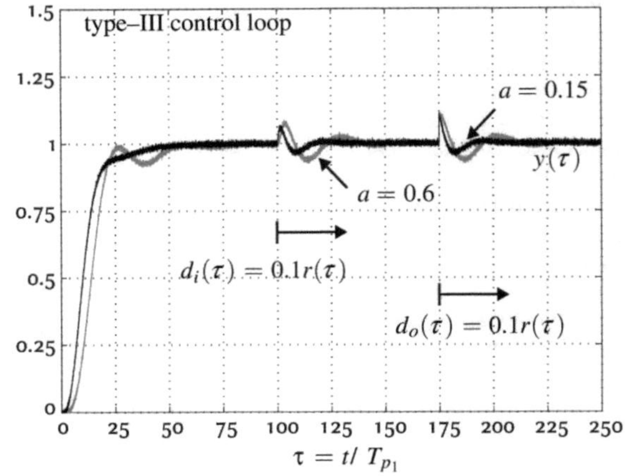

Fig. 5.23 Step response of the PID type-III control loop when $a = 0.15$ and $a = 0.6$ for a process defined by (5.155)

5.4 Summary

Explicit PID tuning rules have been presented towards the design of type-III control loops and regardless of the process complexity in Sect. 5.2.2. The proposed control law is considered feasible for many real world applications since it is of PID type. For the definition of the optimal control law, the powerful principle of the Symmetrical Optimum criterion was adopted. The advantage of type-III control loops compared to type-I, type-II (control of integrating processes) is obvious since the higher the type of the control loop, the faster reference signals can be tracked by the output of the process.

This advantage has been justified through simulation examples for the control of a variety of process models as show in Sect. 5.2.3. It was shown that the conventional PID tuning (type-II control loops, current state of the art) via the Symmetrical Optimum criterion fails to track parabolic reference signals. Even in cases when the conventional tuning is used for the design of a type-III control loop, the performance is still suboptimal especially in cases when the process complexity is increased.

In contrast to this, the proposed PID control law tracks with zero steady state position, velocity and acceleration error step, ramp, and parabolic reference signals regardless of the plant complexity. The robustness of the proposed control law was also tested to parameters variations showing finally promising results, see Sect. 5.2.3.4. To this end, control engineers are capable of designing type-III control loops, firstly on a simulation level before going finally on a real time implementation.

Moreover, the Symmetrical Optimum criterion has been extended for the design of type-p control loop in Sect. 5.3. Based on the design of type-III control loops (design with pole-zero cancellation), the proposed control law was extended for tuning PID type-p control loops so that tracking of faster reference signals is achieved.

The development of the proposed control is carried out in the frequency domain where the transfer function of the process involves the dominant time constants and the plant's unmodeled dynamics, see Sect. 5.3.1. Once more, the proposed theory has been evaluated for the control of representative plants met in many industry applications, see Sect. 5.3.3. The robustness of the proposed control law achieves promising results (see Sect. 5.3.3.3) also for the control of processes with parameters the control law disregards, such as nonminimum phase processes and processes with time delay.

References

1. Åström KJ, Hägglund T (1995) PID controllers: theory, design and tuning, 2nd edn. Instrument Society of America, Research Triangle Park
2. Kessler C (1958) Das symmetrische optimum. Regelungstechnik, pp 395–400 and 432–426
3. Margaris NI (2003) Lectures in applied automatic control (in Greek), 1st edn. Tziolas
4. Oldenbourg RC, Sartorius H (1954) A uniform approach to the optimum adjustment of control loops. Trans ASME 76:1265–1279
5. Papadopoulos KG, Margaris NI (2012) Extending the symmetrical optimum criterion to the design of PID type-p control loops. J Process Control 12(1):11–25
6. Papadopoulos KG, Mermikli K, Margaris NI (2011a) Optimal tuning of PID controllers for integrating processes via the symmetrical optimum criterion. In: 19th mediterranean conference on control & automation (MED), IEEE, Corfu, Greece, pp 1289–1294
7. Papadopoulos KG, Papastefanaki EN, Margaris NI (2011b) Optimal tuning of PID controllers for type-III control loops. In: 19th mediterranean conference on control & automation (MED), IEEE, Corfu, Greece, pp 1295–1300
8. Papadopoulos KG, Papastefanaki EN, Margaris NI (2012a) Automatic tuning of PID type-III control loops via the symmetrical optimum criterion. In: International conference on industrial technology, (ICIT), IEEE, Athens, Greece, pp 881–886
9. Papadopoulos KG, Tselepis ND, Margaris NI (2012b) Revisiting the magnitude optimum criterion for robust tuning of PID type-I control loops. J Process Control 22(6):1063–1078
10. Papadopoulos KG, Papastefanaki EN, Margaris NI (2013) Explicit analytical PID tuning rules for the design of type-III control loops. IEEE Trans Ind Electron 60(10):4650–4664
11. Poulin E, Pomerleau A (1999) PI settings for integrating processes based on ultimate cycle information. IEEE Trans Control Syst Technol 7(4):509–511
12. Preitl S, Precup RE (1999) An extension of tuning relation after symmetrical optimum method for PI and PID controllers. Automatica 35(10):1731–1736

Chapter 6
Sampled Data Systems

Abstract In this chapter, analytical tuning rules for digital PID type-I, type-II, type-III control loops are presented. Controller parameters are determined explicitly as a function of the process parameters and the sampling time T_s of the controller. For developing the proposed theory in type-I, type-II control loops, a generalized single-input single-output stable process model is used consisting of n-poles, m-zeros plus unknown time delay-d. As far as type-III control loops is concerned the principle of pole-zero cancellation according to the method proposed in Sect. 5.2.1, see [3], is followed. The derivation of the proposed PID control law lies in the principle of the Magnitude Optimum criterion and the optimization conditions proved in Appendix A.1 are used for extracting the explicit solution. For all control loop types, a performance comparison is presented in terms of simulation examples. The comparison focuses on the effect of the sampling time T_s to the control loops response both in the time and frequency domain.

6.1 Type-I Control Loops

For presenting the proposed explicit solution, the closed loop system of Fig. 6.1 is considered. The transfer function of the process $G(s)$ is defined by

$$G(s) = k_{\mathrm{p}} \frac{s^m \beta_{\mathrm{m}} + \cdots + s^4 \beta_4 + s^3 \beta_3 + s^2 \beta_2 + s \beta_1 + 1}{\begin{pmatrix} s^n p_n + s^{n-1} p_{n-1} + \cdots + s^5 p_5 + s^4 p_4 + s^3 p_3 \\ +s^2 p_2 + s p_1 + 1 \end{pmatrix}} \mathrm{e}^{-sT_{\mathrm{d}}}, \quad n > m \tag{6.1}$$

and the proposed controller is given by

$$C(s) = C^*(s) C_{\mathrm{ZOH}}(s) = \left(\frac{1 + sX + s^2 Y}{sT_{\mathrm{i}}}\right)^* \left(\frac{1 - \mathrm{e}^{-sT_{\mathrm{s}}}}{sT_{\mathrm{s}}}\right). \tag{6.2}$$

Note in this case that $C^*(s)$ stands for the digital representation of the PID control law and $C_{\mathrm{ZOH}}(s)$ stands for the zero order hold unit. T_s stands for the sampling period of the controller.

K.G. Papadopoulos, *PID Controller Tuning Using the Magnitude Optimum Criterion*,
DOI 10.1007/978-3-319-07263-0_6

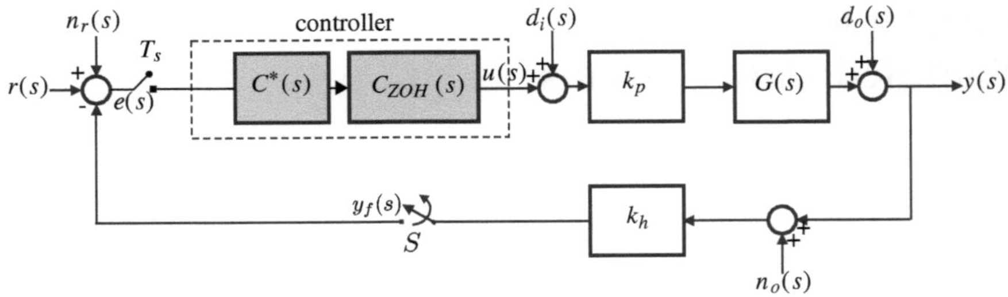

Fig. 6.1 Block diagram of the closed-loop control system. $G(s)$ is the plant transfer function, $C(s)$ is the controller transfer function, $r(s)$ is the reference signal, $y(s)$ is the output of the control loop, $y_f(s)$ is the output signal after k_h, $d_o(s)$ and $d_i(s)$ are the output and input disturbance signals respectively and $n_r(s)$, $n_o(s)$ are the noise signals at the reference input and process output respectively. k_p stands for the plant's dc gain and k_h is the feedback path

Normalizing all time constants in the frequency domain with the sampling period T_s and substituting with $s' = sT_s$ results in

$$G(s') = k_p \frac{\left(s'^m z_m + \cdots + s'^4 z_4 + s'^3 z_3 + s'^2 z_2 + s' z_1 + 1\right)}{\left(\begin{array}{c} s'^n r_n + s'^{n-1} r_{n-1} + \cdots + s'^5 r_5 + s'^4 r_4 + s'^3 r_3 \\ +s'^2 r_2 + s' r_1 + 1 \end{array}\right)} e^{-s'd} \tag{6.3}$$

for the process, and

$$C(s') = C^*(s') C_{ZOH}(s') = \left(\frac{1 + s'x + s'^2 y}{s' t_i}\right)^* \left(\frac{1 - e^{-s'}}{s'}\right) \tag{6.4}$$

for the controller, respectively. Let it be noted that for normalizing both (6.1), (6.2) the substitutions $x = \frac{X}{T_s}$, $y = \frac{Y}{T_s^2}$, $t_i = \frac{T_i}{T_s}$, $d = \frac{T_d}{T_s}$, $r_j = \frac{p_j}{T_s^i}$, $\forall j = 1, \ldots, n$, and $z_i = \frac{\beta_i}{T_s^i}$, $\forall i = 1, \ldots, m$ have been made.

The transition from $\mathcal{L}\{.\}$ to the $\mathcal{Z}\{.\}$ domain takes place according to the transformation

$$s' = \frac{z - 1}{z} = \frac{e^{s'} - 1}{e^{s'}}. \tag{6.5}$$

Since $z = e^{s'}$, the digital PID type controller takes now the form

$$\begin{aligned} C(s') &= C^*(s') C_{ZOH}(s') \\ &= \frac{1}{t_i} \frac{(1 + x + y)e^{2s'} - (x + 2y)e^{s'} + y}{e^{s'}(e^{s'} - 1)} \end{aligned} \tag{6.6}$$

For simplifying the calculation of the final closed loop transfer function, the substitutions

$$x = 2\hat{y} - \hat{x} - 2 \quad \text{and} \quad y = \hat{x} - \hat{y} + 1 \tag{6.7}$$

take place. As a result, (6.6) becomes

$$C(s') = C^*(s')C_{\text{ZOH}}(s') = \frac{1}{t_\text{i}}\frac{(1-\text{e}^{s'})\hat{x} + (\text{e}^{2s'}-1)\hat{y} + 1}{\text{e}^{s'}(\text{e}^{s'}-1)}. \tag{6.8}$$

In addition, the respective open $F_\text{ol}(s')$ and closed loop $T(s')$ transfer functions become

$$F_\text{ol}(s') = k_\text{h}C(s')G(s') \tag{6.9}$$

or

$$F_\text{ol}(s') = k_\text{h}\frac{k_\text{p}}{t_\text{i}}\frac{\left[\begin{pmatrix} s'^m z_\text{m} + \cdots + s'^3 z_3 \\ +s'^2 z_2 + s' z_1 + 1\end{pmatrix}\left[(1-\text{e}^{s'})\hat{x} + (\text{e}^{2s'}-1)\hat{y} + 1\right]\right]}{\left[\begin{pmatrix} s'^n r_n + \cdots + s'^3 r_3 + s'^2 r_2 \\ +s' r_1 + 1\end{pmatrix}\text{e}^{s'(d+1)}(\text{e}^{s'}-1)\right]} \tag{6.10}$$

and

$$T(s') = \frac{C(s')G(s')}{1 + k_\text{h}C(s')G(s')} = \frac{N(s')}{D(s')} = \frac{N(s')}{D_1(s') + k_\text{h}N(s')} \tag{6.11}$$

or

$$T(s') = \frac{k_\text{p}\begin{pmatrix} s'^m z_\text{m} + \cdots + s'^3 z_3 \\ +s'^2 z_2 + s' z_1 + 1\end{pmatrix}\left[(1-\text{e}^{s'})\hat{x} + (\text{e}^{2s'}-1)\hat{y} + 1\right]}{\left[\begin{matrix} t_\text{i}\begin{pmatrix} s'^n r_n + \cdots + s'^3 r_3 \\ +s'^2 r_2 + s' r_1 + 1\end{pmatrix}\text{e}^{s'(d+1)}(\text{e}^{s'}-1) \\ +k_\text{h}k'_\text{p}\begin{pmatrix} s'^m z_\text{m} + \cdots + s'^3 z_3 \\ +s'^2 z_2 + s' z_1 + 1\end{pmatrix}\left[\begin{matrix}(1-\text{e}^{s'})\hat{x} \\ +(\text{e}^{2s'}-1)\hat{y} \\ +1\end{matrix}\right]\end{matrix}\right]}. \tag{6.12}$$

Substituting the time delay constant by the "all pole" series approximation

$$\text{e}^{s'} = 1 + s' + \frac{1}{2!}s'^2 + \frac{1}{3!}s'^3 + \frac{1}{4!}s'^4 + \frac{1}{5!}s'^5 + \frac{1}{6!}s'^6 + \cdots \tag{6.13}$$

yields as proved in the Appendix C.1, that the corresponding polynomials for both the numerator $N(s')$ and denominator $D(s')$ of the closed loop transfer function are given by

$$\begin{aligned} N(s') = \cdots &+ k_\mathrm{p}(z_6 + y_6\hat{y} - x_6\hat{x})s'^6 \\ &+ k_\mathrm{p}(z_5 + y_5\hat{y} - x_5\hat{x})s'^5 + k_\mathrm{p}(z_4 + y_4\hat{y} - x_4\hat{x})s'^4 \\ &+ k_\mathrm{p}(z_3 + y_3\hat{y} - x_3\hat{x})s'^3 + k_\mathrm{p}(z_2 + y_2\hat{y} - x_2\hat{x})s'^2 \\ &+ k_\mathrm{p}(z_1 + 2\hat{y} - \hat{x})s' + k_\mathrm{p} \end{aligned} \tag{6.14}$$

and

$$\begin{aligned} D(s') = D_1(s') + k_\mathrm{h}N(s') = \cdots &+ \left[t_\mathrm{i}q_6 + k_\mathrm{h}k_\mathrm{p}(z_6 + y_6\hat{y} - x_6\hat{x})\right]s'^6 \\ &+ \left[t_\mathrm{i}q_5 s'^5 + k_\mathrm{h}k_\mathrm{p}(z_5 + y_5\hat{y} - x_5\hat{x})\right]s'^5 + \left[t_\mathrm{i}q_4 + k_\mathrm{h}k_\mathrm{p}(z_4 + y_4\hat{y} - x_4\hat{x})\right]s'^4 \\ &+ \left[t_\mathrm{i}q_3 + k_\mathrm{h}k_\mathrm{p}(z_3 + y_3\hat{y} - x_3\hat{x})\right]s'^3 + \left[t_\mathrm{i}q_2 + k_\mathrm{h}k_\mathrm{p}(z_2 + y_2\hat{y} - x_2\hat{x})\right]s'^2 \\ &+ \left[t_\mathrm{i} + k_\mathrm{h}k_\mathrm{p}(z_1 + 2\hat{y} - \hat{x})\right]s' + k_\mathrm{h}k_\mathrm{p}. \end{aligned} \tag{6.15}$$

As it is proved in the Appendix C.1, the final PID control law is defined by

$$k_\mathrm{h} = 1 \tag{6.16}$$

$$t_\mathrm{i} = 2k_\mathrm{h}k_\mathrm{p}\left(r_1 + d - z_1 - x - \frac{1}{2}\right) \tag{6.17}$$

$$\hat{x} - a_1\hat{y} = b_1 \quad \text{and} \quad \hat{x} - a_2\hat{y} = b_2 \tag{6.18}$$

where

$$a_1 = \frac{2(q_2^2 - q_3) - (q_2 y_2 - y_3)}{(q_2^2 - q_3) - (q_2 x_2 - x_3)} \tag{6.19}$$

$$b_1 = \frac{(q_3 z_1 - q_2 z_2 + z_3 - q_4) - (q_2^2 - 2q_3)(q_2 - z_1)}{(q_2^2 - q_3) - (q_2 x_2 - x_3)} \tag{6.20}$$

$$a_2 = \frac{2q_3^2 - 4q_2 q_4 + q_2 y_4 - q_3 y_3 - y_5 + 2q_5 + q_4 y_2}{(q_3 - x_3)q_3 - (q_4 - x_4)q_2 - (q_2 - x_2)q_4 + q_5 - x_5} \tag{6.21}$$

$$b_2 = \frac{-\left(q_3^2 - 2q_2 q_4 + 2q_5\right)(q_2 - z_1) + \begin{pmatrix} q_2 z_4 - q_3 z_3 + q_4 z_2 \\ -q_5 z_1 - z_5 + q_6 \end{pmatrix}}{\begin{bmatrix} (q_3 - x_3)q_3 - (q_4 - x_4)q_2 - (q_2 - x_2)q_4 \\ + (q_5 - x_5) \end{bmatrix}}. \tag{6.22}$$

By solving (6.17), (6.18) parameters $\hat{x}$, $\hat{y}$ are determined by

$$\hat{x} = \frac{a_1 b_2 - a_2 b_1}{a_1 - a_2}, \quad \hat{y} = \frac{b_2 - b_1}{a_1 - a_2}. \tag{6.23}$$

From the definition of the integrator's time constant (6.17) or (C.48) it is critical to point out that

$$\frac{T_\mathrm{i}}{T_\mathrm{s}} = 2k_\mathrm{h}k_\mathrm{p}\left(r_1 + d - z_1 - x - \frac{1}{2}\right) \tag{6.24}$$

or according to (C.6), (C.7)

$$\begin{aligned} T_\mathrm{i} &= 2k_\mathrm{h}k_\mathrm{p}\left(T_\mathrm{s}r_1 + T_\mathrm{s}d - T_\mathrm{s}z_1 - T_\mathrm{s}x - \frac{1}{2}T_\mathrm{s}\right) \\ &= 2k_\mathrm{h}k_\mathrm{p}\left(p_1 + T_\mathrm{d} - \beta_1 - T_\mathrm{s}x - \frac{1}{2}T_\mathrm{s}\right) \\ &= 2k_\mathrm{h}k_\mathrm{p}\left(\sum_{i=1}^{n}(T_{p_\mathrm{i}}) + T_\mathrm{d} - \sum_{i=1}^{m}(T_{z_\mathrm{i}}) - X - \frac{1}{2}T_\mathrm{s}\right). \end{aligned} \tag{6.25}$$

In other words as it was proved in (3.72) and (C.36), integrator's time constant is equal

$$T_{i_\mathrm{dig}} = T_{i_\mathrm{an}} - \underbrace{2k_\mathrm{h}k_\mathrm{p}\frac{1}{2}T_\mathrm{s}}, \tag{6.26}$$

where T_{i_dig} and T_{i_an} are the optimal values for the integrator's time constant regarding the analog and digital design, respectively.

6.1.1 Performance Comparison Between Analog and Digital Design in Type-I Control Loops

For testing the proposed digital control law, a comparison between the revised analog PID tuning presented in Sect. 3.3 and the proposed PID digital control law are presented in this section. As it was shown in Sect. 3.4, the revised analog control action outperforms in comparison with the conventional tuning especially in cases where the complexity of the process is increased, in terms of poles, zeros and time delay. For that reason, it makes more sense to concentrate within this section on the effect of the sampling time T_s to the quality of the proposed PID control action, compared to the optimal analog design.

In the sequel, two curves are plotted within each figure. Given the transfer function of the plant $G(s)$, the response of the output $y(\tau)$ and the command signal $u(\tau)$ are investigated. To do this, both control loops are normalized with the sampling time T_s, the digital controller is implemented according to the relation $s' = sT_s$.

6.1.1.1 Sampling Time Half of the Dominant Time Constant

In this case, the plant is given by

$$G\left(s'\right) = \frac{0.5}{\left(1+2s'\right)\left(1+1.56s'\right)\left(1+1.34s'\right)\left(1+0.67s'\right)\left(1+0.6s'\right)} \tag{6.27}$$

where ratio $\frac{T_{p1}}{T_s} = 2$. From Fig. 6.2b it is clear that output disturbance rejection regarding analog control action outperforms in terms of settling time the response coming from the digital control action. This conclusion holds also for the step response and input disturbance rejection, see Fig 6.2a.

In the case of output disturbance rejection, settling time is $t_{ss} = 12\tau$ and $t_{ss} = 27.5\tau$ regarding analog and digital control respectively.

The response of the output $y(\tau)$ is also reflected by the command signal $u(\tau)$ response, see Fig. 6.3a, b. There, it is shown that the fast input disturbance suppression observed in Fig. 6.2b is achieved by the strong and fast command effort observed in Fig. 6.3a. The same conclusion holds also for the output disturbance rejection applied at $\tau = 76$, see Figs. 6.2a and 6.3b.

6.1.1.2 Sampling Time 20× Less Than the Dominant Time Constant

In this example, the transfer function of the plant is defined by

$$G\left(s'\right) = \frac{0.5}{\left(1+10s'\right)\left(1+7.79s'\right)\left(1+6.73s'\right)\left(1+3.39s'\right)\left(1+2.97s'\right)} \tag{6.28}$$

for which $\frac{T_{p1}}{T_s} = 20$ holds by. The sampling time of the controller has been decreased to twenty times less the dominant time constant of the process. In this case, it is apparent from the response in the time domain, see Fig. 6.4 the reference tracking, input and output disturbance rejection exhibit almost the same behavior, see Fig. 6.4a. This result is also reflected by the response of the command signal at the presence of an input disturbance $d_i(\tau) = 0.25r(\tau)$.

However, the response in the frequency domain both for $|T(ju)|$, $|S(ju)|$, see Fig. 6.5 shows that, the region for which $|T(ju)| \approx 1$ becomes shorter in case when the sampling time of the controller is chosen such that $\frac{T_{p1}}{T_s} = 20$. This result is against the target of the Magnitude Optimum criterion the goal of which it to try to maintain $|T(ju)| \approx 1$ in the widest possible frequency range.

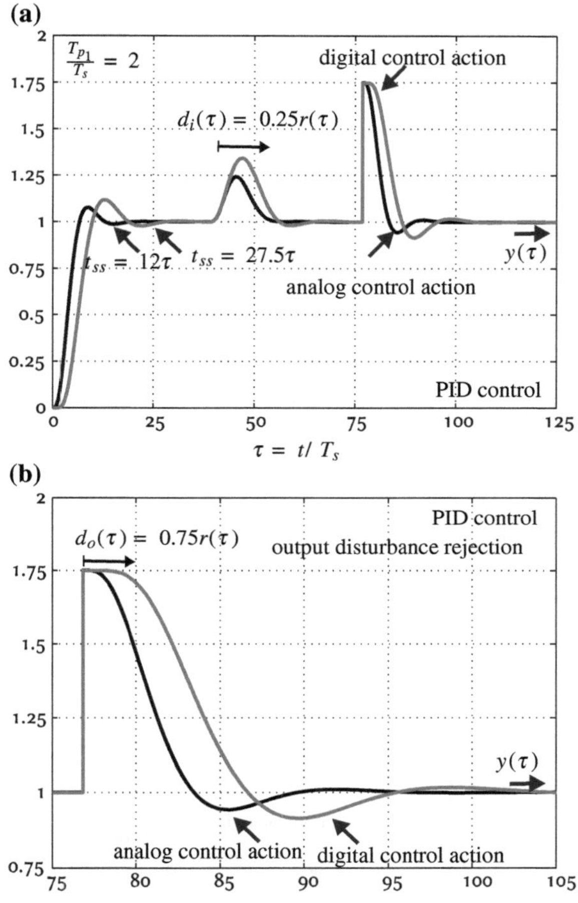

Fig. 6.2 Comparison of the analog and digital control action for the control of the process defined by (6.27). **a** Response of the output $y(\tau)$ in the presence of input $d_i(\tau) = 0.25r(\tau)$ and output $d_o(\tau) = 0.75r(\tau)$ disturbance. **b** Output disturbance rejection

On the other hand, the frequency range for which complementary sensitivity is $|S(ju)| \approx 0$, is shorter in the case where $\frac{T_{p1}}{T_s} = 2$ compared to the region in the case where $\frac{T_{p1}}{T_s} = 20$. As mentioned in Sect. 2.6 this behavior is not desired, since such a control loop becomes more sensitive to possible disturbances in the low frequency range region.

For that reason, control engineers have to find a compromise as far as the choice of the sampling time T_s of the controller is concerned. This is also the goal of this chapter. The introduction of the sampling time T_s in the control action along with

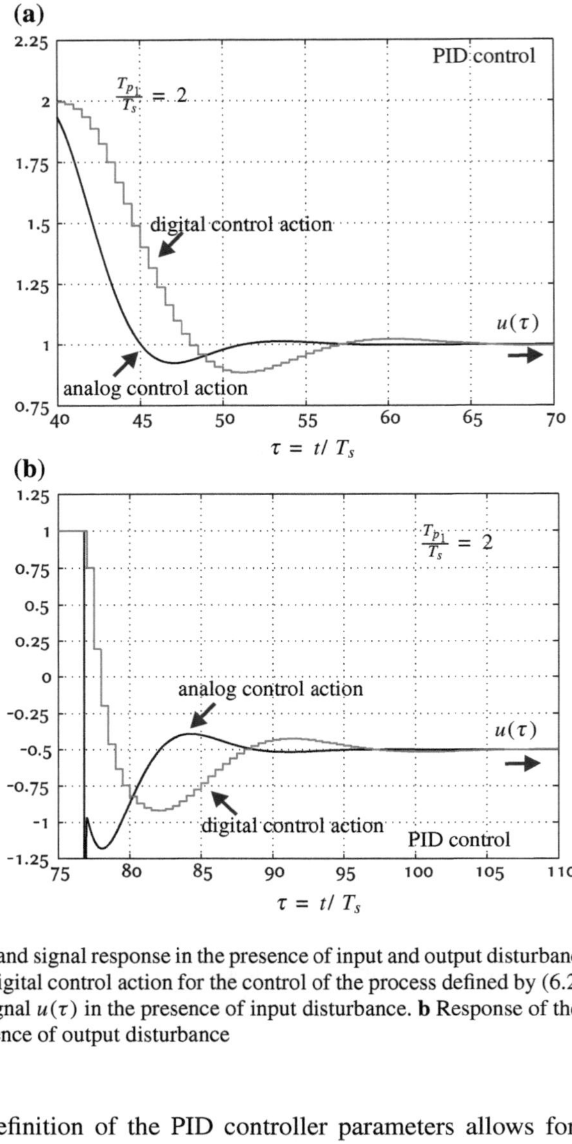

Fig. 6.3 Command signal response in the presence of input and output disturbance. Comparison of the analog and digital control action for the control of the process defined by (6.27). **a** Response of the command signal $u(\tau)$ in the presence of input disturbance. **b** Response of the command signal $u(\tau)$ in the presence of output disturbance

the explicit definition of the PID controller parameters allows for such accurate investigation before the final integration of the control law within a real-time system.

6.1.1.3 Robustness to Model Uncertainties

In this section, the dc gain k_p of the process (or actuator's) gain is violated with an error of the form $k_\text{p}' = k_\text{p}(1+\epsilon)$ while the controller stays tuned with its nominal value k_p. The plant is defined again by

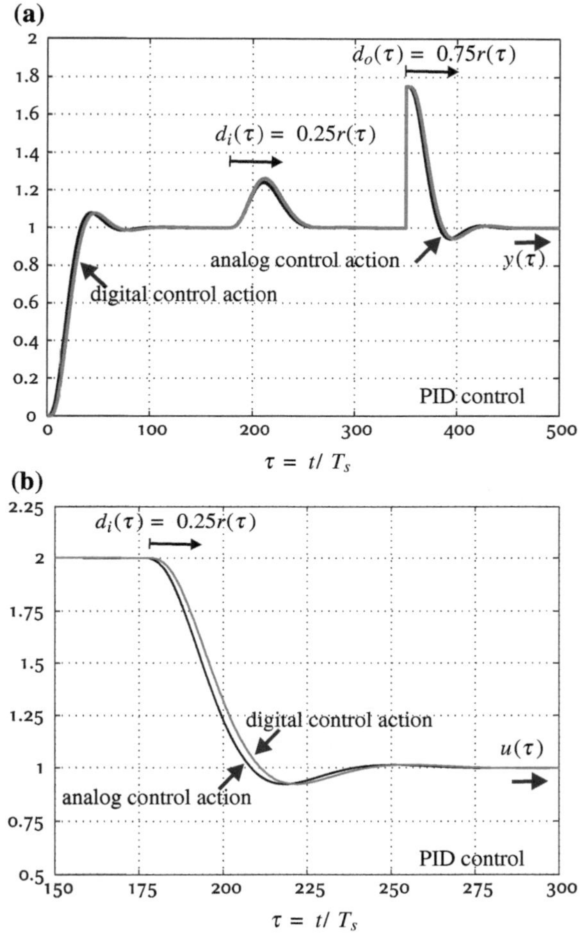

Fig. 6.4 Comparison of the analog and digital control action for the control of the process defined by (6.28). **a** Response of the output $y(\tau)$ in the presence of input $d_{\rm i}(\tau) = 0.25r(\tau)$ and output $d_{\rm o}(\tau) = 0.75r(\tau)$ disturbance. **b** Response of the command signal $u(\tau)$ at the presence of an output disturbance

$$G(s') = \frac{0.5}{(1+2s')(1+1.56s')(1+1.34s')(1+0.67s')(1+0.6s')} \tag{6.29}$$

for which $\frac{T_{p1}}{T_s} = 2$. From Fig. 6.6 it is apparent that in case when the the actuator's gain changes by $+20\,\%$ the overshoot of the step response increases from $6\,\%$ to $21\,\%$.

In the opposite case, when the error is $\epsilon = -20\,\%$ the step response of the control loop exhibits an overshoot of $0\,\%$ while the rise time has been increased. Let it be

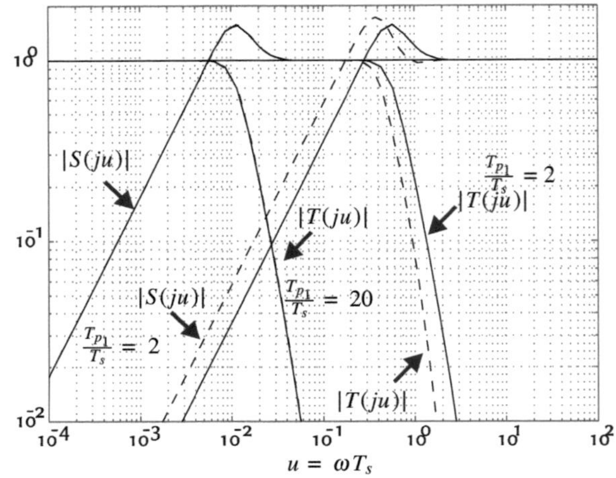

Fig. 6.5 Frequency response of the digital control loop when for the controlled of the same process the sampling time of the digital PID controller is chosen such that $\frac{T_{p1}}{T_s} = 2$ and $\frac{T_{p1}}{T_s} = 20$. Decrease of the sampling time T_s of the controller decreases the bandwidth of $|T(ju)|$ for which $|T(ju)| \approx 1$

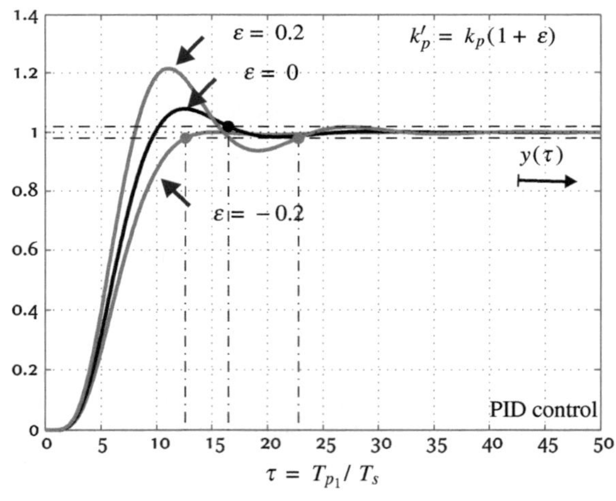

Fig. 6.6 Effect on the step response of the closed loop control system due to changes on the plant's dc gain k_p

noted that such an investigation is critical in the field of electric motor drives, since the gain k_p in the frequency domain stands for the modulation policy followed in vector-controlled electrical drives.

6.2 Type-II Control Loops

In similar fashion with the analog design and for presenting the explicit solution for digital PID controllers in type-II control loops, the transfer function of

$$G(s) = k_\mathrm{p} \frac{\left(s^m \beta_\mathrm{m} + \cdots + s^4 \beta_4 + s^3 \beta_3 + s^2 \beta_2 + s \beta_1 + 1\right)}{\left(\begin{array}{l} s^n p_n + s^{n-1} p_{n-1} + \cdots + s^5 p_5 + s^4 p_4 + s^3 p_3 \\ + s^2 p_2 + s p_1 + 1 \end{array}\right)} \mathrm{e}^{-sT_\mathrm{d}} \tag{6.30}$$

is introduced where $n > m$. Note that since the control loop is of type-II, two pure integrators must be included in the open loop $F_\mathrm{ol}(s)$ transfer function. As a result, the proposed PID type controller is given by

$$C(s) = C^*(s) C_\mathrm{ZOH}(s) = \left(\frac{1 + sX + s^2 Y}{s^2 T_\mathrm{i}}\right)^* \left(\frac{1 - \mathrm{e}^{-sT_\mathrm{s}}}{s}\right) \tag{6.31}$$

for which the second integrator is introduced by the control action since it is of I-PID. Let it be noted that the same analysis holds for the control of integrating processes since one integrator is introduced by the process itself and one more by the controller from the PID control action. For calculating the closed loop transfer function $T(s)$, both the controller and the process are normalized with the sampling period T_s of the zero order hold. In that after substituting with $s' = sT_\mathrm{s}$, relations (6.30), (6.31) become

$$G(s') = k_\mathrm{p} \frac{\left(s'^m z_\mathrm{m} + \cdots + s'^4 z_4 + s'^3 z_3 + s'^2 z_2 + s' z_1 + 1\right)}{\left(\begin{array}{l} s'^n r_n + s'^{n-1} r_{n-1} + \cdots + s'^5 r_5 + s'^4 r_4 + s'^3 r_3 \\ + s'^2 r_2 + s' r_1 + 1 \end{array}\right)} \mathrm{e}^{-s'd} \tag{6.32}$$

and

$$C(s') = C^*(s') C_\mathrm{ZOH}(s') = T_\mathrm{s} \left(\frac{1 + s'x + s'^2 y}{s'^2 t_\mathrm{i}^2}\right)^* \frac{1 - \mathrm{e}^{-s'}}{s'} \tag{6.33}$$

for which $x = \frac{X}{T_\mathrm{s}}$, $y = \frac{Y}{T_\mathrm{s}^2}$, $t_\mathrm{i} = \frac{T_\mathrm{i}}{T_\mathrm{s}}$, $d = \frac{T_\mathrm{d}}{T_\mathrm{s}}$ and $r_j = \frac{p_j}{T_\mathrm{s}^j}$, $\forall j = 1, \ldots, n$, $z_\mathrm{i} = \frac{\beta_\mathrm{i}}{T_\mathrm{s}^i}$, $\forall i = 1, \ldots, m$, has been set.

Once more, the transition from $\mathcal{L}\{.\}$ to the $\mathcal{Z}\{.\}$ domain, takes place according to the transformation

$$s' = \frac{z - 1}{z} = \frac{\mathrm{e}^{s'} - 1}{\mathrm{e}^{s'}} \tag{6.34a}$$

$$\frac{1}{s'^2} = \frac{T_s z'}{(z' - 1)^2} = \frac{T_\mathrm{s} \mathrm{e}^{s'}}{(\mathrm{e}^{s'} - 1)^2} \tag{6.34b}$$

Since $z = \mathrm{e}^{s'}$, the digital PID type controller takes the form

$$C(s') = C^*(s')C_{\mathrm{ZOH}}(s') = \frac{T_\mathrm{s}}{t_\mathrm{i}^2}\left(\frac{1}{s'^2} + \frac{x}{s'} + y\right)^* \left(\frac{1 - \mathrm{e}^{-s'}}{s'}\right) \tag{6.35}$$

or

$$C(s') = \frac{T_\mathrm{s}}{t_\mathrm{i}^2}\left[\frac{(x+y)\mathrm{e}^{2s'} - (x + 2y - T_\mathrm{s})\mathrm{e}^{s'} + y}{(\mathrm{e}^{s'} - 1)^2}\right] \tag{6.36}$$

or finally

$$C(s') = \frac{T_\mathrm{s}^2}{t_\mathrm{i}^2}\left[\frac{\left(\frac{x}{T_\mathrm{s}} + \frac{y}{T_\mathrm{s}}\right)\mathrm{e}^{2s'} - \left(\frac{x}{T_\mathrm{s}} + 2\frac{y}{T_\mathrm{s}} - 1\right)\mathrm{e}^{s'} + \frac{y}{T_\mathrm{s}}}{(\mathrm{e}^{s'} - 1)^2}\right]. \tag{6.37}$$

For simplifying the calculations following in the sequel, the substitution

$$\hat{x} = \frac{x}{T_\mathrm{s}} + 2\frac{y}{T_\mathrm{s}} - 1 \tag{6.38}$$

and

$$\hat{y} = \frac{x}{T_\mathrm{s}} + \frac{y}{T_\mathrm{s}} \tag{6.39}$$

takes place. This results in

$$\frac{x}{T_\mathrm{s}} = 2\hat{y} - \hat{x} - 1 \tag{6.40}$$

and

$$\frac{y}{T_\mathrm{s}} = \hat{x} - \hat{y} + 1. \tag{6.41}$$

By substituting equations (6.38)–(6.39), (6.36) takes the form

$$C(s') = \frac{T_\mathrm{s}^2}{t_\mathrm{i}^2}\left[\frac{(1 - \mathrm{e}^{s'})\hat{x} + (\mathrm{e}^{2s'} - 1)\hat{y} + 1}{(\mathrm{e}^{s'} - 1)^2}\right]. \tag{6.42}$$

With respect to the above, the corresponding open $F_{\mathrm{ol}}(s')$ and closed loop $T(s')$ transfer functions become

$$F_{\mathrm{ol}}(s') = k_\mathrm{h} C(s') G(s') \tag{6.43}$$

or

$$k_{\rm h}C(s')G(s') = k_{\rm h}\frac{T_{\rm s}^2 k_{\rm p}}{t_{\rm i}^2} \times \frac{\begin{bmatrix} \left(s'^m z_{\rm m} + \cdots + s'^3 z_3 + s'^2 z_2 + s' z_1 + 1\right) \\ \times\left[(1-{\rm e}^{s'})\hat{x} + ({\rm e}^{2s'}-1)\hat{y}+1\right] \end{bmatrix}}{\left(s'^n r_n + \cdots + s'^3 r_3 + s'^2 r_2 + s' r_1 + 1\right){\rm e}^{s'd}\left({\rm e}^{s'}-1\right)^2} \tag{6.44}$$

and

$$T(s') = \frac{C(s')G(s')}{1+k_{\rm h}C(s')G(s')} = \frac{N(s')}{D(s')} = \frac{N(s')}{D_1(s')+k_{\rm h}N(s')} = \frac{\begin{bmatrix} k_{\rm p}'(s'^m z_{\rm m} + \cdots + s'^3 z_3 + s'^2 z_2 + s' z_1 + 1) \\ \times\left[(1-{\rm e}^{s'})\hat{x} + ({\rm e}^{2s'}-1)\hat{y}+1\right] \end{bmatrix}}{\begin{bmatrix} t_{\rm i}^2(s'^n r_n + \cdots + s'^3 r_3 + s'^2 r_2 + s' r_1 + 1){\rm e}^{s'd}({\rm e}^{s'}-1)^2 \\ + \begin{bmatrix} k_{\rm h}k_{\rm p}'(s'^m z_{\rm m} \cdots + s'^3 z_3 + s'^2 z_2 + s' z_1 + 1) \\ \times\left[(1-{\rm e}^{s'})\hat{x} + ({\rm e}^{2s'}-1)\hat{y}+1\right] \end{bmatrix} \end{bmatrix}}. \tag{6.45}$$

Finally, the corresponding polynomials $N(s')$, $D(s')$ for both the numerator and denominator of the closed loop transfer function are given by

$$N(s') = \sum_{j=0}^{m}\left[k_{\rm p}'\left(y_j\hat{y} - x_j\hat{x} + z_j\right)s'^j\right] \tag{6.46}$$

where $y_1 = 2$, $x_1 = 1$, $z_0 = 1$ and

$$D\left(s'\right) = \sum_{i=0}^{n}\left[\left(q_{\rm i}t_{\rm i}^2 + k_{\rm h}k_{\rm p}'\left(y_{\rm i}\hat{y} - x_{\rm i}\hat{x} + z_{\rm i}\right)s'^i\right)\right]. \tag{6.47}$$

From the application of the optimization conditions presented in A.1, the final PID control action as proved in Appendix C.2 is defined by

$$\begin{bmatrix} t_{\rm i}^2 \\ \hat{x} \\ \hat{y} \end{bmatrix} = \begin{bmatrix} 1 & -2k_{\rm h}k_{\rm p}'\left(x_2 - q_3\right) & -2k_{\rm h}k_{\rm p}'\left(2q_3 - y_2\right) \\ 0 & D & E \\ 0 & 0 & -2\frac{[(2D+E)Z+D(AD+BE)]}{(2D+E)^2} \end{bmatrix}^{-1} \times \begin{bmatrix} -\left(z_2 + q_3 z_1 - q_4\right) \\ Z \\ -\frac{D(2BZ+CD)+Z^2}{(2D+E)^2} - \hat{y}^2 \end{bmatrix}. \tag{6.48}$$

It is necessary to mention that all variables within (6.48) apart from $t_\mathrm{i}, \hat{x}, \hat{y}$ are process-dependent as defined in C.2.

6.2.1 Performance Comparison Between Analog and Digital Design in Type-II Control Loops

For justifying the potential of the proposed optimal control law a comparison between the revised analog PID tuning, see Sect. 4.3, and the proposed digital control law, see Sect. 6.2 will be performed when controlling the same process $G(s)$. In both cases all time constants have been normalized with sampling time T_s, $s' = sT_\mathrm{s}$.

Controller unmodeled dynamics have been chosen equal to $T_{\Sigma c} = 0.1T_{p_1}$. Special attention is drawn on the output $y(\tau)$ and the controller's command signal $u(\tau)$ regarding reference tracking $r(\tau)$ and at the presence of input $d_\mathrm{i}(\tau)$ and output $d_\mathrm{o}(\tau)$ disturbances, see Fig. 6.1. For coping with the issue of great overshoot in both cases,[1] as mentioned in Sect. 4.2.3, an external filter $C_\mathrm{ex}(s)$ of the form

$$C_\mathrm{ex}\left(s'\right) = \frac{1}{1 + s' x_\mathrm{ex} + s'^2 y_\mathrm{ex}} \tag{6.49}$$

is added in series after the reference signal $r(\tau)$, where $x,\ y$ are the zeros of the corresponding PID controller. Specifically, once the $x,\ y$ controller parameters are determined by the explicit solution from (4.44), (4.45) for the analog and (6.48) for the digital control law respectively, the external filter is tuned then according to these values, with $x_\mathrm{ex} = x_\mathrm{an},\ y_\mathrm{ex} = y_\mathrm{an}$ and $x_\mathrm{ex} = x_\mathrm{dig},\ y_\mathrm{ex} = y_\mathrm{dig}$.

6.2.1.1 Sampling Time Equal to the Dominant Plant's Time Constant

In this example, the process is defined by

$$G(s') = \frac{0.8147}{(1+s')(1+0.99s')(1+0.69s')(1+0.13s')(1+0.1s')} \mathrm{e}^{-0.6s'}. \tag{6.50}$$

From (6.50) it is apparent that $T_\mathrm{s} = T_{p_1}$. From Fig. 6.7a it is apparent that the digital control action leads to an unsatisfactory step response of the control loop. The digital control loop exhibits an overshoot of around 16 % compared to the analog control loop which is around 0.5 %.

Of course this response can be finely tuned by properly choosing the parameters of the external filter $C_\mathrm{ex}(s)$ in (6.49). From Fig. 6.7b it is clear that spends less effort in terms of overshoot compared to the analog control action at the presence of input

[1] Since the control loop in both cases analog and digital control law is of type-II a high overshoot at the output $y(\tau)$ is expected at step changes on the reference signal $r(\tau)$.

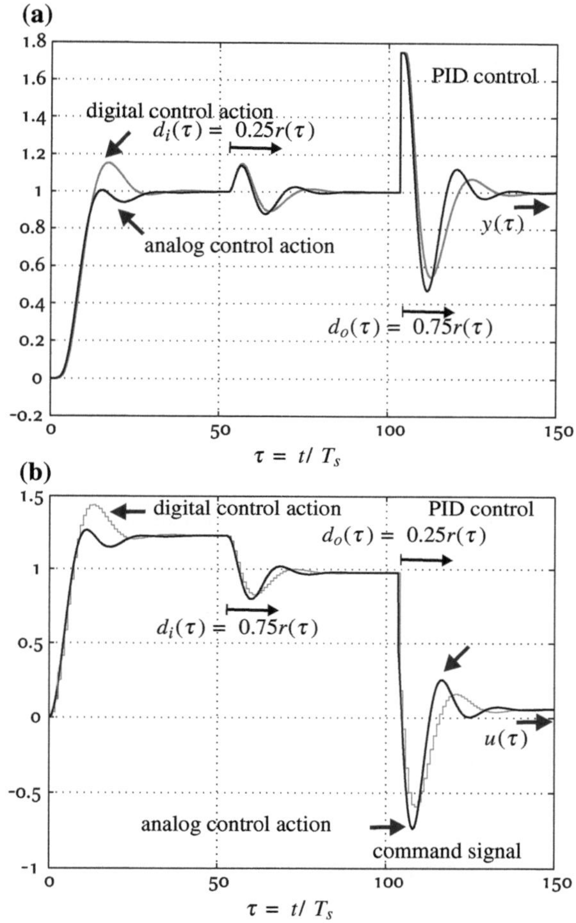

Fig. 6.7 Response of the output $y(\tau)$ and the controller's command signal $u(\tau)$ for the control loop with the plant defined by (6.50). **a** Response of the output $y(\tau)$ in the presence of input $d_i(\tau) = 0.25r(\tau)$ and output $d_o(\tau) = 0.75r(\tau)$ disturbance. **b** Response of the command signal $u(\tau)$ in the presence of input $d_i(\tau) = 0.25r(\tau)$ and output $d_o(\tau) = 0.75r(\tau)$ disturbance

and output disturbance. In Fig. 6.8a, the settling time of the analog control loop's response is faster compared to the digital control loop's response, which is also reflected by the effort spent from the digital controller.

6.2.1.2 Sampling Time 10× Less Than the Plant's Dominant Time Constant

In this case, the sampling time of the controller has been decreased to 10× less the dominant time constant of the process. The plant's transfer function is given by

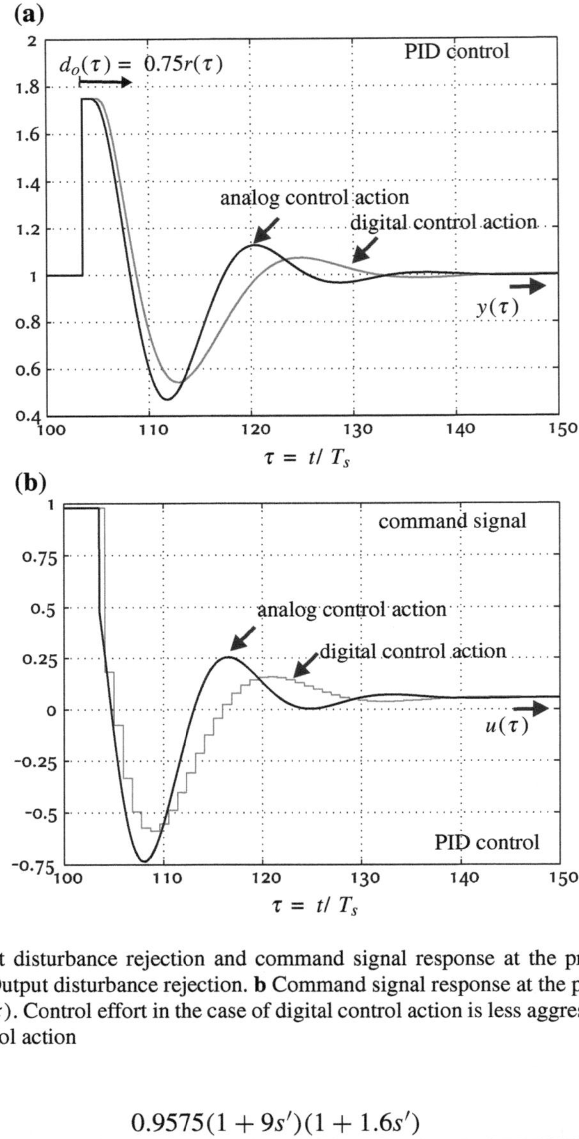

Fig. 6.8 Output disturbance rejection and command signal response at the presence of output disturbance. **a** Output disturbance rejection. **b** Command signal response at the presence of output disturbance $d_o(\tau)$. Control effort in the case of digital control action is less aggressive compared to the analog control action

$$G(s') = \frac{0.9575(1 + 9s')(1 + 1.6s')}{(1 + 10s')(1 + 9.9s')(1 + 9.86s')(1 + 8.2s')(1 + 1.4s')} e^{-9.4s'} \quad (6.51)$$

for which two zeros also exist. The response of both the control loop's output $y(\tau)$ and the controller's command signal is presented in Fig. 6.9. From there it is clear that the digital controller spends less effort, see Fig. 6.9b for achieving almost the same output response in terms of settling time of disturbance rejection, see Fig. 6.9a.

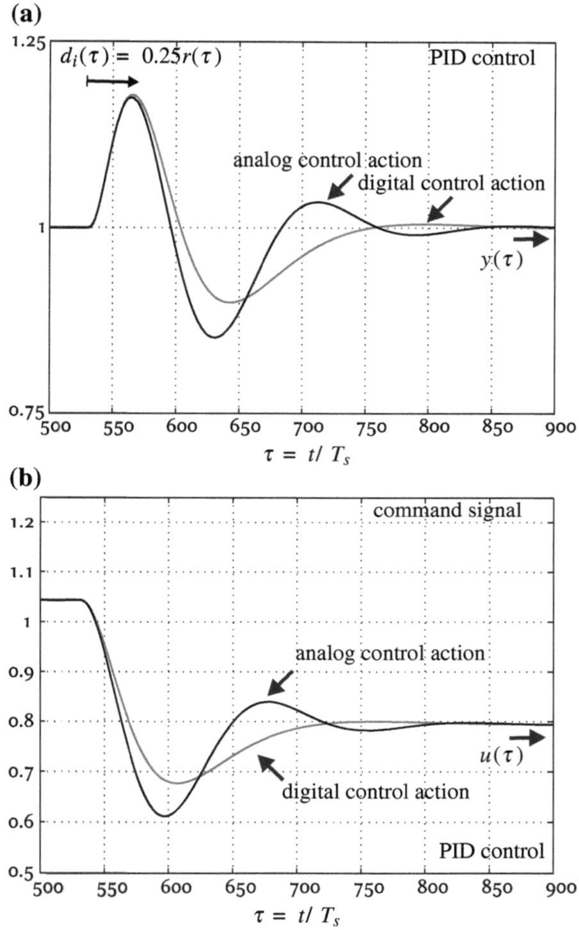

Fig. 6.9 Response of the output $y(\tau)$ and the controller's command signal $u(\tau)$ for the control loop with the plant defined by (6.51). **a** Response of output $y(\tau)$ in the presence of output disturbance. **b** Command signal response at the presence of output disturbance $d_o(\tau)$

6.2.1.3 Robustness to Model Uncertainties

In this case, the process is defined by

$$G(s') = \frac{0.96}{(1+1s')(1+0.91s')(1+0.72s')(1+0.7s')(1+0.03s')} \tag{6.52}$$

for which the nominal dc gain of the process is $k_{\mathrm{p}} = 0.96$. Initially, the digital controller is tuned according to (6.48). Let it be noted that the plant's dc gain is involved

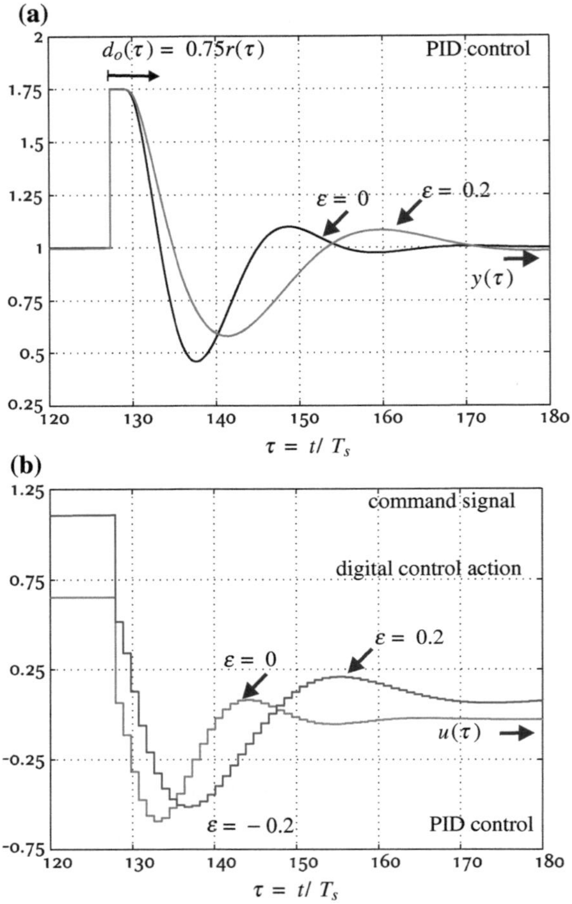

Fig. 6.10 Variation of the plant's dc gain $k_p' = k_p(1+\epsilon)$, $\epsilon = 20\,\%$. **a** Effect of the plant's dc k_p gain variation to the quality of response of $y(\tau)$. **b** Effect of the plant's dc k_p gain variation to the quality of response of $u(\tau)$

only within the integrator's closed form expression, see (C.100) after parameters $x,\ y$ or $\hat{x},\ \hat{y}$ are optimally determined.

To this end, the change on the plant's dc gain affects only the tuning of the integrator's time constant. In this example, the first tuning of the digital PID controller is done based on (6.48) and the nominal measured gain $k_p = 0.96$ whereas in the other case, the controller stays tuned with its initial nominal value and the plant's dc gain changes by 20 %, $k_p' = k_p(1+\epsilon)$.

In Fig. 6.10a, b the response of the output $y(\tau)$ and the command signal $u(\tau)$ is presented. From there it is apparent that variations of the plant's dc gain up to $\epsilon = 20\,\%$ cause a change in the settling time of output disturbance suppression by

30.09 %. Initially, settling time is $t_{ss} = (156 - 127)\tau = 29\tau$ whereas in the case where $\epsilon = 20\,\%$, settling time is $t_{ss} = (171 - 127)\tau = 44\tau$.

6.3 Type-III Control Loops

As mentioned in the abstract, for proving the proposed explicit PID control action, the principle of pole-zero cancellation is followed. For doing this, the integrating process introduced in Sect. 5.2.1 is adopted defined by

$$G(s) = \frac{1}{sT_{\mathrm{m}}(1+sT_{p_1})(1+sT_{\Sigma_{\mathrm{p}}})}. \tag{6.53}$$

The proposed controller is given by

$$C(s) = C^*(s)C_{\mathrm{ZOH}}(s) = \left[\frac{(1+sT_n)(1+sT_v)(1+sT_x)}{s^2T_{\mathrm{i}}(1+sT_{\Sigma c_1})(1+sT_{\Sigma c_2})}\right]^* \frac{(1-\mathrm{e}^{-sT_{\mathrm{s}}})}{sT_{\mathrm{s}}} \tag{6.54}$$

where T_{s} stands for the controller's sampling period. Again all time constants in the control loop are normalized in the frequency domain with the sampling period T_{s} and the substitution $s' = sT_{\mathrm{s}}$ takes place. In that, (6.53) and (6.54) become

$$G(s') = \frac{1}{s't_{\mathrm{m}}(1+s't_{p_1})(1+s't_{\Sigma_{\mathrm{p}}})} \tag{6.55}$$

and

$$C(s') = C^*(s')C_{\mathrm{ZOH}}(s') = T_{\mathrm{s}}\left[\frac{(1+s't_n)(1+s't_v)(1+s't_x)}{s'^2t_{\mathrm{i}}(1+s't_{\Sigma c_1})(1+s't_{\Sigma c_2})}\right]^* \frac{(1-\mathrm{e}^{-s'})}{s'} \tag{6.56}$$

for which $t_{\mathrm{i}} = \frac{T_{\mathrm{i}}}{T_{\mathrm{s}}}$, $t_{\Sigma_{c_1}} = \frac{T_{\Sigma_{c_1}}}{T_{\mathrm{s}}}$, $t_{\Sigma_{c_2}} = \frac{T_{\Sigma_{c_2}}}{T_{\mathrm{s}}}$, $t_n = \frac{T_n}{T_{\mathrm{s}}}$, $t_v = \frac{T_v}{T_{\mathrm{s}}}$, $t_x = \frac{T_x}{T_{\mathrm{s}}}$, $t_{\mathrm{m}} = \frac{T_{\mathrm{m}}}{T_{\mathrm{s}}}$, $t_{p_1} = \frac{T_{p_1}}{T_{\mathrm{s}}}$, $t_{\Sigma_{\mathrm{p}}} = \frac{T_{\Sigma_{\mathrm{p}}}}{T_{\mathrm{s}}}$ has been set.

In similar fashion with the analog design procedure in Section B.3, the open loop transfer function $F_{\mathrm{ol}}(s')$ is given by

$$\begin{aligned} F_{\mathrm{ol}}(s') &= k_{\mathrm{p}}k_{\mathrm{h}}C(s')G(s') \\ &= T_{\mathrm{s}}\left[\frac{\left(1+s't_n\right)\left(1+s't_v\right)\left(1+s't_x\right)}{s'^2t_{\mathrm{i}}\left(1+s't_{\Sigma c_1}\right)\left(1+s't_{\Sigma c_2}\right)}\right]^*\left[\frac{(1-\mathrm{e}^{-s'})}{s'}\right] \\ &\quad\times \frac{k_{\mathrm{p}}k_{\mathrm{h}}}{\left[s't_{\mathrm{m}}(1+s't_{p_1})(1+s't_{\Sigma_{\mathrm{p}}})\right]}. \end{aligned} \tag{6.57}$$

For moving from the $\mathcal{L}\{.\}$ to the $\mathcal{Z}\{.\}$ domain, the substitutions below are considered

$$\frac{1}{s'} = \frac{z'}{z'-1} = \frac{e^{s'}}{e^{s'}-1}, \tag{6.58}$$

$$\frac{1}{s'^2} = \frac{T_s z'}{(z'-1)^2} = \frac{T_s e^{s'}}{(e^{s'}-1)^2}. \tag{6.59}$$

To this end and since $z' = e^{s'}$, $F_{ol}(s')$ becomes finally equal to

$$F_{ol}(s') = \frac{k_p k_h T_s^2}{s' t_m (1+s' t_{p_1})(1+s' t_{\Sigma_p})} \left[\frac{\left(1+s' t_n\right)\left(1+s' t_v\right)\left(1+s' t_x\right)}{t_i \left(1+s' t_{\Sigma c_1}\right)\left(1+s' t_{\Sigma c_2}\right)} \right] \times \frac{e^{s'}}{\left(e^{s'}-1\right)^2}. \tag{6.60}$$

Assuming that the dominant time constant is accurately measured, as mentioned in Sect. 6.3, pole-zero cancellation takes place for determining parameter t_x. Therefore

$$t_x = t_{p_1} \tag{6.61}$$

is set. This results in

$$F_{ol}(s') = \frac{k_p' k_h e^{s'} (1+s' t_n)(1+s' t_v)}{s' t_m t_i (1+s' t_{\Sigma_p})(1+s' t_{\Sigma c_1})(1+s' t_{\Sigma c_2})(e^{s'}-1)^2} \tag{6.62}$$

and after setting $k_p' = k_p T_s^2$. In similar fashion with the analog design it is set $t_{\Sigma_{c_1}} t_{\Sigma_{c_2}} \approx 0$ and $t_{\Sigma_c} = t_{\Sigma_{c_1}} + t_{\Sigma_{c_2}}$.

This results in $(1+s' t_{\Sigma_p})(1+s' t_{\Sigma_{c_1}})(1+s' t_{\Sigma_{c_2}}) = (1+s' t_{\Sigma_p})(1+s' t_{\Sigma_c})$. Moreover if $t_{\Sigma_c} t_{\Sigma_p} \approx 0$ and $t_\Sigma = t_{\Sigma_c} + t_{\Sigma_p}$ then (6.62) becomes equal to

$$F_{ol}(s') = \frac{k_p' k_h e^{s'} (1+s' t_n)(1+s' t_v)}{s' t_m t_i (1+s' t_\Sigma)(e^{s'}-1)^2}. \tag{6.63}$$

Finally the closed loop transfer function becomes equal to

$$T(s') = \frac{k_p C(s') G(s')}{1 + k_p k_h C(s') G(s')} = \frac{\dfrac{k_p' e^{s'} (1+s' t_n)(1+s' t_v)}{s' t_m t_i (1+s' t_\Sigma)(e^{s'}-1)^2}}{1 + k_h \dfrac{k_p' e^{s'} (1+s' t_n)(1+s' t_v)}{s' t_m t_i (1+s' t_\Sigma)(e^{s'}-1)^2}}$$

$$= \frac{k_p' e^{s'} (1+s' t_n)(1+s' t_v)}{s' t_m t_i (1+s' t_\Sigma)(e^{s'}-1)^2 + k_h k_p' e^{s'} (1+s' t_n)(1+s' t_v)}. \tag{6.64}$$

Since (6.64) is in the form of (A.1), the optimization conditions (A.9)–(A.12) can be applied for determining the optimal digital PID control law.

In Appendix C.3, it is proved that parameters $k_{\mathrm{h}}, t_x, t_n, t_v, t_{\mathrm{i}}$ are determined finally by

$$\begin{bmatrix} k_{\mathrm{h}} \\ t_x \\ t_n \\ t_v \\ t_{\mathrm{i}} \end{bmatrix} = \begin{bmatrix} \dfrac{1}{t_{p_1}} \\ \dfrac{-[(n-1)t_\Sigma(4nt_\Sigma^2 - 4(2B-1))] - \sqrt{\Delta}}{2[nt_\Sigma^2(4-n) - 2(2B-1)]} \\ nt_\Sigma \\ \dfrac{k_{\mathrm{h}}k_{\mathrm{p}}T_{\mathrm{s}}^2[2t_\Sigma t_n t_v + (2B-1)(t_\Sigma - t_n - t_v)]}{t_{\mathrm{m}}} \end{bmatrix} \tag{6.65}$$

where variables B, Δ are also process dependent parameters.

6.3.1 Performance Comparison Between Analog and Digital Design in Type-III Control Loops

In this section, three benchmark process models are controlled both by the analog and digital PID control action and the choice of sampling time T_{s} compared to the dominant time constant T_{p_1} is investigated, see ratio $\frac{T_{p_1}}{T_{\mathrm{s}}}$. Input $d_i(\tau)$ and output $d_{\mathrm{o}}(\tau)$ step disturbances are applied at the locations shown in Figs. 3.1 and 6.1 at the presence of the reference signal $r(s)$.

Normalization of the control loop in both cases (analog and digital control design) has been made according to the substitution $s' = sT_{\mathrm{s}}$. In all control actions, design parameter n has been set equal to a value such that $n > 4$, i.e., $(n = 7.46)$, see [1] whereas for filtering the reference signal $r(\tau)$ the first-order filter

$$C_{\mathrm{ex_{an}}} = \frac{1}{1 + \gamma t_{v_{\mathrm{an}}} s} \tag{6.66}$$

$$C_{\mathrm{ex_{dig}}} = \frac{1}{1 + \gamma t_{v_{\mathrm{dig}}} s} \tag{6.67}$$

is utilized.[2] The presence of the external filter is necessary for higher than type-I control loops to avoid high overshoot on the output $y(\tau)$ when step changes on $r(\tau)$ occur, see Sects. 4.2.3, 5.2.3 also [2, 3], where the 2DoF (two Degree of Freedom controller) is described.

[2] Parameter γ is chosen so that the overshoot of $y(\tau)$ satisfies a certain value (depending on the application) when step changes on the reference signal $r(\tau)$ occur.

6.3.1.1 Process with Dominant Time Constants

The process defined by

$$G_1(s') = \frac{0.1}{0.8s'(1+2s')(1+1.6s')} \tag{6.68}$$

is introduced in this example. The calculated digital and analog controllers according to the theory presented in Sects. 5.2.1, 6.3 are defined by

$$C_{\text{an}}(s') = \frac{(1+s't_n)(1+s't_v)(1+s't_x)}{s'^2 t_i(1+s't_{\Sigma_{c_1}})(1+s't_{\Sigma_{c_2}})} = \frac{(1+s'12.68)(1+s'12.69)(1+s'2)}{s'^2 171(1+s'0.1)(1+s'0.1)} \tag{6.69}$$

$$\begin{aligned} C_{\text{dig}}(s') &= T_s \frac{(1+s't_n)(1+s't_v)(1+s't_x)}{s'^2 t_i(1+s't_{\Sigma c_1})(1+s't_{\Sigma c_2})} \\ &= 0.2 \frac{(1+s'9.83)(1+s'12.6)(1+2s')}{s'^2 328.43(1+s'0.1)(1+s'0.1)}. \end{aligned} \tag{6.70}$$

From Fig. 6.11a, it is apparent that when the ratio $\frac{T_{p1}}{T_s} = 2$, the response of the digital control action is oscillatory compared to the analog control action and exhibits an undesired overshoot of ≈22 %. The same unsatisfactory behavior is observed also as far as output disturbance rejection is concerned, see Fig. 6.11a.

This is the result of the oscillating command signal which comes out of the digital controller, see Fig. 6.12a. In the frequency domain, the response of sensitivity $S|(ju)|$ and complementary sensitivity $|T(ju)|$ is shown in Fig. 6.12b where the two systems exhibit almost the same behavior.

In Fig. 6.13a the sampling time of the control loop has been increased, see ratio $\frac{T_{p1}}{T_s} = 500$. In this case, the output disturbance rejection has been significantly improved, see Fig. 6.13b but on the contrary, the region where the magnitude of complementary sensitivity $|T(ju)| \approx 1$ has been also reduced.

This contradicts of course with the principle of the Magnitude Optimum criterion, for which the closed loop control system is designed such that $|T(j\omega)| \simeq 1$ in the widest possible frequency range.

6.3.1.2 Process with Long Time Delay

In this example, a process with time delay half of the process's dominant time constant is introduced. Although the time delay t_d is not considered as a parameter in the proposed control law (analog and digital control action), in this example the robustness of the PID controller to parameter uncertainties is also investigated. The process is defined by

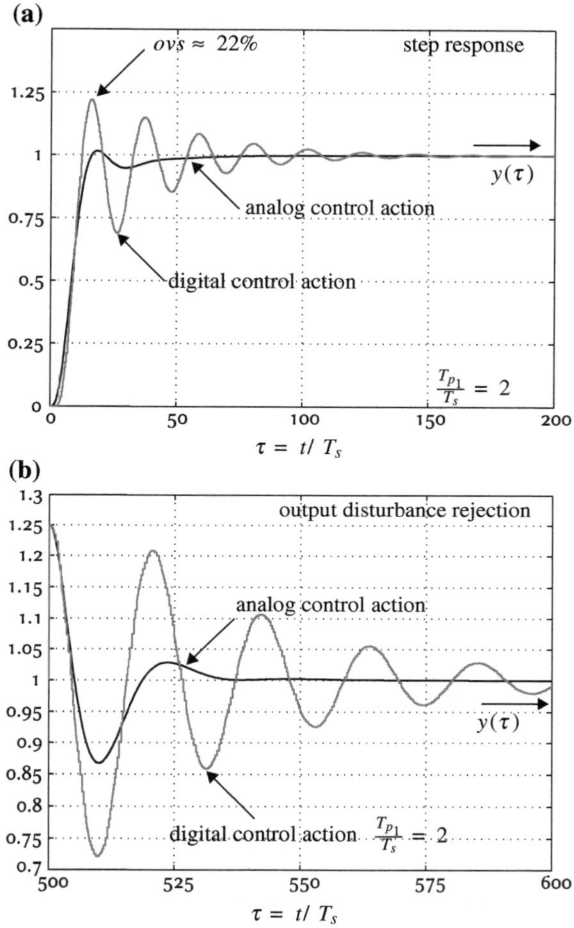

Fig. 6.11 Control of an integrating process defined by (6.68). **a** Step response of the analog and digital control action. **b** The control loop's output $y(\tau)$ in the presence of input $r(\tau)$ and $d_o(\tau)$= $0.25r(\tau)$ output disturbance at $\tau = 500$

$$G_2(s') = \frac{0.1}{s'(1+5s')(1+4.5s')}\mathrm{e}^{-2.5s'} \tag{6.71}$$

and the calculated analog and digital PID control actions are given by

$$\begin{aligned} C_{\text{an}}(s') &= \frac{(1+s't_n)(1+s't_v)(1+s't_x)}{s'^2 t_i(1+s't_{\Sigma_{c1}})(1+s't_{\Sigma_{c2}})} \\ &= \frac{(1+35.43s')(1+35.47s')(1+5s')}{s'^2 119.41(1+s'0.1)(1+s'0.1)} \end{aligned} \tag{6.72}$$

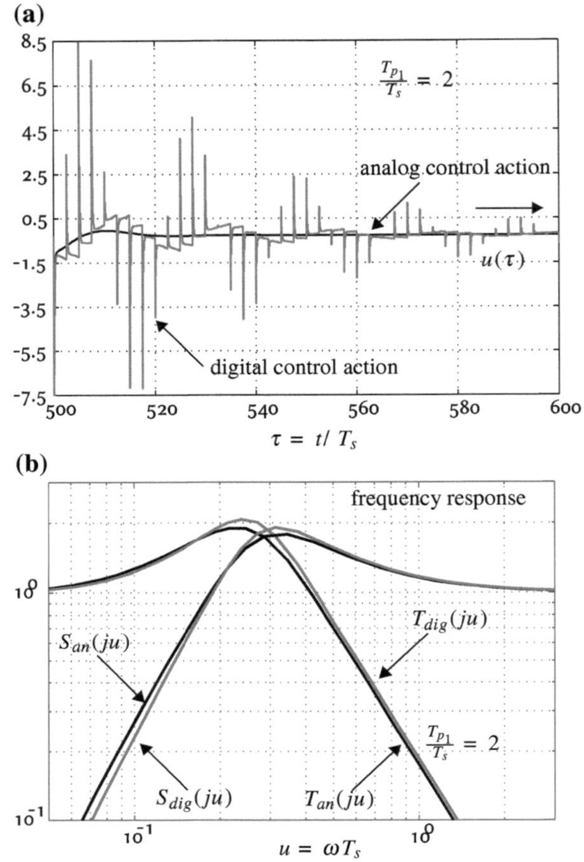

Fig. 6.12 Control of an integrating process defined by (6.68). Analog and digital control loop: ratio $\frac{T_{p_1}}{T_s} = 2$. An output disturbance $d_o(\tau) = 0.25r(\tau)$ is applied at $\tau = 500$ at the presence of $r(\tau)$. **a** Step response of the command signal. **b** Frequency response of sensitivity $|S(ju)|$ and complementary sensitivity $|T(ju)|$

and

$$\begin{aligned} C_{\text{dig}}(s') &= T_s \frac{(1+s't_n)(1+s't_v)(1+s't_x)}{s'^2 t_i (1+s't_{\Sigma c_1})(1+s't_{\Sigma c_2})} \\ &= 0.1 \frac{(1+s'21.2)(1+s'35.43)(1+5s')}{s'^2 7.1(1+s'0.1)(1+s'0.1)}. \end{aligned} \tag{6.73}$$

The performance of the aforementioned control actions is presented in Fig. 6.14a, b. From there it is apparent that settling time of the analog controller is faster than the digital control loop, $t_{ss} = 108\tau$ compared to $t_{ss} = 248\tau$.

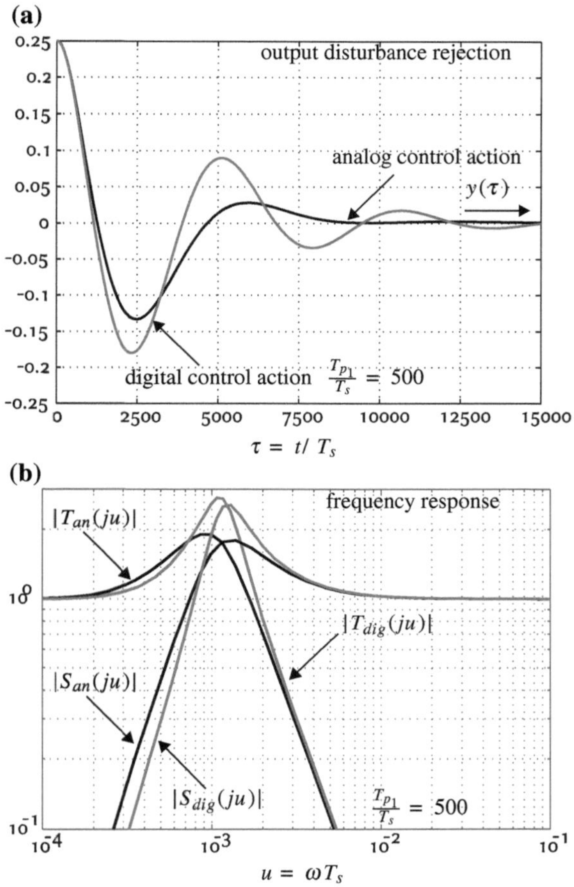

Fig. 6.13 Control of an integrating process defined by (6.68). Ratio $\frac{T_{p1}}{T_s} = 500$. **a** An output disturbance $d_o = 0.25r(\tau)$ is applied at $\tau = 0$. **b** Increase of the sampling time $T(s)$ has improved disturbance rejection in the time domain but reduced the region for which $|T(ju)| \approx 1$ is satisfied

Note that in this example the ratio $\frac{T_{p1}}{T_s}$ has been chosen equal to $\frac{T_{p1}}{T_s} = 5$. Within the digital control action Fig. 6.14b, control effort has a more oscillatory behavior than the analog control action. In Sect. 6.3.2 the choice of the sampling time is discussed so that such behavior is avoided.

6.3.1.3 A Nonminimum Phase Process

In this example, the nonminimum phase process described by

$$G_3(s') = \frac{0.1(1 - 10s')}{10s'(1 + 50s')(1 + 40s')} \tag{6.74}$$

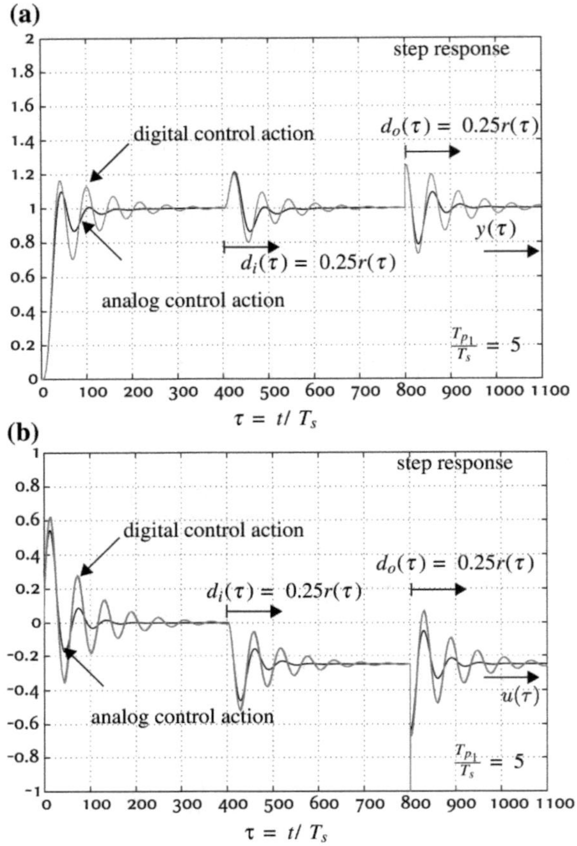

Fig. 6.14 Control of an integrating process with time delay half of the dominant time constant defined by (6.71). Response of the output $y(\tau)$ and the command signal $u(\tau)$ in the presence of input $d_o(\tau) = 0.25r(\tau)$ and output disturbance $d_i(\tau) = 0.25r(\tau)$. **a** Step response of the analog and digital control action of $y(\tau)$. **b** Step response of the analog and digital control action of $u(\tau)$

is controlled both by the analog and digital PID control action which are described by

$$\begin{aligned} C_{an}(s') &= \frac{(1+s't_n)(1+s't_v)(1+s't_x)}{s'^2 t_i(1+s't_{\Sigma_{c_1}})(1+s't_{\Sigma_{c_2}})} \\ &= \frac{(1+317s')(1+317.4s')(1+50s')}{s'^2 85536(1+s'2.5)(1+s'2.5)} \end{aligned} \tag{6.75}$$

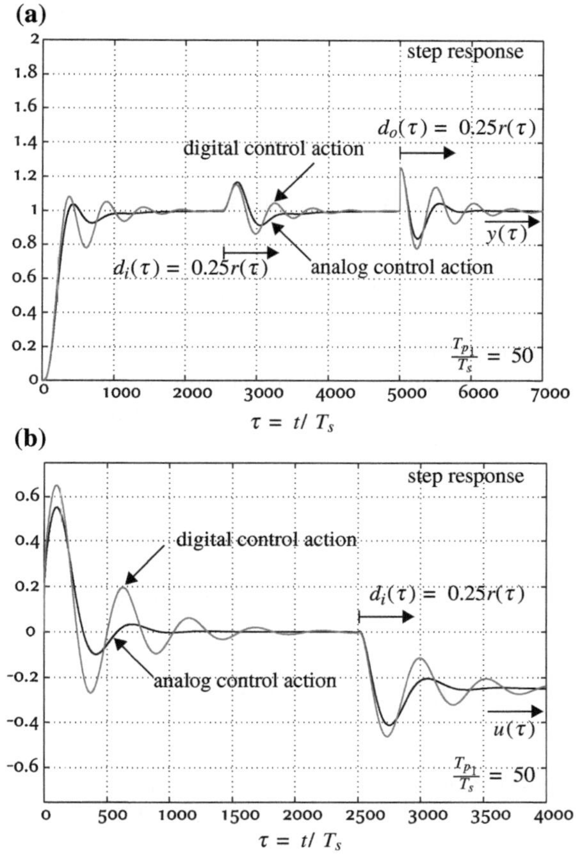

Fig. 6.15 Control of an integrating nonminimum phase process defined by (6.74). Response of the output $y(\tau)$ and the command signal $u(\tau)$ in the presence of input $d_o(\tau) = 0.25r(\tau)$ and output disturbance $d_i(\tau) = 0.25r(\tau)$. **a** Step response of the analog and digital control action of $y(\tau)$. **b** Step response of the analog and digital control action of $u(\tau)$

and

$$C_{\text{dig}}(s') = T_s \frac{(1+s't_n)(1+s't_v)(1+s't_x)}{s'^2 t_i(1+s't_{\Sigma c_1})(1+s't_{\Sigma c_2})} = \frac{(1+s'162.2)(1+s'317)(1+50s')}{s'^2 43730.22(1+s'2.5)(1+s'2.5)} \tag{6.76}$$

respectively. The robustness of the proposed controller is also investigated in this example since for the derivation of the proposed control law no zeros in the model of the process have been considered, see (6.53).

From Fig. 6.15a it is apparent that disturbance rejection is not suppressed that fast as in the analog design. This is also apparent in the command signal Fig. 6.15b where the control effort $u(\tau)$ is oscillating compared to the analog command signal of the PID controller.

6.3.1.4 Robustness to Model Uncertainties

In this case, the process to be controlled is defined by

$$G(s') = \frac{0.1}{2s'(1+10s')(1+9.5s')}. \tag{6.77}$$

For testing the robustness to model uncertainties a change in $k_{p_{er}} = (1+\epsilon)k_\mathrm{p}$ is provoked in the process model, while the controller stays tuned with its initial nominal value k_p. Therefore, in this case, the product between the plant $G(s)$ along with the plant's dc gain k_p, is given by

$$k_{\mathrm{p_{er}}} G(s') = k_\mathrm{p}(1+a)\frac{1}{s't_\mathrm{m}(1+s't_{p_1})(1+s't_{\Sigma_\mathrm{p}})} \tag{6.78}$$

but the integrator's time constant t_i in (6.54) stays still tuned according to (6.54) which is equal to

$$t_\mathrm{i} = \frac{k_\mathrm{h} k_\mathrm{p} T_\mathrm{s}^2[2t_\Sigma t_n t_v + (2B-1)(t_\Sigma - t_n - t_v)]}{t_\mathrm{m}}. \tag{6.79}$$

In Fig. 6.16 changes to ϵ are forced, which are equal to $\epsilon = \pm 0.2$. From the step response Fig. 6.16a and the output disturbance rejection Fig. 6.16b it is apparent that a non significant change is caused in the settling time and the overshoot of the output $y(\tau)$ of the control loop.

6.3.2 Sampling Time Effect Investigation in Type-III Control Loops

In this section the normalized plant transfer function defined by

$$G_1(s') = \frac{k_\mathrm{p}^{'}}{s't_\mathrm{m}(1+t_{p_1}s')(1+t_{\Sigma_\mathrm{p}}s')} \tag{6.80}$$

for which $t_m = \frac{T_m}{T_\mathrm{s}}$, $t_{p_1} = \frac{T_{p_1}}{T_\mathrm{s}}$, $t_{\Sigma_\mathrm{p}} = \frac{T_{\Sigma\mathrm{p}}}{T_\mathrm{s}}$ and $k_\mathrm{p}^{'} = k_\mathrm{p} T_\mathrm{s}$ as shown in Sect. 6.3. Three different ratios of $\frac{T_{p_1}}{T_\mathrm{s}}$ are investigated regarding the performance of the digital control action compared to the analog control law, both in the time and frequency domain.

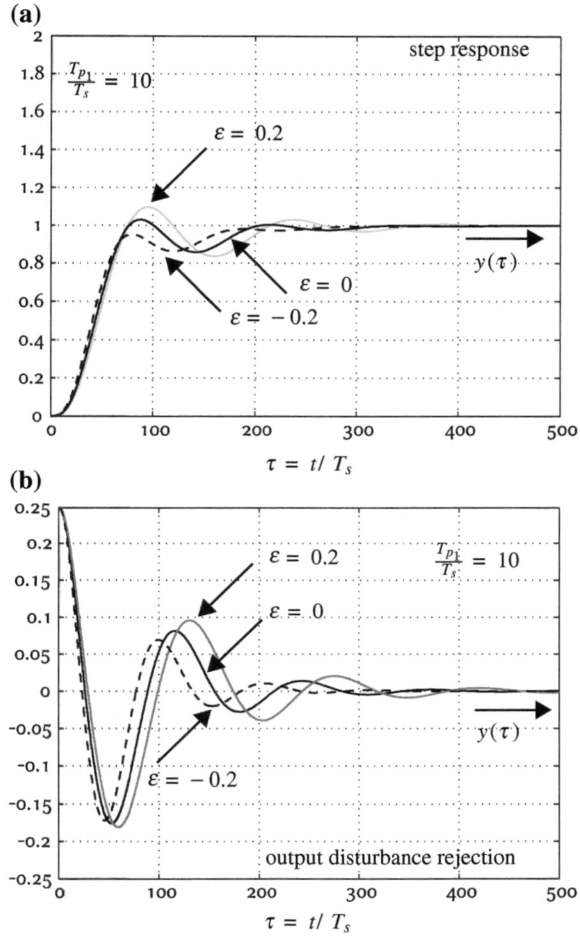

Fig. 6.16 Robustness of the proposed digital control law to model uncertainties. A change in the dc gain of the process k_p is provoked of the form $k_{p_{er}} = k_p(1+\epsilon)$ while the integrator's time constant of the controller stays tuned with its initial nominal value k_p. **a** Step response of the closed loop control system. **b** Output disturbance rejection

6.3.2.1 Sampling Time 2× Less Than the Plant's Dominant Time Constant

Process (6.80) is now defined by

$$G_1(s') = \frac{0.1}{0.4s'(1+2s')(1+1.8s')}. \tag{6.81}$$

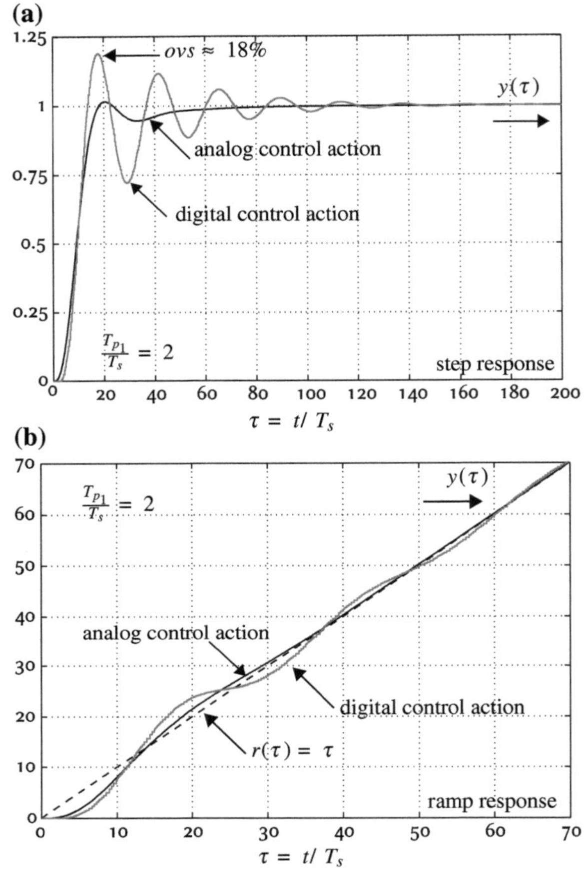

Fig. 6.17 Step and ramp response of the digital and analog control loop when $\frac{T_{p1}}{T_s} = 2$ for the process defined by (6.81). **a** Step response of the digital and analog control loop when $\frac{T_{p1}}{T_s} = 2$. **b** Ramp response of the digital and analog control loop when $\frac{T_{p1}}{T_s} = 2$

In Fig. 6.17 the step (Fig. 6.17a), ramp (Fig. 6.17b) response is presented along with the frequency response and output disturbance rejection. For avoiding the great overshoot at the output of the control loop, a first-order reference filter has been added of the form (6.66) and (6.67) for the analog and digital controller.

Parameter γ has been chosen equal to $\gamma = 0.75$ and parameters $t_{v_{an}} = t_v$ in (6.66), $t_{v_{dig}} = t_v$ in (6.67) are coming from the optimal control law (5.10) and (6.65), respectively.

From Fig. 6.18b it becomes apparent that the frequency response of the closed loop control system is almost the same both for the analog and the digital implementation.

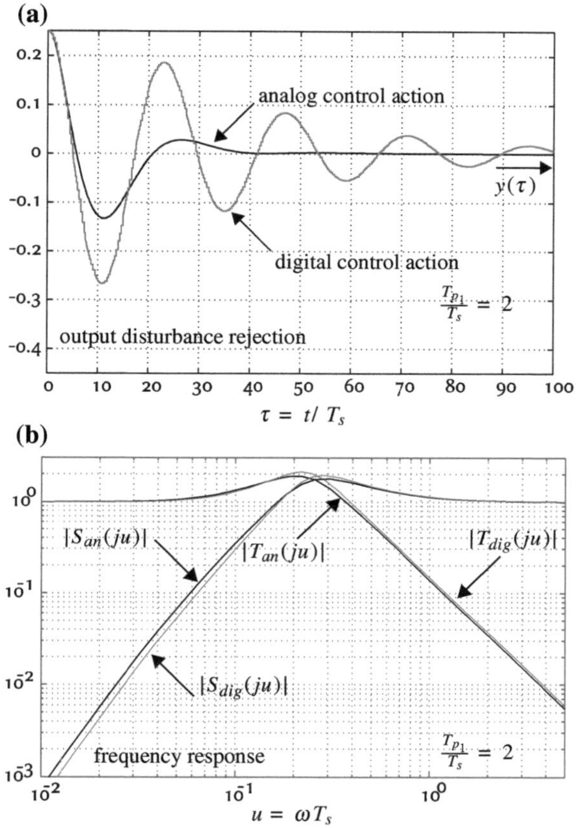

Fig. 6.18 Output disturbance rejection and frequency response of sensitivity S and complementary sensitivity T for the analog and digital control action when the plant is defined by (6.81). **a** Output disturbance rejection for the analog and digital control action. **b** Frequency response of sensitivity S and complementary sensitivity T for the analog and digital control action

On the contrary, since $\frac{T_{p_1}}{T_s} = 2$ output disturbance rejection of the digital control action is poor compared to the analog control loop, see Fig. 6.18a.

6.3.2.2 Sampling Time 10× Less Than the Plant's Dominant Time Constant

In this case, the process (6.80) is defined by

$$G_2(s') = \frac{0.1}{2s'(1 + 10s')(1 + 9.5s')} \tag{6.82}$$

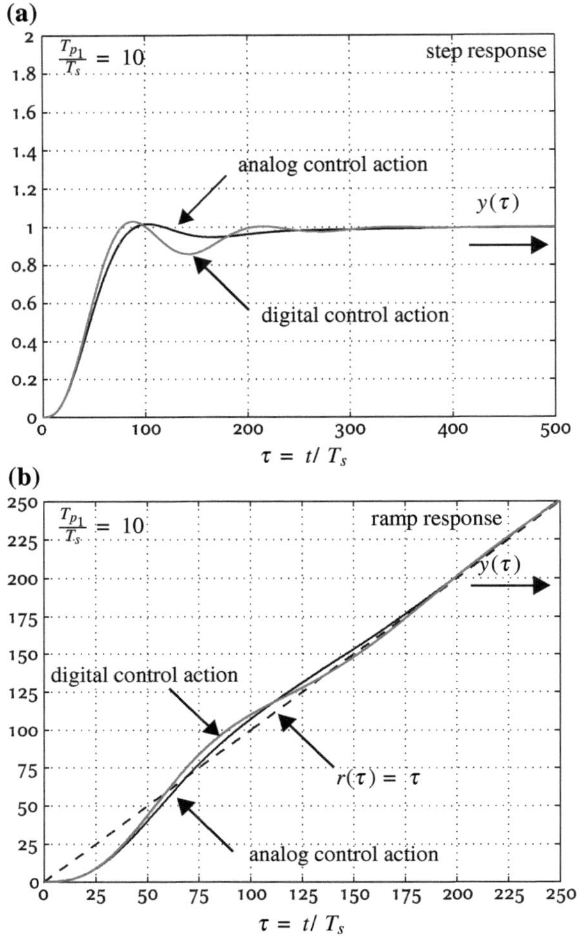

Fig. 6.19 Step and ramp response of the analog and digital control loop for the plant defined by (6.82). Ratio $\frac{T_{p1}}{T_s} = 10$. **a** Step response of the analog and digital control loop. **b** Ramp response of the analog and digital control loop

from which it is apparent that $\frac{T_{p1}}{T_s} = 10$. The time domain performance of the digital controller has been significantly improved, see Figs. 6.19a, b and 6.20a. On the contrary, the magnitude of $|T(ju)|$ is equal to 0.707 at $u = 0.09$, see Fig. 6.20b whereas in Fig. 6.18b this takes place at $u = 0.47$.

6.3.2.3 Sampling Time 100× Less Than the Plant's Dominant Time Constant

The results from the previous example are also confirmed in the following case for which the process defined by

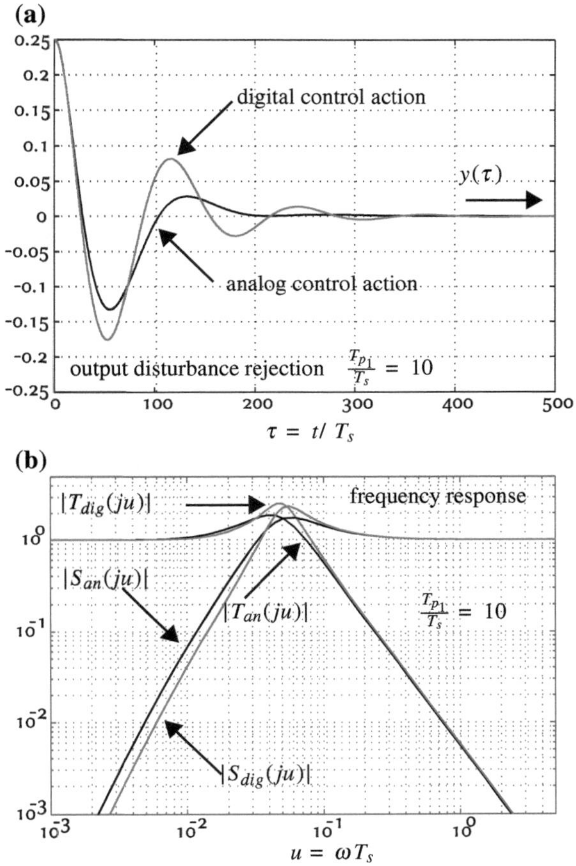

Fig. 6.20 Output disturbance rejection and frequency response of sensitivity S and complementary sensitivity T for the analog and digital control action when the plant is defined by (6.82). Bandwidth of T has been decreased compared to Fig. 6.18 but time domain performance has been significantly improved compared to Fig. 6.18. Ratio $\frac{T_{p1}}{T_s} = 10$. **a** Output disturbance rejection for the analog and digital control action. **b** Frequency response of sensitivity S and complementary sensitivity T for the analog and digital control action

$$G_3(s') = \frac{0.1}{s'(1 + 2.5s')(1 + 2.25s')} \tag{6.83}$$

has been considered. The region of u for which $|T(ju)| \approx 1$ has been reduced even more while the performance in the time domain delivers similar satisfactory results with the previous example at which $\frac{T_{p1}}{T_s} = 10$. The frequency of $|T(ju)|$ at which $|T(ju)| \approx 0.707$ is equal to $u = 0.01$.

From the above analysis, it is apparent that in the case where the $\frac{T_{p1}}{T_s}$ decreases (i.e., $\frac{T_{p1}}{T_s} = 10,\ 100$), the frequency region where the magnitude of $|T(ju)|$ remains

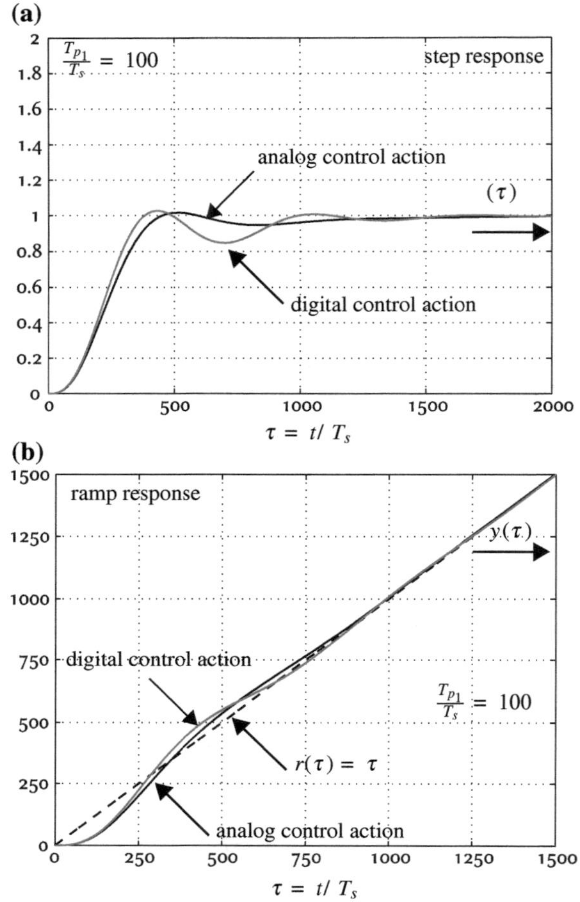

Fig. 6.21 Step and ramp response of the analog and digital control loop for the plant defined by (6.83). Ratio $\frac{T_{p1}}{T_s} = 100$. **a** Step response of the analog and digital control loop. **b** Ramp response of the analog and digital control loop

equal to one, $|T(ju)| \approx 1$ becomes smaller compared to magnitude where the ratio of the control loop $\frac{T_{p1}}{T_s}$ is equal to $\frac{T_{p1}}{T_s} = 2$. This feature contradicts with the principle of the Magnitude Optimum criterion, see Section A.1 which requires that $|T(ju)|$ must be maintained equal to the unity in the widest possible frequency range.

As a result, the digital control design has to satisfy all requirements both in the time and frequency domain. Therefore it has to satisfy an acceptable behavior in the time domain (step response) and comply with the principle $|T(ju)| \approx 1$ in the widest possible frequency range. Therefore, sampling time T_s has to be chosen such that the command signal is noise free and the magnitude of the closed loop transfer function is $|T(ju)| \approx 1$ in the widest possible frequency range. The latter feature is

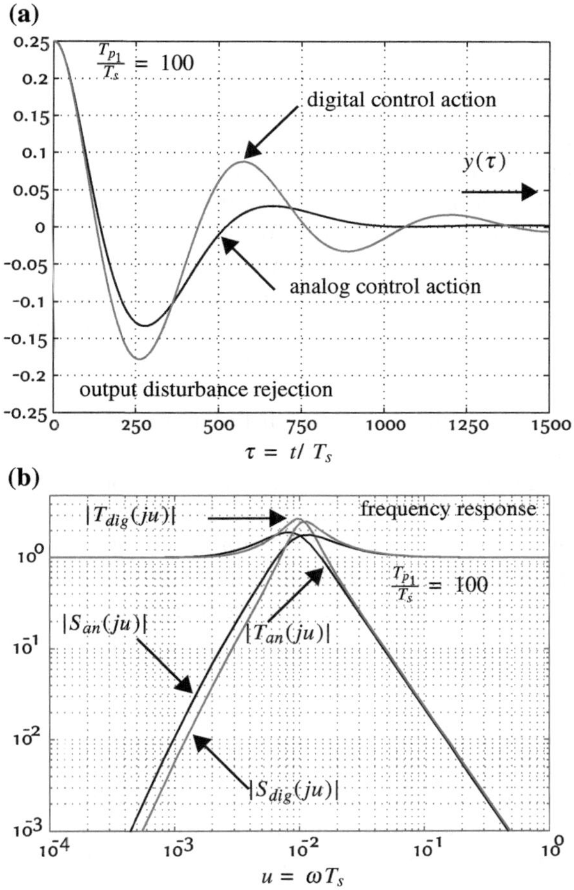

Fig. 6.22 Output disturbance rejection and frequency response of sensitivity S and complementary sensitivity T for the analog and digital control action when the plant is defined by (6.83). Bandwidth of T has been decreased compared to Figs. 6.18 and 6.20 but time domain performance has been significantly improved compared to Fig. 6.17. Ratio $\frac{T_{p1}}{T_s} = 100$. **a** Output disturbance rejection for the analog and digital control action. **b** Frequency response of sensitivity S and complementary sensitivity T for the analog and digital control action

highly desired, since it forces the amplitude of the sensitivity function

$$S(s) = 1 - T(s) = \frac{y(s)}{d_o(s)} = \frac{1}{1 + k_p k_h C(s) G(s)} \tag{6.84}$$

to be equal to zero in the widest possible frequency range starting from the low frequency region (Fig. 6.21).

6.4 Summary

Analytical expressions for the digital PID controller tuning have been presented regarding the control of type-I, type-II, type-III control loops. The explicit control law takes into account all modeled process parameters (model of n poles, m zeros plus unknown time delay d) plus the controller's sampling time T_s. Basis of the proposed theory is the Magnitude Optimum criterion and the proof of each one of the control actions for type-I, type-II, type-III control loops is presented in Appendix C. One big advantage of the proposed theory is the introduction of the sampling time T_s within the explicit closed form expressions regarding the determination of the PID parameters. This idea gives the benefit to control engineers to investigate the effect of the sampling time T_s to the control loops performance both in the time and the frequency domain (Fig. 6.22).

For that reason, during the comparison between the analog and the digital control loop design, all time constants within the control loop have been normalized according to the relation $s' = sT_s$. One interesting result observed in type-I control loops is the trade of the control engineer is faced with, regarding the choice of the sampling time against the control loop's performance. Specifically, it was shown that the higher the ratio $\frac{T_{p1}}{T_s} \to \infty$ is, the more the analog response $y(\tau)$ is identical to the digital as far as the time domain is concerned. However, the decrease of the sampling time versus the dominant time constant affects the bandwidth of $|T(j\omega)|$ in the frequency domain.

For two different sampling times T_{s_1} and T_{s_2} for which $T_{s_1} < T_{s_2}$, it was shown that the frequency range BW for which $|T(ju)| \approx 1$, is decreased in the case where the controller has been designed with sampling time $T_s = T_{s1}$ compared to the controller designed with sampling time T_{s2}. This is a feature against the principle of the Magnitude Optimum criterion, for which the controller is designed such that $|T(ju)| \approx 1$ in the widest possible frequency range. Therefore, control engineers have to find a compromise between the desired bandwidth of T and the desired response in the time domain so that these two basic requirements of the design are satisfied.

References

1. Papadopoulos KG, Margaris NI (2012) Extending the symmetrical optimum criterion to the design of PID type-p control loops. J Process Control 12(1):11–25
2. Papadopoulos KG, Papastefanaki EN, Margaris NI (2013) Explicit analytical PID tuning rules for the design of type-III control loops. IEEE Trans Ind Electron 60(10):4650–4664
3. Papadopoulos KG, Tselepis ND, Margaris NI (2013) Type III control loops-digital PID controller design. J Process Control 23(10):1401–1414

Part III
Automatic Tuning of the PID Controller

Chapter 7
Automatic Tuning of PID Regulators for Type-I Control Loops

Abstract A systematic automatic tuning method for PID-type controllers in Single Input–Single Output processes is proposed. The method is inspired from the Magnitude Optimum design criterion and (1) considers the existence of a poor process model and (2) requires only access to the output of the process and not to its states (3) requires an open-loop experiment on the plant itself for initializing the algorithm. The application of the Magnitude Optimum criterion for tuning the PID controller in the case of a known single input–single output linear process model and regardless of its complexity shows that the step response of the control loop exhibits a certain performance in terms of overshoot (4.4 %), settling and rise time as it was already shown in Chap. 3 and Sect. 3.2. The proposed method exploits this feature and tunes the PID controller parameters, so that the aforementioned performance is achieved. Since the proposed control law is not restricted to specific plants regarding their complexity, a performance comparison in Sects. 7.3 and 7.4.3 discusses the closed-loop frequency response when the controller is tuned optimally according to Sect. 3.3 and when the controller is tuned automatically according to Sect. 7.2.

7.1 Why Automatic Tuning?

The problem of tuning a PID controller involves two sides of the same coin. The first side deals with the problem of tuning the PID parameters based on a known process model. In this case, the transfer function of the process model is often acquired through experimental data along with the use of system identification techniques and therefore controller parameters are tuned based on the modeled time constants of the process. In such cases, this kind of tuning involves an explicit solution regarding the PID controller's parameters which is often expressed as a function of the plant's known dynamics, see part II of this book and also [21, 24, 25].

The second side deals with the problem of the PID controller's tuning when there is almost little or no a priori knowledge regarding the model of the process. This kind of tuning is often called "tuning on demand", "one shot tuning", or "automatic tuning", see [4, 6, 7, 9, 10, 13, 15]. Roughly speaking, as stated in [3], by automatic

K.G. Papadopoulos, *PID Controller Tuning Using the Magnitude Optimum Criterion*,
DOI 10.1007/978-3-319-07263-0_7

tuning, we mean a method where a controller is tuned automatically on demand from a user. In this case, the user typically either pushes a button or sends a command to the controller.

The problem of automatic tuning of PID-type controllers seems to have been treated thoroughly enough according to the number of patents reported in [1]. However, as stated in [2], a vast majority of the PID controllers in the industry are still tuned manually by control or commissioning engineers and operators. A typical example of such a case is the tuning of the PI controller (speed, current, or flux controllers) in vector controlled medium voltage drives where commissioning engineers on site, carry out the tuning of the controller based on past experiences and heuristics. The reason for this is basically owed to the lack of knowledge of the process model itself.

Concrete examples of such lack of knowledge of the process model itself are as follows. The nonlinear behavior of the modulator along with frequent changes in the motor model all over the motor's operating range may lead often to unstable control loops. The reason for instability stems from the fact that the involved PID controllers treat the modulator itself as a linear gain along with an inaccurate time delay constant in series most of the times, and remain tuned with this specific set of parameters all over the whole operating range (various loads, various frequencies). For this reason and in order to avoid nonlinear phenomena, control engineers spend much effort on achieving a linear behavior within the modulator itself, as far as modulation index and modulation angle is concerned, see [27]. This problem becomes especially challenging when the modulator is required to operate also in the overmodulation region where the problem of nonlinearity becomes strongly apparent, see [5, 8, 16, 17].

A second reason responsible for the poor tuning of PID-type controllers is met in cases where the model of the process is of second order and its behavior is strongly oscillating. Such plants are often modeled by a transfer function with complex conjugate poles if modeling in the frequency domain is followed. Examples of this case are met in the field of AC/DC and DC/DC power converters, see [11, 12, 18, 26] where the transfer function of the process model is characterized by the damping ratio ζ and the resonance frequency ω_{n}. The problem in this case lies in the fact that control engineers often apply PID tuning methods which have been developed for the well-known First Order Plus Dead Time (FOPLDT) model, see [19] which in this case is most of the times inappropriate.

Last but not least, the use of the derivative D term when the control law is of PID, still remains an open topic, see [8]. Many are the cases where the addition of the D term is often avoided since its addition to the control law is said to cause amplification of the noise in the error term which often is blamed to lead to an unstable control loop. With respect to the above, the development of a systematic automatic PID tuning procedure has to solve three issues.

Firstly, such a tuning procedure has to decide the optimal PID-type controller for the process. In that, it has to decide whether the process needs I or PI control and if the D part has to be added or omitted.

Secondly, it is necessary for such a tuning procedure to end up in a control loop which achieves a robust performance in terms of satisfactory reference tracking and output disturbance rejection. The latter is of great importance especially in the field of electric motor drives where demanding requirements are often met regarding the speed, current and flux PI controllers (SFOC, RFOC).[1]

Last but not least, such a method should consider an adaptive behavior of the controller in case the process model changes rather frequently. In other words, the controller should have the benefit of retuning its parameters in cases when variations of the plant parameters occur.

In order to develop such a tuning technique able to satisfy all the aforementioned requirements, the advantages introduced by the Magnitude Optimum criterion are exploited throughout this chapter, see [20, 28]. The Magnitude Optimum criterion, introduced by Sartorius and Oldenbourg is based on the idea of designing a controller which renders the magnitude of the closed-loop frequency response as close as possible to unity in the widest possible frequency range [20]. The conventional tuning of the PID controller based on this principle has been thoroughly discussed in Chap. 3 (see Sect. 3.2) where a revised PID control law has also been presented, see Sect. 3.3 and [20].

One important feature of both the conventional and the revised PID control law presented in Sect. 3.2 [20] is the 'preservation' of the shape of the step[2] and frequency response of the final closed-loop control system regardless of the process complexity. The 'preservation' of the shape means that the output of the control loop exhibits a specific overshoot (4.4 %), settling and rise time in the time domain, whereas the amplitude of the closed-loop transfer function remains as close as possible to unity in the widest possible frequency range.

The second important feature of both the conventional and the revised PID control law is the coupling analytical relation between the PID control parameters when the principle of the Magnitude Optimum criterion is followed. This coupling relation gives the flexibility to express all three control parameters based on one, and therefore by tuning only one parameter (zero of the PID controller) all two other parameters of the PID controller are tuned automatically. To this end, target of the proposed method is to tune automatically only one parameter of the PID controller (all others are tuned automatically) by achieving the prescribed aforementioned performance of the step response in terms of overshoot (4.4 %), settling and rise time.

For the sake of a clear presentation of this chapter, in Sect. 7.2, the direct tuning of the conventional PID tuning is presented. There it is shown how the step and frequency response are preserved when the plant is controlled under I, PI, and PID control via the Magnitude Optimum criterion. In Sect. 7.2.5, the proposed method is presented. In Sects. 7.3 and 7.4.3, evaluation results demonstrate the potential of the proposed automatic tuning method where a comparison between the explicit solution presented in Sect. 3.3 and the solution provided by automatically tuned PID controller takes place.

[1] Stator or rotor field oriented vector control.

[2] Overshoot, settling and rise time remain unaltered.

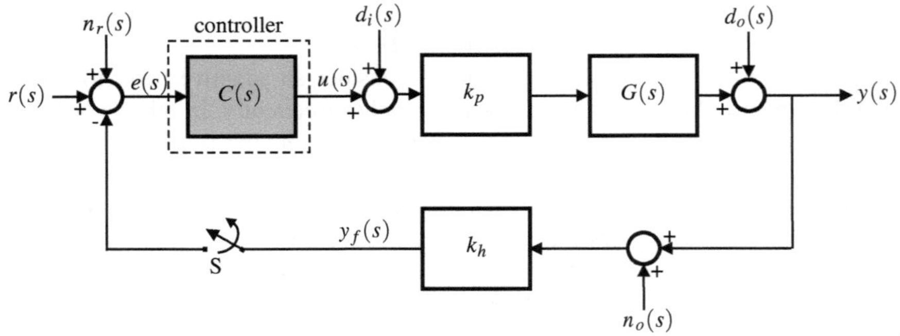

Fig. 7.1 Block diagram of the closed-loop control system. $G(s)$ is the plant transfer function, $C(s)$ is the controller transfer function, $r(s)$ is the reference signal, $y(s)$ is the output of the control loop, $y_f(s)$ is the output signal after k_h, $d_o(s)$ and $d_i(s)$ are the output and input disturbance signals, respectively, and $n_r(s)$, $n_o(s)$ are the noise signals at the reference input and process output, respectively. k_p stands for the plant's dc gain and k_h is the feedback path

7.2 The Algorithm of Automatic Tuning of PID Regulators

The closed-loop system of Fig. 7.1 is again considered, where $r(s)$, $e(s)$, $u(s)$, $y(s)$, $d_o(s)$ and $d_i(s)$ are the reference input, the control error, the input and output of the plant, the output and the input disturbances, respectively. In addition, the real process is described by

$$G(s) = \frac{1}{(1 + sT_{p_1})(1 + sT_{p_2}) \cdots (1 + sT_{p_n})}, \tag{7.1}$$

where $T_{p_1} > T_{p_2} > \cdots > T_{p_n}$. This type of modeling is not restrictive since it is shown in Sect. 7.2 that the proposed method can be applied in processes with time delay or right half plane zeros. Parameter k_p stands for the plant's dc gain. In vector controlled medium voltage drives, for example, k_p stands for the pulse width modulator's linear gain k_{PWM} which is assumed to remain linear over the whole operating range of the motor.

Supposing that little information about the process is available, it is conceived as a first order one defined by the approximation

$$\widetilde{G}(s) = \frac{1}{1 + sT_{\Sigma_p}}, \tag{7.2}$$

where $T_{\Sigma_p} = \sum_{i=1}^{n} T_{p_i}$ is the equivalent sum time constant of the plant. When the information about the plant is limited, the control that can consciously be applied is limited to integral control, so that the system exhibits at least zero steady state position error.

7.2.1 Integral Control of the Approximate Plant

By applying integral action given by

$$C(s) = \frac{1}{sT_{\mathrm{i_I}}(1 + sT_{\Sigma_\mathrm{c}})}, \tag{7.3}$$

to the approximate plant (7.2), the resulting closed-loop transfer function

$$T(s) = \frac{k_\mathrm{p}C(s)G(s)}{1 + k_\mathrm{p}k_\mathrm{h}C(s)G(s)} \tag{7.4}$$

takes the form

$$\begin{aligned}\widetilde{T}(s) &= \frac{k_\mathrm{p}}{sT_{\mathrm{i_I}}(1 + sT_{\Sigma_\mathrm{c}})(1 + sT_{\Sigma_\mathrm{p}}) + k_\mathrm{h}k_\mathrm{p}} \\ &\approx \frac{k_\mathrm{p}}{s^2T_{\mathrm{i_I}}T_\Sigma + sT_{\mathrm{i_I}} + k_\mathrm{p}k_\mathrm{h}}\end{aligned} \tag{7.5}$$

for which $T_{\Sigma_\mathrm{c}}T_{\Sigma_\mathrm{p}} \approx 0$[3] and $T_\Sigma = T_{\Sigma_\mathrm{c}} + T_{\Sigma_\mathrm{p}}$ has been considered. Note that T_{Σ_c} stands for the controller's unmodeled dynamics arising from its implementation. According to the conventional design via the Magnitude Optimum principle see Sect. 3.2.1, the integration time constant $T_{\mathrm{i_I}}$ along with the feedback path k_h prove to be equal to

$$k_\mathrm{h} = 1 \quad \text{and} \quad T_{\mathrm{i_I}} = 2k_\mathrm{p}k_\mathrm{h}T_\Sigma. \tag{7.6}$$

Condition $k_\mathrm{h} = 1$ implies that the closed-loop system has zero steady state position error. Substituting (7.6) into (7.5), leads to

$$\widetilde{T}(s) = \frac{1}{2T_\Sigma^2 s^2 + 2sT_\Sigma + 1}. \tag{7.7}$$

Normalizing the time by setting $s' = sT_\Sigma$ leads to

$$\widetilde{T}(s') = \frac{1}{2s'^2 + 2s' + 1}. \tag{7.8}$$

At this point, it is necessary to declare that by using only the integration time constant $T_{\mathrm{i_I}}$ and if $k_\mathrm{h} = 1$, which results in the above closed-loop dynamic behavior, the sum time constant of the closed-loop system T_Σ can be estimated by the relation

[3] The controller's unmodeled dynamics T_{Σ_c} are negligible compared to the plant's unmodeled dynamics T_{Σ_p}, $T_{\Sigma_\mathrm{c}} \ll T_{\Sigma_\mathrm{p}}$.

$$T_{\Sigma_{\mathrm{est}}} = \frac{T_{\mathrm{i_I}}}{2k_{\mathrm{p}}k_{\mathrm{h}}} = \frac{T_{\mathrm{i_I}}}{2k_{\mathrm{p}}}. \tag{7.9}$$

7.2.2 Integral Control of the Real Plant

If the same control law, (7.6), is applied to the real plant (7.1), the resulting closed-loop transfer function is given by

$$T(s') = \frac{1}{\begin{bmatrix} s'^{n+1}\left(\frac{2}{T_\Sigma^n}\prod_{j=1}^{n}T_{\mathrm{p}_j}\right) + \cdots + s'^3\left(\frac{2}{T_\Sigma^2}\sum_{i\neq j=1}^{n}T_{\mathrm{p}_i}T_{\mathrm{p}_j}\right) \\ +2s'^2 + 2s' + 1 \end{bmatrix}} \tag{7.10}$$

as it was proved Sect. 3.2.2. There, it was shown that depending on the ratio $\rho = \frac{T_{\mathrm{p}1}}{T_\Sigma}$ the step and frequency response exhibits certain performance characterized by

- Mean rise time $t_{\mathrm{r}} = 4.40T_\Sigma$ ($4.7T_\Sigma$ for $\rho \geq 0.9$ and $4.1T_\Sigma$ for $\rho = 0.3$).
- Mean settling time $t_{\mathrm{ss}} = 7.86T_\Sigma$ ($8.40T_\Sigma$ for $\rho \geq 0.9$ and $7.32T_\Sigma$ for $\rho = 0.3$).
- Mean overshoot 4.47 % (4.32 % for $\rho \geq 0.9$ and 4.62 % for $\rho = 0.3$).
- Gain margin $\alpha_{\mathrm{m}} = 205$ db.
- Phase margin $\phi_{\mathrm{m}} = 65.27°$.

7.2.3 Proportional-Integral Control

If the dominant time constant T_{p_1} of the plant is evaluated (conventional design method via the Magnitude Optimum criterion), the transfer function process model can be defined by

$$\widetilde{G}(s) = \frac{1}{(1 + sT_{\mathrm{p}_1})(1 + sT_{\Sigma_{1\mathrm{p}}})}, \tag{7.11}$$

where $T_{\Sigma_{1\mathrm{p}}} = \sum_{i=2}^{n} T_{\mathrm{p_i}}$ is the parasitic time constant of the plant. Since the plant has a dominant time constant, PI control of the form

$$C(s) = \frac{1 + sT_{\mathrm{n}}}{sT_{\mathrm{i_{PI}}}(1 + sT_{\Sigma_{\mathrm{c}}})}, \tag{7.12}$$

is imposed to (7.11). The resulting closed-loop transfer function is again defined by

$$\widetilde{T}(s) = \frac{1}{2T_{\Sigma_1}^2 s^2 + 2T_{\Sigma_1}s + 1}. \tag{7.13}$$

Setting again $s' = sT_{\Sigma_1}$ leads to

$$\widetilde{T}(s') = \frac{1}{2s'^2 + 2s' + 1}, \tag{7.14}$$

for which the optimal PI control action has been proved in Sect. 3.2.3 to be given by

$$k_h = 1, \tag{7.15}$$

$$T_n = T_{p_1}, \tag{7.16}$$

$$T_{iPI} = 2k_p k_h T_{\Sigma_1} = 2k_p k_h (T_\Sigma - T_{p1}) = 2k_p k_h (T_\Sigma - T_n). \tag{7.17}$$

Comparing (7.14) with (7.8), it is concluded that with the application of PI control via the conventional design of the Magnitude Optimum criterion, a closed-loop system with time and frequency response of the same shape results.

However, the response of (7.14) is faster, because the timescale is smaller ($T_{\Sigma_1} < T_\Sigma$). In other words, the compensation of the dominant time constant T_{p_1} has left the shape (performance features) of the system time and frequency responses unaltered and produced only a change both in the time and frequency scale, respectively. In addition, through the new integration time constant T_{iPI}, with which a step response with mean overshoot 4.47 % is achieved, the 'parasitic' time constant of the closed-loop system can be estimated through the relation

$$T_{\Sigma_{1est}} = \frac{T_{iPI}}{2k_p k_h} = \frac{T_{iPI}}{2k_p}. \tag{7.18}$$

7.2.4 Proportional-Integral-Derivative Control

If two dominant time constants T_{p_1}, T_{p_2} of the plant are measured accurately, the transfer function of the process can be approximated by

$$\widetilde{G}(s) = \frac{1}{(1 + sT_{p_1})(1 + sT_{p_2})(1 + sT_{\Sigma_{2p}})}, \tag{7.19}$$

where again $T_{\Sigma_{2p}} = \sum_{i=3}^{n} T_{\Sigma_{p_i}}$ stands for the parasitic time constant of the plant. Since the plant has two dominant time constants, PID control defined by

$$C(s) = \frac{(1 + sT_n)(1 + sT_v)}{sT_{iPID}(1 + sT_{\Sigma_c})} \tag{7.20}$$

is imposed to (7.19). In similar fashion, in Sect. 3.2.3, it was proved that the end closed-loop transfer function is given by

$$\widetilde{T}(s) = \frac{1}{2T_{\Sigma_2}^2 s^2 + 2T_{\Sigma_2} s + 1}. \tag{7.21}$$

Normalizing the time by setting $s' = sT_{\Sigma_2}$ leads to

$$\widetilde{T}(s') = \frac{1}{2s'^2 + 2s' + 1}. \tag{7.22}$$

Condition $|\widetilde{T}(j\omega)| \simeq 1$ is now satisfied when

$$k_\mathrm{h} = 1, \tag{7.23}$$

$$T_\mathrm{n} = T_{\mathrm{p}_1}, \tag{7.24}$$

$$T_\mathrm{v} = T_{\mathrm{p}_2}, \tag{7.25}$$

$$\begin{aligned} T_\mathrm{iPID} &= 2k_\mathrm{p}k_\mathrm{h}T_{\Sigma 2} = 2k_\mathrm{p}k_\mathrm{h}(T_\Sigma - T_{\mathrm{p}_1} - T_{\mathrm{p}_2}) \\ &= 2k_\mathrm{p}k_\mathrm{h}(T_\Sigma - T_\mathrm{n} - T_\mathrm{v}). \end{aligned} \tag{7.26}$$

Comparing (7.22) with (7.14) and (7.8), it becomes evident that with the application of PID control, we end up again, in a closed-loop system with time and frequency responses of the same shape (performance features), but with even smaller timescale ($T_{\Sigma_2} < T_{\Sigma_1} < T_\Sigma$) and consequently even faster.

Moreover, with the integration time constant T_iPID, with which we achieve a step response with 4.47 % mean overshoot, we can estimate the new 'parasitic' time constant of the closed-loop system using the relation

$$T_{\Sigma 2\mathrm{est}} = \frac{T_\mathrm{iPID}}{2k_\mathrm{p}k_\mathrm{h}} = \frac{T_\mathrm{iPID}}{2k_\mathrm{p}}. \tag{7.27}$$

7.2.5 The Tuning Process

The conventional Magnitude Optimum design criterion, presented in Sect. 3.2, leads effortlessly to the automatic tuning procedure of the controller parameters. The procedure follows the next steps:

Step 1: *Determination of the gain* k_p. The gain k_p is determined from the step response of the plant at steady state, Fig. 7.2a.

$$\lim_{t\to\infty} y(t) = \lim_{s\to 0} sG(s)u(s) = k_\mathrm{p} \tag{7.28}$$

and if the process $G(s)$ is stable. In vector controlled induction motor drives, k_p stands for the pulse width modulator gain which is a priori known for the whole operating range of the motor. Moreover, an estimation of the sum time constant T_{Σ_p} of the plant can be derived from the step response according to

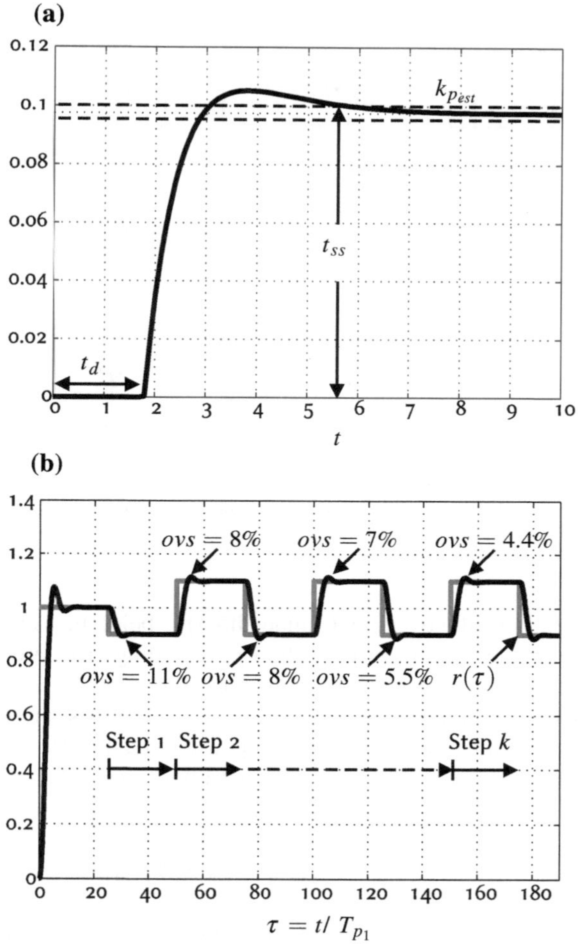

Fig. 7.2 Typical step response after an open-loop experiment of the process and screen shots of the automatic tuning procedure. **a** Typical step response of the process. **b** A series of small step variations of the reference input with alternating sign are imposed for tuning the PID controller's parameters

$$T_{\Sigma_{\text{Pest}}} \approx \frac{t_{\text{ss}}}{4}, \tag{7.29}$$

where t_{ss} is the settling time of the step response.

Then, an auxiliary loop (gray shaded) is placed in the closed-loop system of Fig. 7.1, as shown in Fig. 7.3. The purpose of this loop is the tuning of the controller $C_x(s)$. The operation of the auxiliary loop is the following.

A series of small step variations of the reference input with alternating sign are imposed, so that the plant does not diverge far from its operating point, Fig. 7.2b.

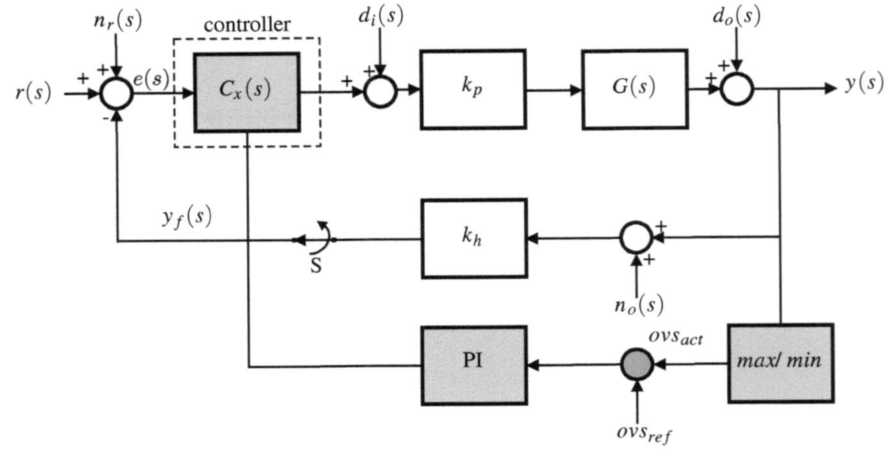

Fig. 7.3 Block diagram of the closed-loop control system and the tuning loop. k_p is the plant's dc gain and k_h stands for the feedback path. C_x stands for the automatically tuned controller. ovs_{act} is the measured overshoot of $y(s)$ and ovs_{ref} is set equal to 4.47 %

During these variations, the overshoot (undershoot) is being measured and is compared with the reference overshoot (undershoot). According to the preceding analysis in Sect. 3.2, the absolute value of the reference overshoot is 0.0447. The error is fed into a PI controller, which tunes the controller $C_x(s)$ in succession, so that the overshoot (undershoot) of the closed-loop step response to be 4.47 %. According to the analysis presented in Sects. 3.2 and 3.2.4, the controller $C_x(s)$ is being given the form

$$C_x(s) = \frac{(1 + sT_{n_x})(1 + sT_{v_x})}{(2k_p k_h T_{\Sigma_x} - 2k_p k_h T_{n_x} - 2k_p k_h T_{v_x})s(1 + sT_{\Sigma_c})}, \tag{7.30}$$

where T_{Σ_x}, T_{n_x} and T_{v_x} are time constants that must be determined automatically.

Step 2: Determination of the time constant T_{Σ_x}. In (7.30) $T_{n_x} = T_{v_x} = 0$ is set. In succession, a series of step variations on the reference input is imposed and time constant T_{Σ_x} is tuned such, so that the overshoot (undershoot) is 4.47 %.

According to Sect. 3.2.1, this occurs when $T_{\Sigma_x} \approx T_{\Sigma}$. Tuning of T_{Σ_x}, (or T_{i_x}) is described in Fig. 7.4

Step 3: Determination of the time constant T_{n_x}. With the value of T_{Σ_x} given from Step 2, $T_{v_x} = 0$ is set in (7.30).

$$C_x(s) = \frac{1 + sT_{n_x}}{(2k_p k_h T_{\Sigma_x} - 2k_p k_h T_{n_x})s(1 + sT_{\Sigma_c})}. \tag{7.31}$$

A series of step variations of the reference input is again imposed and T_{n_x} is tuned, so that the overshoot (undershoot) becomes again 4.47 %. As shown in Fig. 7.4a, this occurs when $T_{n_x} \approx T_{p_1}$, Sect. 3.2.3, PI control. If the 'parasitic' time constant

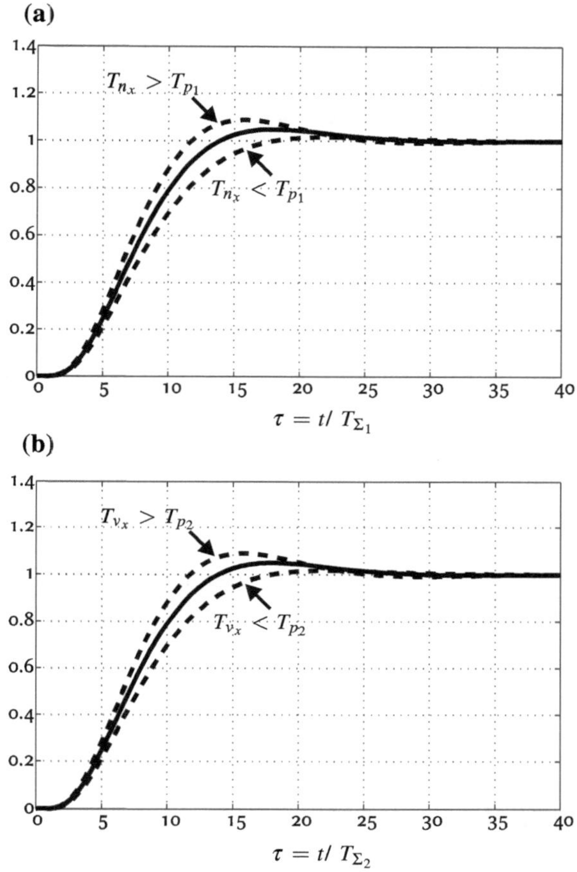

Fig. 7.4 Tuning of the PID controller. According to step 2: let at step (k) a series of step pulses is applied at the reference input. If $\text{ovs}_{\text{act}} < \text{ovs}_{\text{ref}}$ then at $(k+1)$ step $T_{\text{i}_x}(k+1) < T_{\text{i}_x}(k)$. The amount of this change is based on the parameters of the PI controller (gray box) which is tuned heuristically. The PI controller takes the error between ovs_{act}, ovs_{ref} at step k and returns the $T_{\text{i}_x}(k+1)$ for the next step. Note that at step 2, $T_{\text{n}_x} = T_{\text{v}_x} = 0$. At step 3, let at step (k), a series of step pulses is applied at the reference input. If $\text{ovs}_{\text{act}} < \text{ovs}_{\text{ref}}$ then at $(k+1)$ step $T_{\text{n}_x}(k+1) > T_{\text{n}_x}(k)$. Since T_{i_x} is tuned automatically (see (7.30)) while all other parameters remain constant, ovs_{act} is controlled only by tuning T_{n_x}. Since T_{n_x} is the zero of the open-loop transfer function, if $T_{\text{n}_x}(k) > T_{\text{n}_x}(k+1)$ then $\text{ovs}_{\text{act}}(k+1) < \text{ovs}_{\text{act}}(k)$. The same tuning procedure stands for T_{v_x}. **a** Tuning of parameter T_{n_x}. **b** Tuning of parameter T_{v_x}

$T_{\Sigma_1} = T_{\Sigma_x} - T_{\text{n}_x}$ is relatively large, the procedure can be continued by attempting step 4. If the 'parasitic' time constant is sufficiently small, PI control is retained.

Step 4: *Determination of the time constant* T_{v_x}. Given the values of T_{Σ_x} and T_{n_x}, T_{v_x} is tuned in such a way, so that the overshoot is again 4.47 %, by imposing again a series of step variations on the reference input. As shown in Fig. 7.4b, this occurs when $T_{v_x} \approx T_{p_2}$. If the 'parasitic' time constant $T_{\Sigma_2} = T_{\Sigma_x} - T_{n_x} - T_{v_x}$ still remains relatively large, a fact that shows that other relatively large time constants may exist, the tuning procedure can be continued incorporating in cascade with the controller $C_x(s)$ the necessary number of high-pass stages of the form

$$C_{\mathrm{h}}(s) = \frac{1 + sT_a}{1 + sT_b}, \tag{7.32}$$

where $T_a > T_b$. For one additional high-pass stage, the controller $C_x(s)$ must take the form

$$\begin{aligned} C_x(s) &= \frac{(1 + sT_{n_x})(1 + sT_{v_x})(1 + sT_a)}{2k_{\mathrm{p}}s(T_{\Sigma_x} - T_{n_x} - T_{v_x} - T_a)(1 + sT_{\Sigma c})(1 + sT_b)} \\ &\approx \frac{(1 + sT_{n_x})(1 + sT_{v_x})(1 + sT_a)}{2k_{\mathrm{p}}s(T_{\Sigma_x} - T_{n_x} - T_{v_x} - T_a)(1 + sT_{\Sigma bc})} \end{aligned} \tag{7.33}$$

where $T_{\Sigma_{bc}} = T_{\Sigma_c} + T_b$ is considered as the new 'parasitic' time constant of the controller.

However, this can only occur if the noise level, that accompanies the controlled physical quantities, allows it. If this is not possible, but the design of a faster closed-loop system is required, then different control techniques should be followed, as cascade control, for example [14]. Obviously, the controllers' tuning of the inner loops can be achieved using the same procedure.

7.2.6 Starting up the Procedure

Essentially, no information regarding the plant is required for starting up the suggested tuning procedure. Consequently, step 1 is not entirely necessary. However, if the gain k_{p} and an estimation of the sum time constant $T_{\Sigma_{\mathrm{p}}}$ are known, the first application of the method is accelerated significantly. Moreover, if the gain k_{p} is not known, while the tuning of the controller is being carried out normally, the knowledge of the exact values of the plant time constants is not possible.

If in step 2 the initial estimation of T_{Σ_x} is smaller than T_Σ, then a significant overshoot occurs. Since a large overshoot is in general undesirable, the procedure must start with an overestimation of T_Σ. For the initiation of steps 3 and 4, the initial values of T_{n_x} and T_{v_x} are set equal to

$$\frac{T_{\Sigma_x}}{2} \quad \text{and} \quad \frac{T_{\Sigma_x} - T_{n_x}}{2} \tag{7.34}$$

respectively. Specifically, the convergence of the tuning procedure is faster when initially it is $T_{n_x} > T_{p_1}$ and $T_{v_x} > T_{p_2}$.

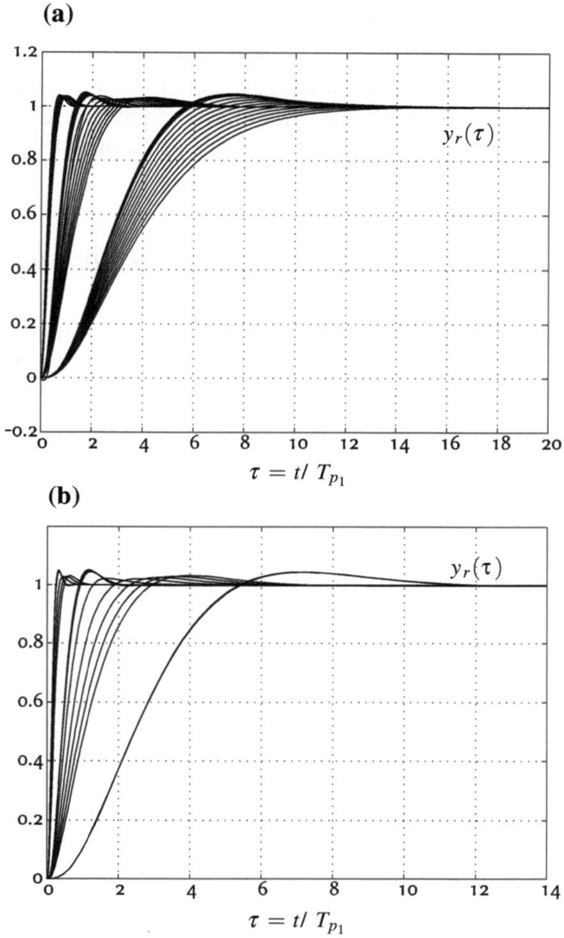

Fig. 7.5 Automatic tuning procedure. **a** Initial conditions differ significantly from the nominal ones. **b** Initial conditions differ by 10 % from the nominal ones

At this point, it should be noted that, as long as the controller parameters are determined, every repetition of the procedure is faster. In Fig. 7.5, it is shown the application of the suggested tuning procedure when the starting conditions are quite different from the nominal ones.

7.3 Simulation Examples

In this section, a performance comparison takes place between (1) the method[4] that tunes automatically the PID controller's parameters as proposed in Sect. 7.2 and (2) the method[5] that tunes explicitly the PID controller parameters proposed in Sect. 3.3.

7.3.1 Plant with One Dominant Time Constant

In this example, the plant exhibits one dominant time constant and its transfer function is defined by

$$G(s') = \frac{1.31}{(1+s')(1+0.2s')(1+0.1s')(1+0.05s')(1+0.02s')}. \tag{7.35}$$

The automatically tuned PI controller according to Sect. 7.2

$$C(s') = \frac{1+s't_{\mathrm{n}}}{s't_{\mathrm{i}}(1+s't_{\mathrm{sc}})} \tag{7.36}$$

is given finally by

$$C_{\mathrm{PIaut}}(s') = \frac{1+s'0.99}{s'1.243(1+s't_{\mathrm{sc}})} \tag{7.37}$$

However, the optimal PI control action calculated analytically according to Sect. 3.3 is defined by

$$C_{\mathrm{PIopt}}(s') = \frac{1+s'1.02}{s'1.17(1+s't_{\mathrm{sc}})}. \tag{7.38}$$

From (7.36) and (7.38) it is apparent that the parameters calculated from both methods are practically the same. This is also justified by the step response of the closed-loop control system in Fig. 7.6a, b where reference tracking and output disturbance rejection is depicted.

In a similar fashion, by tuning automatically the PID controller of the form

$$C(s') = \frac{(1+s't_{\mathrm{n}})(1+s't_{\mathrm{v}})}{s't_{\mathrm{i}}(1+s't_{\mathrm{sc}})} \tag{7.39}$$

---

[4] In this case, only an open-loop experiment is required to the process for initializing the algorithm and no other information.

[5] In this case, the transfer function is assumed accurately modeled.

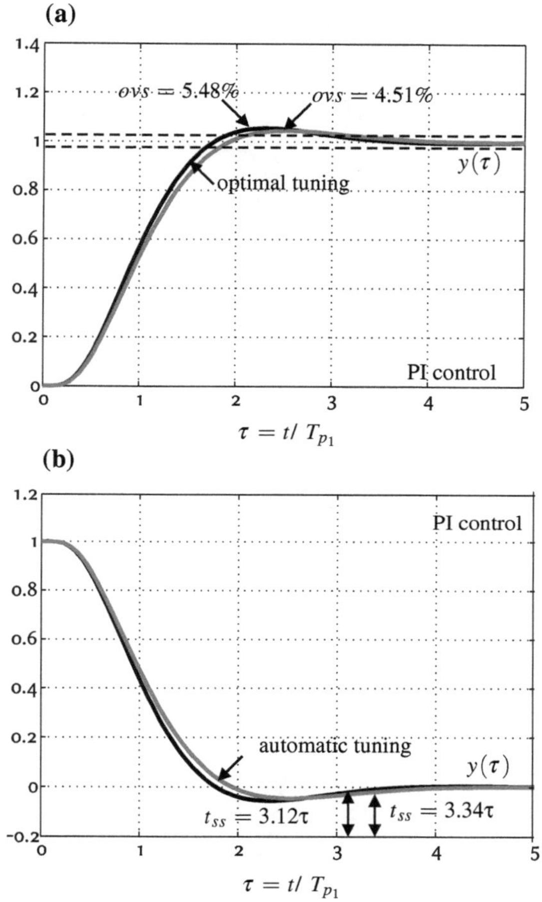

Fig. 7.6 PI control of a plant with one dominant time constant defined by (7.35). **a** Step response of the closed loop control system. **b** Output disturbance rejection

we ended up in

$$C_{\mathrm{PIDaut}}(s') = \frac{(1 + s'0.99)(1 + s'0.209)}{s'0.692(1 + s't_{\mathrm{sc}})} \tag{7.40}$$

whereas the optimal PID control action is given by

$$C_{\mathrm{PIDopt}}(s') = \frac{1 + s'1.252 + s'^2 0.25}{s'0.57(1 + s't_{\mathrm{sc}})}. \tag{7.41}$$

Let it be noted that the zeros of (7.41) are real positive values since

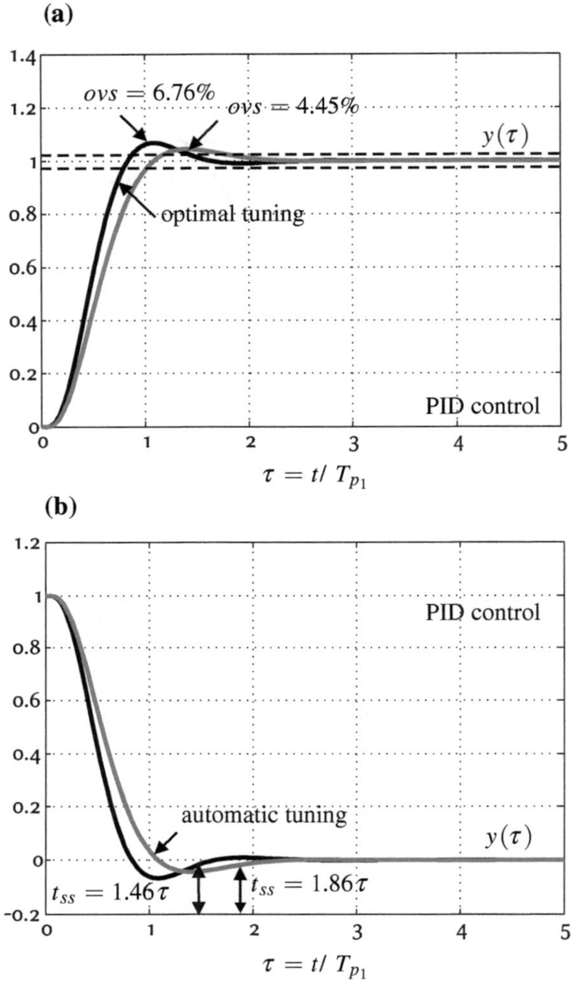

Fig. 7.7 PID control of a plant with one dominant time constant defined by (7.35). **a** Step response of the closed loop control system. **b** Output disturbance rejection

$$C_{\mathrm{PIDopt}}(s') = \frac{(1+s'0.99)(1+s'0.25)}{s'0.57(1+s't_{\mathrm{sc}})} \tag{7.42}$$

After comparing (7.40) with (7.42) it is apparent that both controller's tuning results in almost the same step and frequency response of the final closed control system, see also Fig. 7.7a, b.

7.3.2 Plant with Two Dominant Time Constants

In this example, the process to be controlled exhibits two dominant time constants and its transfer function is defined by

$$G(s') = \frac{0.84}{(1+s')(1+0.9s')(1+0.1s')(1+0.05s')(1+0.02s')}. \tag{7.43}$$

The automatically tuned PI controller defined by

$$C(s') = \frac{1+s't_\mathrm{n}}{s't_\mathrm{i}(1+s't_\mathrm{sc})} \tag{7.44}$$

is finally given by

$$C_\mathrm{PIaut}(s') = \frac{1+s'1.179}{s'1.7(1+s't_\mathrm{sc})} \tag{7.45}$$

whereas the optimal PI controller is given by

$$C_\mathrm{PIopt}(s') = \frac{1+s'1.3}{s'1.46(1+s't_\mathrm{sc})}. \tag{7.46}$$

By tuning automatically the PID controller of the form

$$C(s') = \frac{(1+s't_\mathrm{n})(1+s't_\mathrm{v})}{s't_\mathrm{i}(1+s't_\mathrm{sc})} \tag{7.47}$$

results in

$$C_\mathrm{PIDaut}(s') = \frac{(1+s'1.179)(1+s'0.718)}{s'0.492(1+s't_\mathrm{sc})}, \tag{7.48}$$

whereas the optimal PID controller is given by

$$C_\mathrm{PIDopt}(s') = \frac{1+s'1.91+s'^2 0.91}{s'0.437(1+s't_\mathrm{sc})}, \tag{7.49}$$

the zeros of which are conjugate complex since (7.49) can be rewritten in the form of

$$C_\mathrm{PIDopt}(s') = \frac{[1+s'(0.95+0.017\mathrm{i})][1+s'(0.95-0.017\mathrm{i})]}{s'0.43(1+s't_\mathrm{sc})}. \tag{7.50}$$

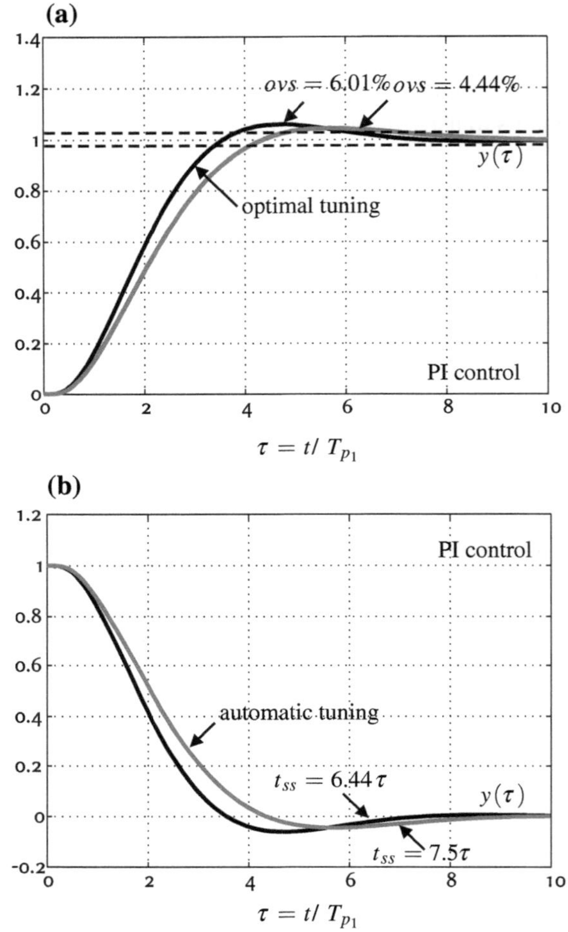

Fig. 7.8 PI control of a plant with two dominant time constants defined by (7.43). **a** Step response of the closed loop control system. **b** Output disturbance rejection

7.3.3 Plant with Dominant Time Constants and Time Delay

In this example, the process exhibits a time delay equal to $t_\mathrm{d} = 1.048$ and its transfer function is given by (Figs. 7.8 and 7.9)

$$G(s') = \frac{1.81}{(1+s')(1+0.69s')(1+0.3s')(1+0.13s')(1+0.1s')} \mathrm{e}^{-1.048s'}. \quad (7.51)$$

The automatically tuned PI controller

$$C(s') = \frac{1+s't_\mathrm{n}}{s't_\mathrm{i}(1+s't_\mathrm{sc})} \quad (7.52)$$

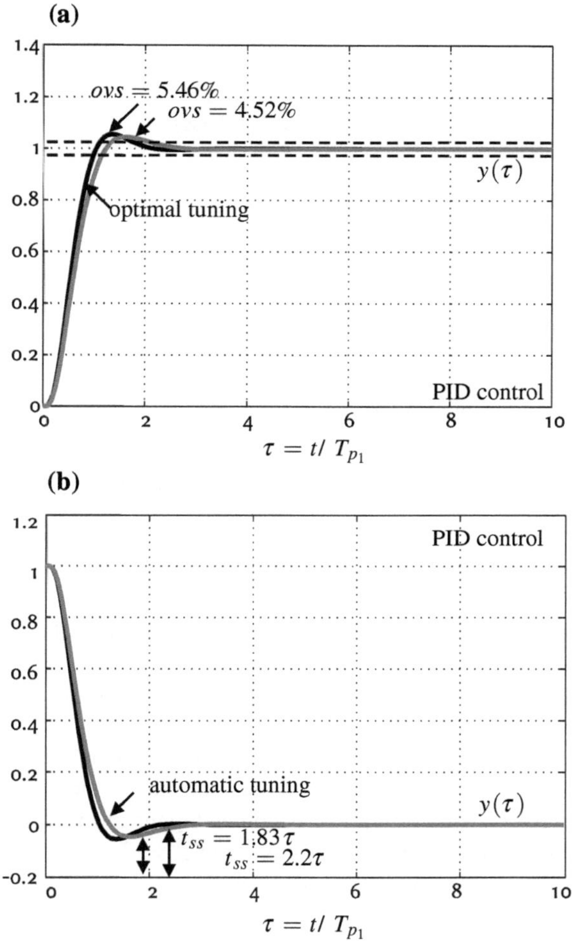

Fig. 7.9 PID control of a plant with two dominant time constants defined by (7.43). **a** Step response of the closed loop control system. **b** Output disturbance rejection

resulted in

$$C_{\mathrm{PIaut}}(s') = \frac{1 + s'1.066}{s'8.3(1 + s't_{\mathrm{sc}})}, \tag{7.53}$$

whereas the optimal PI controller is given by

$$C_{\mathrm{PIopt}}(s') = \frac{1 + s'1.44}{s'6.98(1 + s't_{\mathrm{sc}})}. \tag{7.54}$$

By automatically tuning the PID controller of the form

$$C(s') = \frac{(1 + s't_n)(1 + s't_v)}{s't_i(1 + s't_{sc})} \tag{7.55}$$

we ended up in

$$C_{PIDaut}(s') = \frac{(1 + s'1.066)(1 + s'0.25)}{s'7.38(1 + s't_{sc})} \tag{7.56}$$

whereas the optimal PID controller is given by

$$C_{PIDopt}(s') = \frac{1 + s'2.049 + s'^2 1.16}{s'4.78(1 + s't_{sc})}, \tag{7.57}$$

the zeros of which are conjugate complex since (7.57) can be rewritten as follows

$$C_{PIDopt}(s') = \frac{[1 + s'(1.024 + 0.341\mathrm{i})][1 + s'(1.024 - 0.341\mathrm{i})]}{s'4.78(1 + s't_{sc})}. \tag{7.58}$$

From (7.56) and (7.58), it is apparent significant difference in the value of the integrator's time constant and the zeros of the controller. This difference is depicted also in Figs. 7.10, and 7.11 regarding the step response of the closed-loop control system and disturbance rejection.

Specifically, the settling time t_{ss} of disturbance rejection in the case of PI control is equal to $t_{ss} = 14.7\tau$ and $t_{ss} = 11.3\tau$ when the controller is tuned automatically and optimally, respectively. This difference becomes bigger in the case of PID control where the corresponding settling time is equal to $t_{ss} = 13.1\tau$ and $t_{ss} = 7.05\tau$ when the controller is tuned automatically and optimally.

7.3.4 Plant with Dominant Time Constants, Zeros, and Time Delay

In this example, the process defined by

$$G(s') = 1.31 \frac{(1 + 0.03s')(1 + 0.9s')}{(1 + s')(1 + 0.81s')(1 + 0.79s')(1 + 0.72s')(1 + 0.41s')} \mathrm{e}^{-s'} \tag{7.59}$$

is considered. After the automatically tuned PI controller of the form

$$C(s') = \frac{1 + s't_n}{s't_i(1 + s't_{sc})} \tag{7.60}$$

resulted in

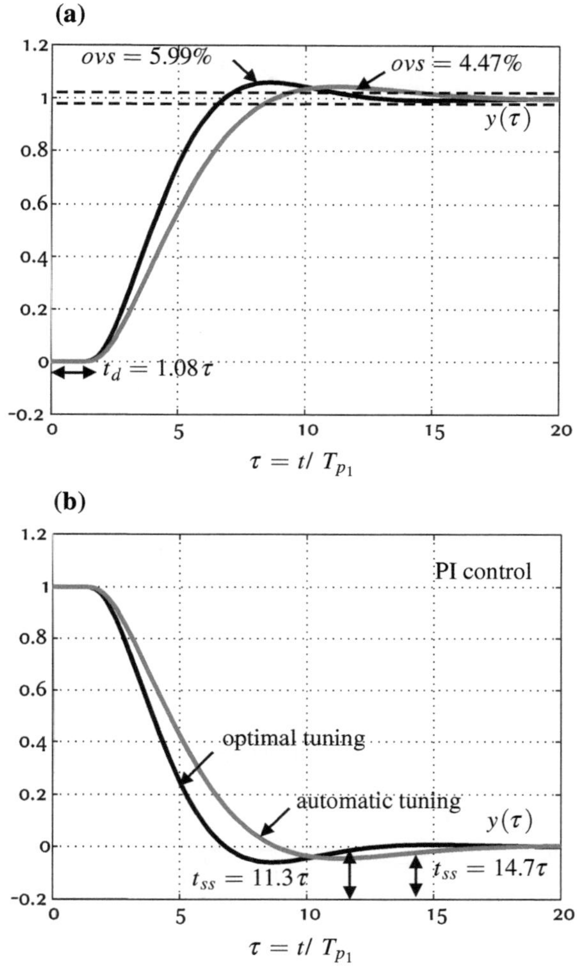

Fig. 7.10 PI control of a plant with dominant time constants and time delay defined by (7.51). **a** Step response of the closed loop control system. **b** Output disturbance rejection

$$C_{\mathrm{PIaut}}(s') = \frac{1 + s'1.06}{s'7.423(1 + s't_{\mathrm{sc}})}, \tag{7.61}$$

whereas the optimal PI controller is given by

$$C_{\mathrm{PIopt}}(s') = \frac{1 + s'1.67}{s'5.84(1 + s't_{\mathrm{sc}})}. \tag{7.62}$$

By automatically tuning the PID controller of the form

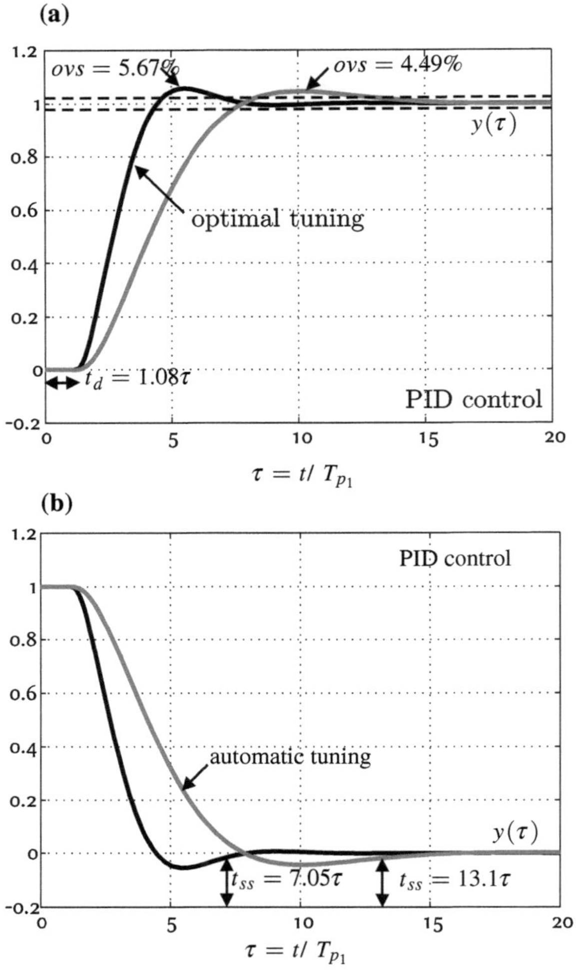

Fig. 7.11 PID control of a plant with dominant time constants and time delay defined by (7.51). **a** Step response of the closed loop control system. **b** Output disturbance rejection

$$C(s') = \frac{(1 + s't_n)(1 + s't_v)}{s't_i(1 + s't_{sc})} \tag{7.63}$$

results in

$$C_{PIDaut}(s') = \frac{(1 + s'1.06)(1 + s'0.31)}{s'6.6(1 + s't_{sc})}, \tag{7.64}$$

whereas the optimal controller is given by

$$C_{\mathrm{PIDopt}}(s') = \frac{1 + s'2.39 + s'^2 1.61}{s'3.94(1 + s't_{\mathrm{sc}})}. \tag{7.65}$$

Note again that the zeros of (7.65) are conjugate complex since its numerator can be rewritten in the form of

$$C_{\mathrm{PIDopt}}(s') = \frac{[1 + s'(1.19 + 0.42\mathrm{i})][1 + s'(1.19 - 0.42\mathrm{i})]}{s'3.95(1 + s't_{\mathrm{sc}})}. \tag{7.66}$$

Note also in this case, the difference in the performance of the final closed-loop control system regarding reference tracking and output disturbance rejection when the PID controller is tuned both automatically and analytically, see Figs. 7.12 and 7.13.

7.3.5 A Nonminimum Phase Plant with Time Delay

In this example, let the nonminimum phase process defined by

$$G(s') = \frac{1.31(1 - 0.03s')(1 - 0.9s')}{(1 + s')(1 + 0.81s')(1 + 0.79s')(1 + 0.72s')(1 + 0.41s')} \mathrm{e}^{-s'}. \tag{7.67}$$

The automatically tuned PI controller

$$C(s') = \frac{1 + s't_{\mathrm{n}}}{s't_{\mathrm{i}}(1 + s't_{\mathrm{sc}})} \tag{7.68}$$

resulted in

$$C_{\mathrm{PIaut}}(s') = \frac{1 + 1.44s'}{s'11.18(1 + s't_{\mathrm{sc}})}, \tag{7.69}$$

whereas the optimal controller is finally given by

$$C_{\mathrm{PIopt}}(s') = \frac{1 + 2.2s'}{9.3s'(1 + s't_{\mathrm{sc}})}. \tag{7.70}$$

After tuning automatically the PID controller of the form

$$C(s') = \frac{(1 + s't_{\mathrm{n}})(1 + s't_{\mathrm{v}})}{s't_{\mathrm{i}}(1 + s't_{\mathrm{sc}})} \tag{7.71}$$

results in

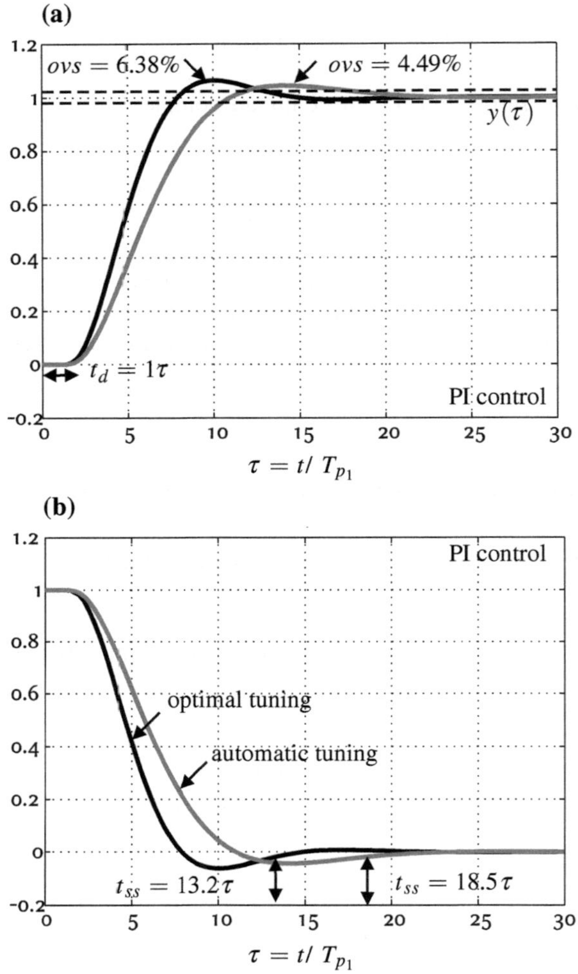

Fig. 7.12 PI control of a plant with dominant time constants, zeros, and time delay defined by (7.59). **a** Step response of the closed loop control system. **b** Output disturbance rejection

$$C_{\text{PIDaut}}(s') = \frac{1 + 1.91s' + 0.68s'^2}{s'9.95(1 + s't_{\text{sc}})}, \tag{7.72}$$

whereas the optimal controller is given by

$$C_{\text{PIDopt}}(s') = \frac{1 + 3.12s' + 2.81s'^2}{s'6.91(1 + s't_{\text{sc}})} \tag{7.73}$$

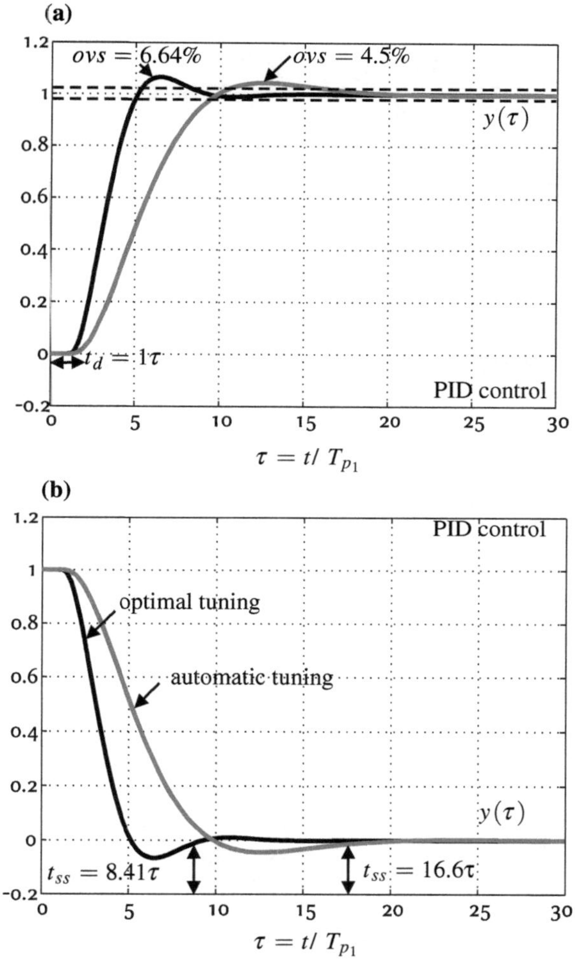

Fig. 7.13 PID control of a plant with dominant time constants, zeros, and time delay defined by (7.59). **a** Step response of the closed loop control system. **b** Output disturbance rejection

Note that in this case, zeros of (7.73) are real values, since the numerator can be rewritten in the form of

$$C_{\text{PIDaut}}(s') = \frac{(1 + 1.44s')(1 + 0.47s')}{s'9.95(1 + s't_{\text{sc}})}. \tag{7.74}$$

In this case, it is apparent that the closed-loop control system with the automatically tuned PI, PID controller exhibits poor performance compared to the optimal PI, PID tuning via the explicit control law, see Figs. 7.14 and 7.15. Specifically, the settling

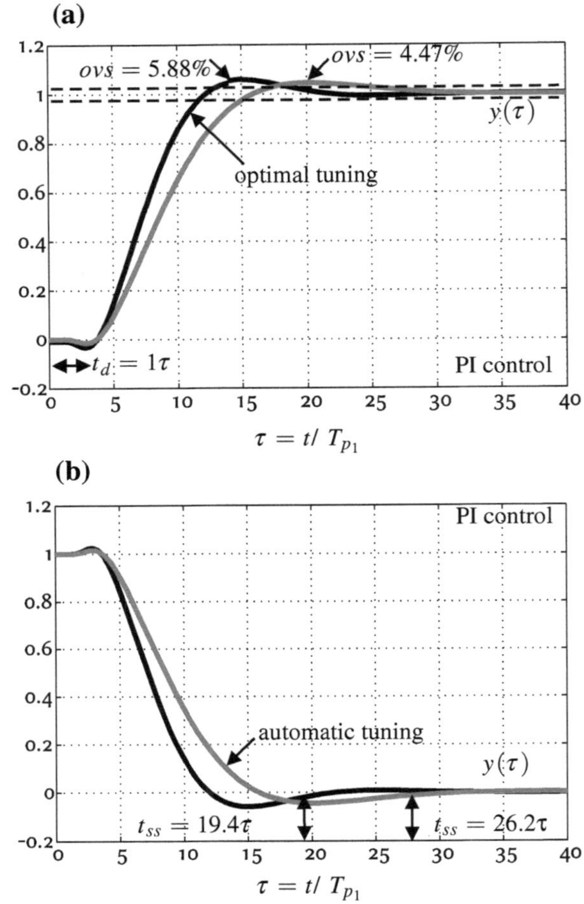

Fig. 7.14 PI control of a nonminimum phase plant defined by (7.67). **a** Step response of the closed loop control system. **b** Output disturbance rejection

time of output disturbance rejection in the case of PID control is $t_{ss} = 23.2\tau$ for the automatically tuned controller and $t_{ss} = 12.9\tau$ for the explicitly tuned controller.

7.4 Automatic Tuning for Processes with Conjugate Complex Poles

In this section, the principle of the Magnitude Optimum criterion is applied to the control of processes with conjugate complex poles, see [22]. It is shown that if applying I-lag control action and PID control action to the process, the same shape

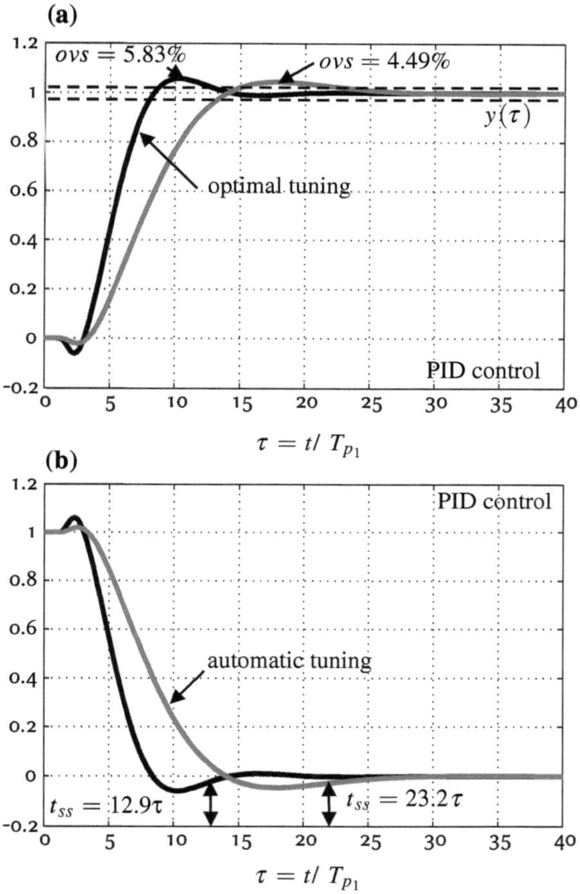

Fig. 7.15 PID control of a nonminimum phase plant defined by (7.67). **a** Step response of the closed loop control system. **b** Output disturbance rejection

of the step response of the control loop is achieved as described in Sect. 7.2. This feature leads effortlessly to the automatic tuning of the PID type controller which is finally presented in Sect. 7.4.2.

7.4.1 Direct Tuning of the PID Controller for Processes with Conjugate Complex Poles

For presenting the proposed method, the oscillatory process of the form

$$G(s) = \frac{1}{(1 + 2\zeta Ts + T^2 s^2)(1 + sT_{\Sigma p})}, \quad \zeta < 1 \tag{7.75}$$

is considered where $\zeta \in (0, 1], \Re$ and $T > 0$. The proposed PID-type controller is defined by

$$C(s) = \frac{1 + sX + s^2 Y}{sT_i(1 + sT_{\Sigma c})} \tag{7.76}$$

allowing its zeros to become conjugate complex if possible. The respective closed-loop transfer function $T(s) = \frac{y(s)}{r(s)}$ according to (2.1) and Fig. 7.1 is given by

$$T(s) = \frac{k_p(1 + sX + s^2 Y)}{\left[\begin{array}{l} T^2 T_\Sigma T_i s^4 + \left(T^2 T_i + 2\zeta T T_\Sigma T_i\right) s^3 + \left(2\zeta T T_i + T_\Sigma T_i + k_h k_p Y\right) s^2 \\ \quad + \left(T_i + k_h k_p X\right) s + k_h k_p \end{array}\right]} \tag{7.77}$$

where $T_\Sigma = T_{\Sigma_c} + T_{\Sigma_p}$ and $T_{\Sigma_c} T_{\Sigma_p} \approx 0$. By normalizing (7.77) with $s' = sT_\Sigma$ results in

$$T(s') = \frac{k_p(1 + s'x + s'^2 y)}{\left[\begin{array}{l} \tau^2 t_i s'^4 + t_i\left(\tau^2 + 2\zeta\tau\right) s'^3 + \left(2\zeta\tau t_i + t_i + k_h k_p y\right) s'^2 \\ \quad + \left(t_i + k_h k_p x\right) s' + k_h k_p \end{array}\right]}. \tag{7.78}$$

By applying I control to the normalized closed-loop transfer function (7.78) thus $x = y = 0$ in (7.78), it is obtained that $|T(j\omega)| \simeq 1$ is preserved in the widest possible frequency range if

$$k_h = 1, \quad t_i = 2k_p k_h\left(1 + 2\zeta\tau\right), \tag{7.79}$$

where $\tau = \frac{T}{T_\Sigma} = \frac{T}{T_{\Sigma c} + T_{\Sigma p}}$ and $t_i = \frac{T_i}{T_\Sigma}$. Integral control law (7.79) is proved as follows. From (7.78), if $x = y = 0$ then

$$T(s') = \frac{k_p}{\tau^2 t_i s'^4 + t_i\left(\tau^2 + 2\zeta\tau\right) s'^3 + t_i\left(1 + 2\zeta\tau\right) s'^2 + t_i s' + k_h k_p}. \tag{7.80}$$

According to A.1 and (7.80), where the principle of the Magnitude Optimum criterion is presented, it is apparent that $k_h k_p = k_p$ or finally

$$k_h = 1. \tag{7.81}$$

The application of (A.10) into (7.80) results in $a_1^2 = 2a_2 a_0$ since the terms b_1, b_2 of (7.80) are $b_1 = b_2 = 0$. Therefore it is apparent that $t_i^2 = 2k_p k_h t_i\left(1 + 2\zeta\tau\right)$ or

$$t_i = 2k_p k_h\left(1 + 2\zeta\tau\right). \tag{7.82}$$

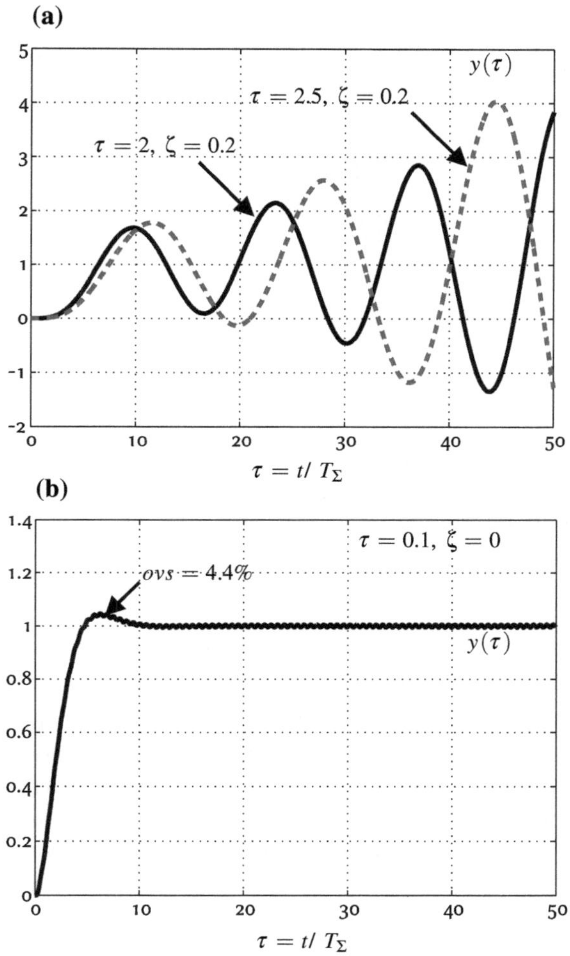

Fig. 7.16 I control—step response of the closed-loop control system for a second-order process with conjugate complex poles for various values of parameters $x,\ \zeta$. The final transfer function of the control loop is defined by (7.83). **a** Unstable step response of the closed loop control system for a second order process with conjugate complex poles. **b** Stable step response of the closed loop control system for a second order process with conjugate complex poles

In that case after substituting (7.81), (7.82) into (7.80) results in

$$T(s') = \frac{1}{\begin{array}{c}2\tau^2(1+2\zeta\tau)s'^4 + 2\tau(1+2\zeta\tau)(\tau+2\zeta)s'^3 \\ +2(1+2\zeta\tau)^2 s'^2 + 2(1+2\zeta\tau)s' + 1\end{array}}. \tag{7.83}$$

From Fig. 7.16a it is apparent that the final closed-loop control system is not stable $\forall\zeta,\ \tau$. However, (7.83) becomes stable $\forall\zeta$ if τ is forced $\tau \to 0$, Fig. 7.16b. In this case (7.83) becomes equal to

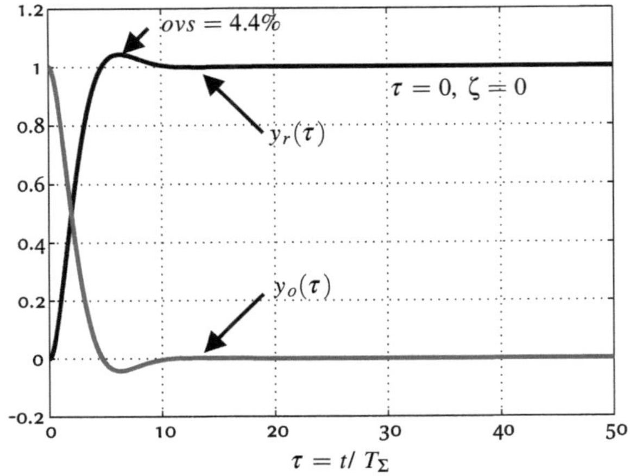

Fig. 7.17 I control—step response of the closed-loop control system for a second-order process with conjugate complex poles $\forall\zeta$ and if $\tau \to 0$. The final transfer function of the control loop is defined by (7.84)

$$T(s) = \frac{1}{2s'^2 + 2s' + 1} \tag{7.84}$$

which is equivalent to (3.10), (3.25) and (3.41) presented in Sect. 3.2. Therefore, according to Sect. 3.2, the step response of the closed-loop control system exhibits overshoot 4.4 %, see Fig. 7.17. From this point and based on the analysis in Sects. 7.2 and 7.4.1 and the determination of the integrator's time constant, see (7.82), a method for the automatic tuning of the PID controller's parameters is proposed in the sequel.

7.4.2 Automatic Tuning of the PID Controller for Processes with Conjugate Complex Poles

Purpose of the proposed method is to tune the PID-type controller's parameters, so that the output $y(s)$ of the control loop exhibits the aforementioned performance of (7.84). For presenting the proposed method, the PID controller of the form

$$C(s) = \frac{1 + sX_x + s^2 Y_x}{sT_{\mathrm{i}_x}(1 + sT_{\Sigma\mathrm{c}})} \tag{7.85}$$

is proposed. The problem is to tune automatically parameters T_{i_x}, X_x, Y_x by having access only to the output of the process $y(s)$, Fig. 7.1. According to the preceding analysis Sect. 7.4.1, in order to force $\tau \to 0$, $T_{\Sigma\mathrm{c}} + T_{\Sigma\mathrm{p}} \gg T$ must hold by since

$$\tau = \frac{T}{T_{\Sigma c} + T_{\Sigma p}}. \tag{7.86}$$

To do this, controller (7.85) is set with $X_x = Y_x = 0$ and the resulting I controller is turned into I-lag control of the form

$$C_x(s) = \frac{1}{sT_{i_1}\,(1 + sT_x)\,(1 + sT_{\Sigma c})}. \tag{7.87}$$

T_x is a known and sufficiently large time constant[6] chosen such

$$\tau = \frac{T}{T_x + T_{\Sigma p} + T_{\Sigma c}} = \frac{T}{T_{\Sigma x}} \ll 1 \tag{7.88}$$

where $T_{\Sigma_x} = T_x + T_{\Sigma_p} + T_{\Sigma_c}$ is the equivalent sum time constant of the closed loop. Again, as mentioned in Sect. 7.4.1, it is assumed in our analysis that $T_{\Sigma_c} T_{\Sigma_p} \approx 0$, $T_{\Sigma_c} T_{\Sigma_p} T_x \approx 0$ and $T_x (T_{\Sigma_c} + T_{\Sigma_p}) \approx 0$. In that case and according to (7.77) ($X = Y = 0$), the respective closed-loop transfer function is equal to

$$T(s) = \frac{F_{ol}(s)}{1 + k_h F_{ol}(s)} = \frac{k_p C_x(s) G(s)}{1 + k_p k_h C_x(s) G(s)}$$
$$= \frac{k_p \dfrac{1}{sT_{i_1}\,(1 + sT_x)\,(1 + sT_{\Sigma c})} \dfrac{1}{(1 + 2\zeta Ts + T^2 s^2)(1 + sT_{\Sigma p})}}{1 + k_h k_p \dfrac{1}{sT_{i_1}\,(1 + sT_x)\,(1 + sT_{\Sigma c})} \dfrac{1}{(1 + 2\zeta Ts + T^2 s^2)(1 + sT_{\Sigma p})}} \tag{7.89}$$

or finally

$$T\,(s) = \frac{k_p}{T^2 T_{i_1} T_{\Sigma_x} s^4 + T_{i_1} T \left(T + 2\zeta T_{\Sigma_x}\right) s^3 + \left(T_{\Sigma_x} + 2\zeta T\right) T_{i_1} + T_{i_1} s + k_p k_h} \tag{7.90}$$

for which we have set $T_{\Sigma_x} = T_x + T_{\Sigma p} + T_{\Sigma c}$ and

$$(1 + sT_{\Sigma p})(1 + sT_{\Sigma c})(1 + sT_x) = 1 + s\left(T_x + T_{\Sigma p} + T_{\Sigma c}\right) \approx 1 + T_{\Sigma_x}. \tag{7.91}$$

Since T_x is known,[7] T_{i_1} is tuned such, so that the overshoot of the closed-loop control system becomes equal to 4.4 %. The tuning of T_{i_1} is made as follows.

Step 1: *Determination of the gain* k_p. Initially, the gain k_p is determined from the step response of the plant at steady state, Fig. 7.20. Therefore,

---

[6] T_x is a design parameter.

[7] This time constant was chosen sufficiently large, so that $\tau \to 0$.

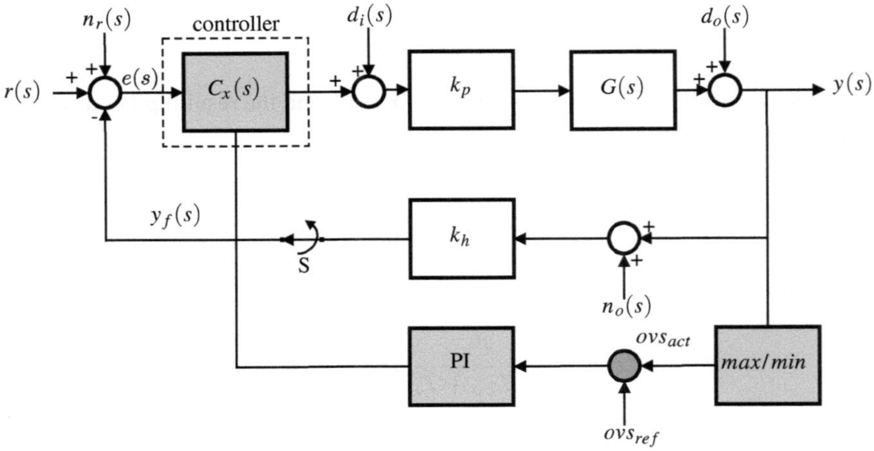

Fig. 7.18 Block diagram of the closed-loop control system and the tuning loop in the frequency domain. k_p is the plant's dc gain and k_h stands for the feedback path. C_x stands for the automatically tuned controller. $\mathrm{ovs}_\mathrm{act}$ is the measured overshoot of $y(t)$ and $\mathrm{ovs}_\mathrm{ref}$ is set equal to 4.4 %

$$\lim_{t\to\infty} y(t) = \lim_{s\to 0} sG(s)\,u(s) = k_\mathrm{p}. \tag{7.92}$$

If k_p is known from the implementation this step can be skipped.

Step 2: *Tuning of the integrator's time constant T_{i_1} and determination of the overall control loop's parasitic time constant.* The control loop of Fig. 7.1 is turned into the control loop of Fig. 7.18. Purpose of this loop is to tune initially parameter T_{i_1}. For that reason, a series of step pulses[8] of alternate sign is imposed in $r(s)$ around the closed loop's operating point, Fig. 7.19b. During this series of step pulses, the overshoot of the output $\mathrm{ovs}_\mathrm{act}$ is measured and compared with $\mathrm{ovs}_\mathrm{ref} = 4.4\,\%$. The comparison is carried out by the |max/min| comparator circuit, which detects the peak overshoot and compares it with the reference. If $\mathrm{ovs}_\mathrm{act} < \mathrm{ovs}_\mathrm{ref}$ then at $(k+1)$ step $T_{\mathrm{i}_x}(k+1) < T_{\mathrm{i}_x}(k)$. From the definition of the open-loop transfer function see (7.89)

$$F_\mathrm{ol}(s) = \frac{k_\mathrm{p}}{sT_{\mathrm{i}_1}(1+sT_x)(1+sT_{\Sigma\mathrm{c}})}\frac{1}{(1+2\zeta Ts+T^2s^2)(1+sT_{\Sigma\mathrm{p}})} \tag{7.93}$$

it is easily seen that the $\mathrm{ovs}_\mathrm{act}$ at the next step increases and the rise time decreases, if the change at the T_{i_1} is done such that $T_{\mathrm{i}_x}(k+1) < T_{\mathrm{i}_x}(k)$. The amount of this change is based on the parameters of the PI controller (gray box), the tuning of

[8] The amplitude of these pulses is small enough, so that the output of the control loop $y(t)$ does not diverge far from its operating point.

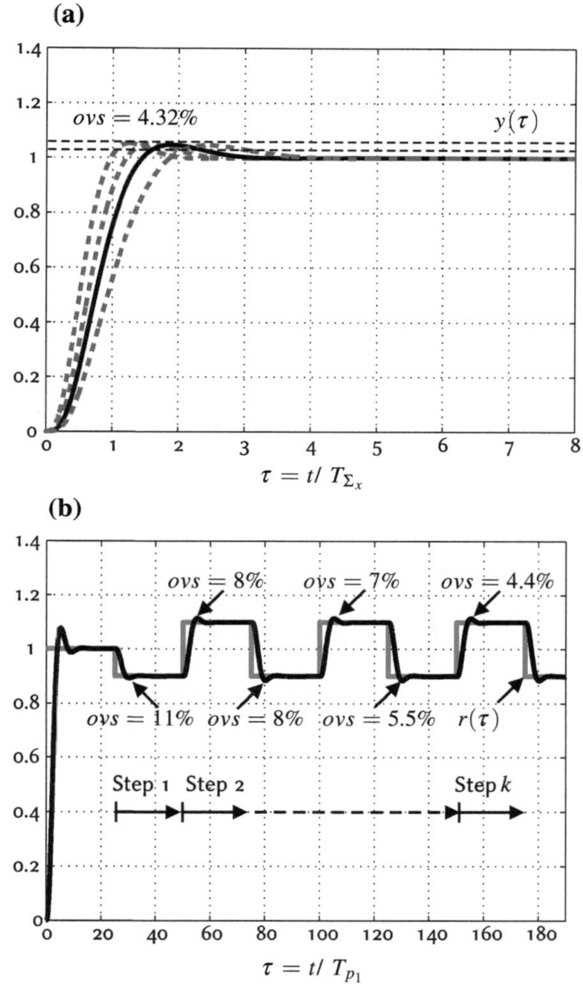

Fig. 7.19 Determination and automatic tuning of the T_{i_1} time constant during I-lag control action. **a** Tuning of the integrator's time constant T_{i_1} so that the overall parasitic time constant T_Σ of the closed loop is determined. **b** series of small step variations of the reference input with alternating sign are imposed for tuning the I-lag controller and the PID controller's parameters

which is heuristic and trivial,[9] [23]. The PI controller is fed with the error between ovs_{act}, ovs_{ref} at step k and returns the $T_{i_x}(k+1)$ for the next step.

Scope of this tuning is the determination of the overall parasitic time constant $T_\Sigma = T_{\Sigma_p} + T_{\Sigma_c}$ of the closed loop. When the overshoot of the closed loop becomes

[9] The PI controller can be avoided and a simple bang-bang control with a hysteresis band in the output overshoot reference can be introduced.

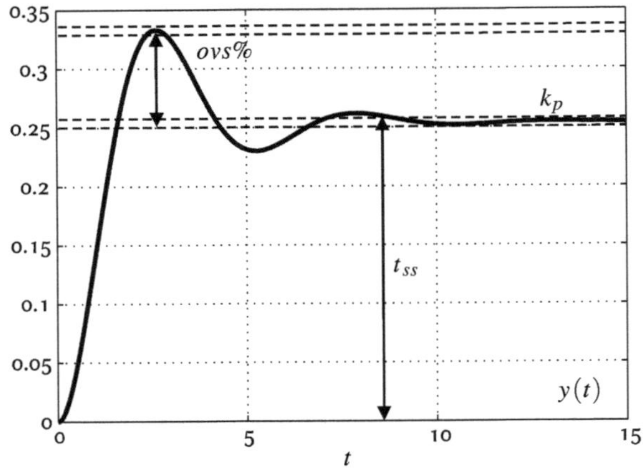

Fig. 7.20 Typical step response of the approximate second-order process with conjugate complex poles

equal to $\text{ovs}_{\text{ref}} = 4.4\,\%$, then according to (7.82), T_{i_1} is equal to

$$T_{\text{i}_1} = 2k_{\text{p}}k_{\text{h}}\left(2\zeta T + T_{\Sigma_x}\right). \tag{7.94}$$

Note that after that step T_{i_1} is known. Thus, for determining T_{Σ_x} through (7.94) a measurement of k_{p}, ζ via an open-loop experiment to the process, Fig. 7.20 is required.

From Fig. 7.20 it is apparent that

$$k_{\text{p}} = y_{\text{r}_{\text{ss}}} = y_{\text{r}}(\infty), \tag{7.95}$$

$$M = \text{e}^{-\frac{\pi\zeta}{\sqrt{1-\zeta^2}}}. \tag{7.96}$$

An accurate estimation of the overshoot Fig. 7.20 is related to the damping ratio ζ through

$$\zeta_{\text{est}} \approx \sqrt{\frac{\ell n^2 M}{\pi^2 + \ell n^2 M}}, \tag{7.97}$$

where

$$M = \frac{\max\left(y_{\text{r}}(t)\right)}{k_{\text{p}}} - 1. \tag{7.98}$$

Moreover, an accurate estimation of T_{est} can be obtained through

$$t_{ss_t} \approx \frac{4}{\zeta\omega_n} = \frac{4}{\zeta}T, \quad T_{est} \approx \frac{t_{ss_{est}}}{4}\zeta_{est}. \tag{7.99}$$

Since a reasonable estimation of ζ_{est}, T_{est} is available, it is obtained through (7.94) and (7.99) that $T_{\Sigma x} = T_\Sigma + T_x$ and $T_\Sigma + T_x = \frac{T_{i_1}}{2k_p k_h} - 2\zeta_{est}T_{est}$ or finally

$$T_\Sigma = \frac{T_{i_1}}{2k_p k_h} - T_x - 2\zeta_{est}T_{est}. \tag{7.100}$$

Note that $k_h = 1$, k_p is measured from (7.95), T_x is known and ζ_{est}, T_{est} are measured from (7.97) and (7.99) respectively. As a result, $C_x(s)$ in (7.87) is finally replaced by the PID-type controller

$$C_z(s) = \frac{1 + 2\zeta_{est}T_{est}s + T_{est}^2 s^2}{sT_{i_2}\left(1 + sT_{\Sigma c}\right)}. \tag{7.101}$$

In that case, the closed-loop transfer function is given by

$$\widetilde{T}(s) = \frac{k_p\left(1 + 2\zeta_{est}T_{est}s + T_{est}^2 s^2\right)}{sT_{i_2}\left(1 + sT_\Sigma\right)\left(1 + 2\zeta Ts + T^2 s^2\right) + k_h k_p\left(1 + 2\zeta_{est}T_{est}s + T_{est}^2 s^2\right)}. \tag{7.102}$$

If $\zeta_{est} \approx \zeta_{real}$ and $\tau_{est} \approx \tau_{real}$ and since $k_h = 1$ then

$$\widetilde{T}(s) \approx \frac{k_p}{sT_{i_2}\left(1 + sT_\Sigma\right) + k_h k_p} \approx \frac{k_p}{s^2 T_{i_2}T_\Sigma + sT_{i_2} + k_h k_p}. \tag{7.103}$$

Therefore, T_{i_2} is tuned exactly as T_{i_1} so that the overshoot of the closed-loop control system becomes equal to 4.4 %. In this case, T_{i_2} is then equal to

$$T_{i_2} \approx 2k_h k_p T_\Sigma. \tag{7.104}$$

Substituting T_{i_2} into (7.103) results in

$$\begin{aligned}\widetilde{T}(s) &\approx \frac{k_p}{s^2 T_{i_2}T_\Sigma + sT_{i_2} + k_h k_p} \\ &\approx \frac{k_p}{2k_p T_\Sigma T_\Sigma s^2 + 2k_p T_\Sigma s + k_h k_p} = \frac{1}{2T_\Sigma^2 s^2 + 2T_\Sigma s + 1}.\end{aligned} \tag{7.105}$$

To this end, $\widetilde{T}(s)$ at (7.102) is approximately equal to (7.84) while the step response of the closed loop has the shape of Fig. 7.17.

7.4.3 Simulation Examples

For verifying the proposed method, we have assumed a process of the form (7.106) is employed.

7.4.3.1 Plant with $\zeta = 0.15$

Nominal parameters of the process are $k_p = 0.0975$, $\zeta = 0.1576$, $T = 0.2785$, $T_{\Sigma p} = 0.0547$.

$$G(s) = \frac{1}{(1 + 2\zeta Ts + T^2s^2)(1 + sT_{\Sigma p})} = \frac{0.0975}{(1 + 0.0878s + 0.0216s^2)(1 + 0.054s)} \tag{7.106}$$

During the open-loop experiment, an estimation of the plant parameters is carried out for parameters k_p, ζ, T, according to (7.92), (7.97) and (7.99) respectively. In Fig. 7.21a, b the step response and frequency after the open-loop experiment of the process is presented.

Furthermore and according to the proposed method presented in Sect. 7.4.2, in Fig. 7.22a, b the tuning of the I-lag controller and the PID controller is presented. The automatic tuning of the I-lag and the PID controller led to $T_{I_1} = 0.1224$, $T_x = 0.2367$ and $T_{I_2} = 0.0746$ respectively. In both cases, the integrator's time constant T_{i_1} and T_{i_2} is tuned accordingly as described in Sect. 7.4.2. In Fig. 7.23 the step response of the control loop's output $y(t)$ and response of the command signal $u(t)$ is presented in the presence of output and input disturbance.

7.4.3.2 Plant with $\zeta = 0.55$

In this example, the transfer function of the process is defined by

$$G(s) = \frac{1}{(1 + 2\zeta Ts + T^2s^2)(1 + sT_{\Sigma p})} = \frac{2.3}{(1 + 0.5576s + 0.25s^2)(1 + 0.05s)}. \tag{7.107}$$

The step and frequency response of the process is shown in Fig. 7.24a, b respectively. In Fig. 7.25a, b the tuning of the I-lag controller (T_{i_1}) and the PID controller (T_{i_2}) is presented.

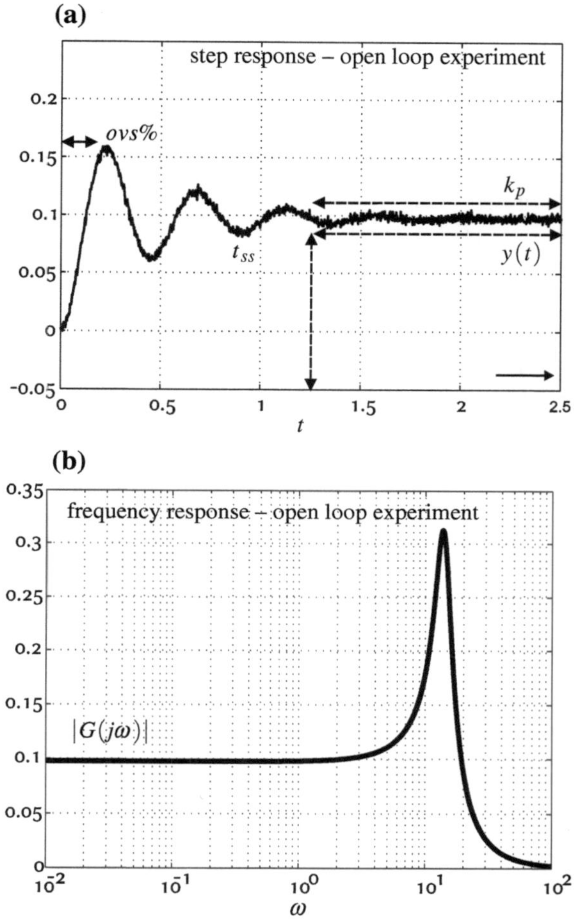

Fig. 7.21 Responses in the time and frequency domain after an open-loop experiment of the process G defined by (7.106). **a** Step response of the process $G(s)$. **b** Frequency response of the process $G(s)$

It is critical to mention that poor initialization of the I-lag controller, Fig. 7.25a can lead to a high overshoot at the output of the control loop. For that reason, initial values both when tuning T_{i_1} and T_{i_2} have to lead to at least 0 % overshoot of the control loop. In this case and according to the I-lag controller tuning, T_x is initialized with $T_x = T_{\text{est}}$ which is measured from the open-loop experiment of the process.

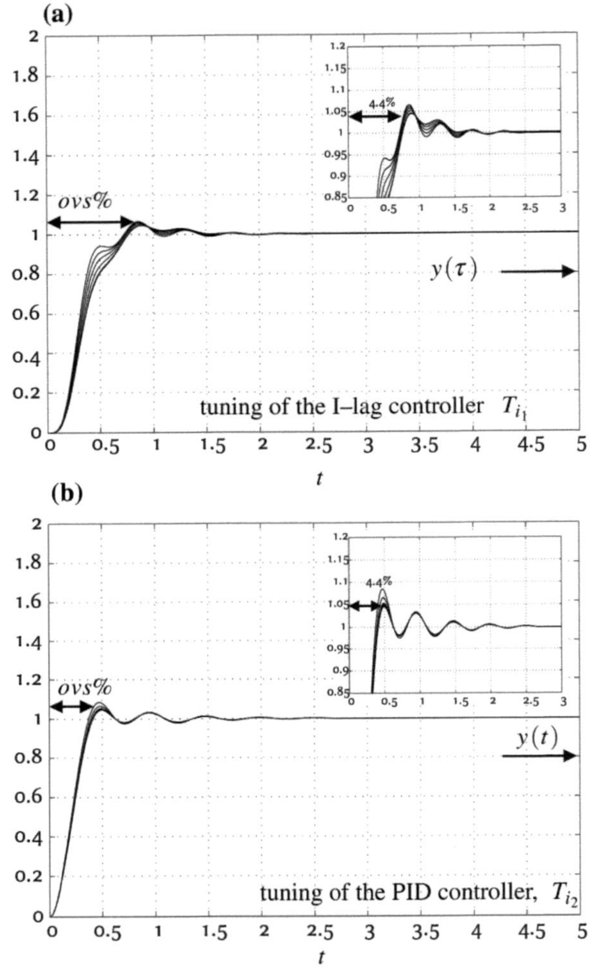

Fig. 7.22 Tuning of the I-lag controller and the PID controller. Steps of the tuning. **a** Tuning of the I-lag controller, T_{i_1} parameter. **b** Tuning of the PID controller, T_{i_2} parameter

7.5 Summary

In this chapter, an automatic tuning algorithm for the PID controller's parameters has been presented. The method requires only measurements from an open-loop experiment of the process, which serves for initializing the proposed algorithm. The method assumes access to the output of the process and not to the states as it

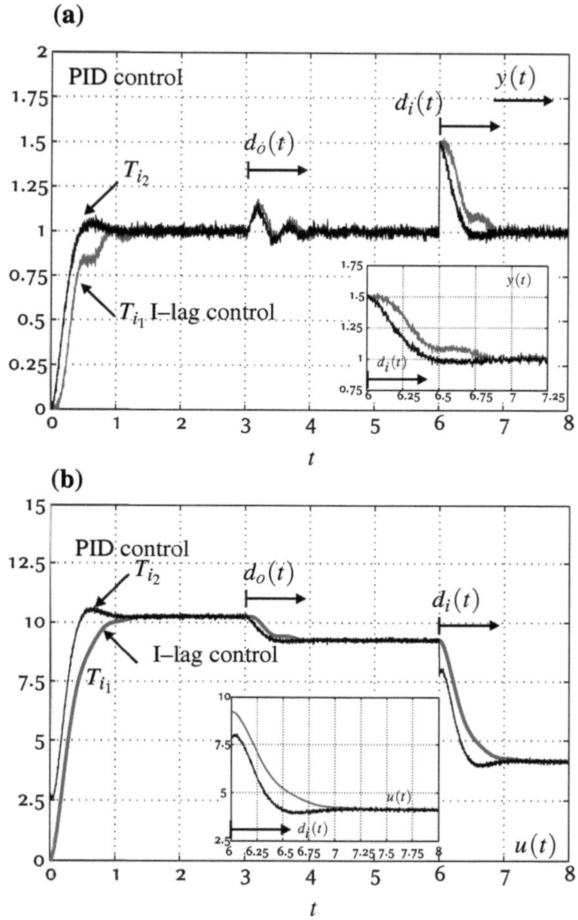

Fig. 7.23 Step response of the control loop. Output $d_o(\tau) = (\tau)$ and input $d_i(\tau) = 0.5r(\tau)$ disturbance is applied at $t = 3$ and $t = 6$ respectively, where $r(s) = \frac{1}{s}$. **a** Response of the output $y(t)$ in the presence of output $d_o(t) = r(t)$ and input $d_i(t) = 0.5r(t)$ disturbance. I-lag control and PID control. **b** Response of the command signal $u(t)$ in the presence of output $d_o(t) = r(t)$ and input $d_i(t) = 0.5r(t)$ disturbance

frequently happens in many industry applications. The method is inspired from an attractive property the direct tuning of the PID controller via the Magnitude Optimum criterion exhibits (Fig. 7.26).

This property is related to the preservation of the shape of the step and frequency response of the final closed-loop control system when the PID controller is tuned through the conventional way. Based on the aforementioned property along with the closed relation which exists between the controller parameters, it is possible to tune only one parameter while all other parameters are tuned automatically. The potential

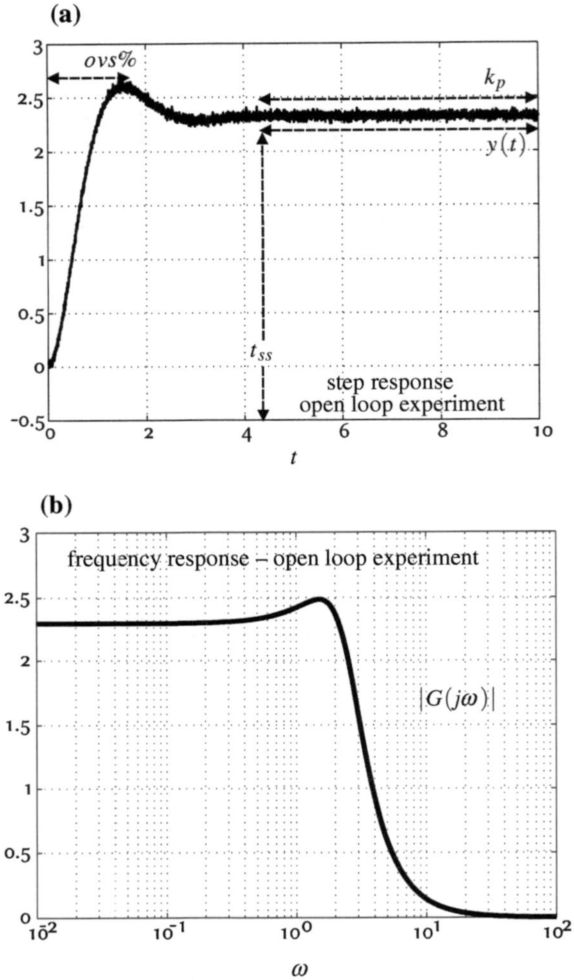

Fig. 7.24 Responses in the time and frequency domain after an open-loop experiment of the process G defined by (7.107). **a** Step response of the process $G(s)$. **b** Frequency response of the process $G(s)$

of the proposed method was evaluated via simulation examples. An extensive simulation test batch was presented in Sect. 7.3 comparing the control action resulting from the proposed method, (very little knowledge of the process) see Sect. 7.2.5, with the control action resulting from the explicit solution (exact knowledge of the process model) presented in Sect. 3.3.

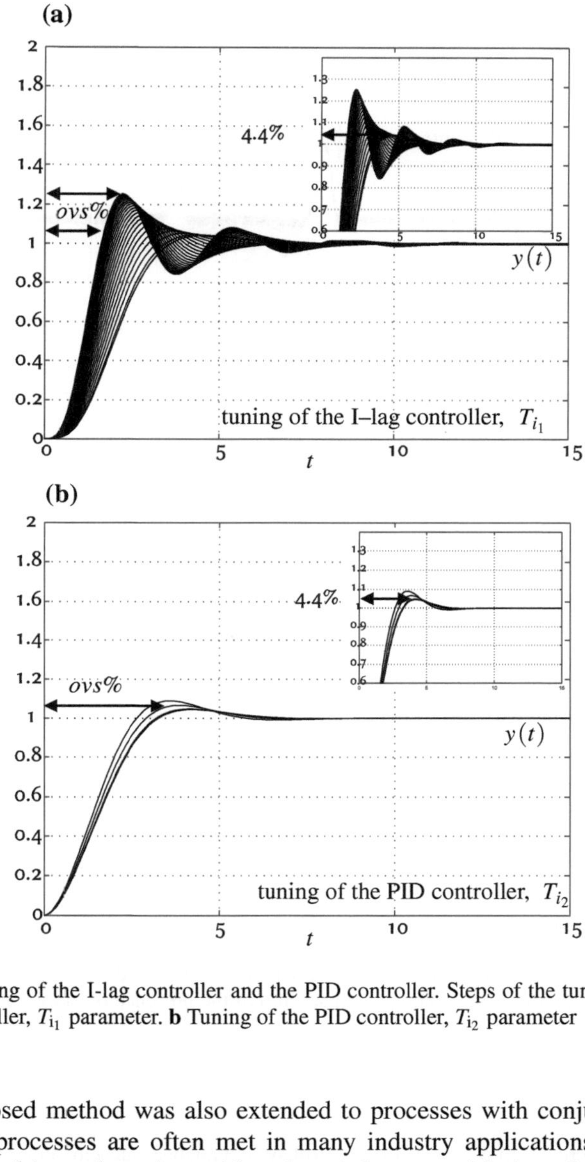

Fig. 7.25 Tuning of the I-lag controller and the PID controller. Steps of the tuning. **a** Tuning of the I-lag controller, T_{i_1} parameter. **b** Tuning of the PID controller, T_{i_2} parameter

The proposed method was also extended to processes with conjugate complex poles. Such processes are often met in many industry applications, i.e., field of electric motor drives where the problem there is known as "design of active damping regulators". The method requires an open-loop experiment of the process, so that basic information (overshoot and the time constant of the process) is measured, which serves for initializing the proposed algorithm. The method assumes access only to the output of the process and not to the states. The proposed method was tested at processes with damping ratio very close to zero achieving promising results.

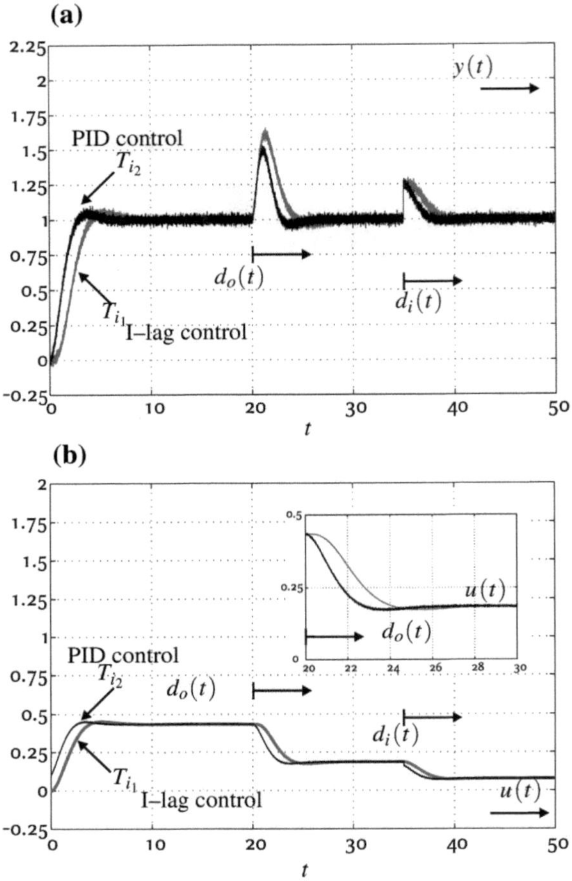

Fig. 7.26 Step response of the control loop. Output $d_o(\tau) = (\tau)$ and input $d_i(\tau) = 0.5r(\tau)$ disturbance is applied at $t = 3$ and $t = 6$ respectively, where $r(s) = \frac{1}{s}$. **a** Response of the output $y(t)$ in the presence of output $d_o(t) = r(t)$ and input $d_i(t) = 0.5r(t)$ disturbance. I-lag control and PID control. **b** Response of the command signal $u(t)$ in the presence of output $d_o(t) = r(t)$ and input $d_i(t) = 0.5r(t)$ disturbance

References

1. Ang KH, Chong G, Li Y (2005) PID control system analysis, design, and technology. IEEE Trans Control Syst Technol 13(4):559–576
2. Åströ KJ, Hägglund T (1995) PID controllers: theory, design and tuning, 2nd edn. Instrument Society of America, North Carolina
3. Åström KJ, Wittenmark B (1973) On self tuning regulators. Automatica 9(2):185–199
4. Åström KJ, Hägglund T, Hang CC, Ho WK (1993) Automatic tuning and adaptation for PID controllers—a survey. Control Eng Pract 1(4):699–714
5. Bakhshai AR, Joos G, Jain PK, Hua J (2000) Incorporating the overmodulation range in space vector pattern generators using a classification algorithm. IEEE Trans Power Electron 15(1):83–94

6. Bi Q, Cai WJ, Wang QC, Hang CC, Lee EL, Sun Y, Liu KD, Zhang Y, Zou B (2000) Advanced controller auto-tuning and its application in HVAC systems. Control Eng Pract 8(6):633–644
7. Cameron F, Seborg DE (1982) A self tuning controller with a PID structure. Int J Control 38(2):401–417
8. Chan FY, Moallem M, Wang W (2007) Design and implementation of modular FPGA-based PID controllers. IEEE Trans Ind Electron 54(4):1898–1906
9. Chen CL (1989) A simple method for on-line identification and controller tuning. AIChE J 35(12):2037–2039
10. Cox CS, Daniel PR, Lowdon A (1997) Quicktune: a reliable automatic strategy for determining PI and PPI controller parameters using FOPDT model. Control Eng Pract 5(10):1463–1472
11. Dannehl J, Fuchs FW, Hansen S, Thogersen PB (2009) Investigation of active damping approaches for PI-based current control of grid-connected pulse width modulation converters with LCL filters. IEEE Trans Ind Appl 46(4):1509–1517
12. Dannehl J, Liserre M, Fuchs FW (2011) Filter-based active damping of voltage source converters with LCL filter. IEEE Trans Ind Electron 58(8):3623–3633
13. Friman M (1997) Automatic retuning of PI controllers in oscillating control loops. Ind Eng Chem Res 36(10):4255–4263
14. Fröhr F, Orttenburger F (1982) Introduction to electronic control engineering. Siemens, Berlin
15. Ho WK, Hang CC, Zhou J (1997) Self-tuning PID control of a plant with under-damped response with specifications on gain and phase margins. IEEE Trans Control Syst Technol 5(4):446–452
16. Kerkman DRJ, Leggate Seibel BJ (1996) Operation of PWM voltage source-inverters in the overmodulation region. IEEE Trans Ind Electron 43(1):132–141
17. Lee DC, Lee GM (1998) A novel overmodulation technique for space-vector PWM inverters. IEEE Trans Power Electron 13(6):1144–1151
18. Liu F, Wu B, Zargari NR, Pande M (2011) An active damping method using inductor-current feedback control for high-power PWM current-source rectifier. IEEE Trans Power Electron 26(9):2580–2587
19. O' Dwyer A (2003) Handbook of PI and PID controller tuning rules, 1st edn. Imperial College Press, London
20. Oldenbourg RC, Sartorius H (1954) A uniform approach to the optimum adjustment of control loops. Trans ASME 76:1265–1279
21. Papadopoulos KG, Margaris NI (2012) Extending the symmetrical optimum criterion to the design of PID type-p control loops. J Process Control 12(1):11–25
22. Papadopoulos KG, Margaris NI (2013) Optimal automatic tuning of active damping PID regulators. J Process Control 23(6):905–915
23. Papadopoulos KG, Tselepis ND, Margaris NI (2012a) On the automatic tuning of PID type controllers via the magnitude optimum criterion. In: International conference on industrial technology (ICIT), IEEE, Athens, Greece, pp 869–874
24. Papadopoulos KG, Tselepis ND, Margaris NI (2012b) Revisiting the magnitude optimum criterion for robust tuning of PID type-I control loops. J Process Control 22(6):1063–1078
25. Papadopoulos KG, Papastefanaki EN, Margaris NI (2013) Explicit analytical PID tuning rules for the design of type-III control loops. IEEE Trans Ind Electron 60(10):4650–4664
26. Rahimi AR, Syberg BM, Emadi A (2009) Active damping in DC/DC power electronic converters: a novel method to overcome the problems of constant power loads. IEEE Trans Ind Electron 56(5):1428–1439
27. Saeedifard M, Bakhshai A (2007) Neuro-computing vector classification SVM schemes to integrate the overmodulation region in neutral point clamped (NPC) converters. IEEE Trans Power Electron 22(3):995–1004
28. Umland WJ, Safiuddin M (1990) Magnitude and symmetric optimum criterion for the design of linear control systems: what is it and how does it compare with the others? IEEE Trans Ind Appl 26(3):489–497

Chapter 8
Changes on the Current State of the Art

Abstract In this chapter, a summary of the book's contribution to the current state of the art is presented. The summary concentrates on the contribution of the book regarding both the direct and the automatic tuning procedure for the PID controller via the Magnitude Optimum criterion. Open issues regarding both tuning approaches are presented in Sect. 8.2.

8.1 The Magnitude Optimum Criterion—Present and Future of PID Control

The main purpose of the book is to present a general principle regarding the tuning of the PID controller based on the Magnitude Optimum criterion. Basic requirement of this principle is to design the controller, such that the magnitude of the closed loop transfer function $|T(j\omega)|$ is equal to the unity in the widest possible frequency range. Since scope of this book is to present a general theory for any process model met within the industry sector (i.e., chemical, electrical engineering), a general transfer function $T(s)$ is adopted in the frequency domain for modeling the closed loop control system.

For presenting the proposed theory and for forcing the magnitude of $|T(j\omega)|$ to be equal to the unity in the widest possible frequency range, certain optimization conditions are presented in Sect. 2.7, A.1. These conditions comprise the basis for the development of the proposed theory. These conditions are used for the design of analog and digital PID control action and for all types of control loops presented within this book, type-I, type-II, type-III ... type-p.

As already mentioned in Chap. 2, the big advantage the principle of the Magnitude Optimum criterion offers, is related to the design of higher order type control loops. Let it be reminded that the higher the type of the control loop is, the faster reference signals the output variable $y(\tau)$ can track. This ability is considered fundamental within the control systems theory, since such kind of loops can track fast reference signals (i.e., ramps, parabolic inputs...) achieving zero steady state position, velocity, acceleration ... error.

Up to now, the Magnitude Optimum criterion was used for the design of type-I control loops. In similar fashion, the Symmetrical Optimum criterion was used for

K.G. Papadopoulos, *PID Controller Tuning Using the Magnitude Optimum Criterion*,
DOI 10.1007/978-3-319-07263-0_8

the design of type-II control loops. For tuning the PID controller, the line of pole-zero cancellation is followed in both tuning methods. This line is considered in this book as the "conventional" way of tuning, which finally proves to lead at suboptimal or even sometimes unstable control loops.

To cope with this issues, and regarding the design of type-I control loops, the so called "revised theory" is proposed. The revised theory does not consider pole-zero cancellation between the plant's poles and the controller's zeros. On the contrary, it determines analytically the PID controller's parameters as a function of all plant's time constants (poles, zeros, delay). In other words, the proposed revised method tunes the controller's gains with all the available information coming from the plant. This was not the case regarding the conventional tuning.

The same line is also followed for the design of type-II control loops where the tuning of the PID controller via the Symmetrical Optimum criterion follows the line of pole-zero cancellation between the process's dominant time constant and the controller's zero, see Sect. 4.2.3. In this case, the conventional tuning proves to lead to unstable closed loops, especially in cases where the plant contains dominant time constants, right half plane zeros or long time delays.

Summarizing the aforementioned state of the art, one can argue that the conventional tuning

- requires that the plant's poles are canceled by the controller's zeros,
- restricts the PID controller's zeros to be tuned with real values because of the pole-zero cancellation method,
- has been tested to simple process models,
- leads to unstable control loops or control loops with unacceptable performance when the complexity of the process is increased,
- no tuning rules or guidelines are presented relevant to the choice of the sampling time T_s, in cases where the controller is implemented digitally.

In contrast to these open points, the revised theory comes to fill this gap by optimally tuning the controller since it

- does not require pole-zero cancellation between the process's poles and the controller's zeros. Therefore, it determines the PID controller's parameters as a function of all time constants coming from the process. This explicit solution is defined by closed form expressions, all of which are proved in the appendix.
- allows the controller's zeros to be tuned with conjugate complex values if needed.
- outperforms the conventional tuning regarding the closed loop's system response, both in time and frequency domain.
- is able to handle the design of higher order control loops (type-II, type-III, type-IV, type-V) and even when the complexity of the process is increased.
- introduces the sampling time T_s of the controller within the closed form expressions, which determine the controller's gains. This allows for accurate investigation of the effect of the sampling time on the control loop's performance both in the time and frequency domain.

In particular, in Part II of the book, the explicit PID tuning solution is proposed for type-I (see Chap. 3), type-II (see Chap. 4), type-III (see Chap. 5) control systems. In Chap. 3, the current state of the art and the so called "conventional" PID tuning procedure is presented in Sect. 3.2. In Sect. 3.2.5 all drawbacks of the conventional tuning are summarized. The proposed revised tuning that follows is Sect. 3.3, the proof of which is presented in Appendix B.1, is compared with the conventional method in a series of simulation examples in Sect. 3.4. There it is shown that when the controlled plant consists of one or two dominant time constants, then both methods lead to the same performance. In any other case, the revised method outperforms the conventional way of tuning both in the time and frequency domain.

Specifically, given a certain plant, the settling time of output disturbance rejection in the control loop can be reduced up to 45 %, when this plant is controlled via the revised method compared to the settling time coming from the conventional control action. Moreover, the revised control loop is less sensitive to input and output disturbances, since the range for which $|T(j\omega)| \approx 1$ is greater compared to the range coming from the conventional tuning.

Furthermore, the proposed revised control action is able to control plants with large zeros. The crystal clear definition of the integrator's time constant, see (4.42) allows the control engineer to understand and decide when the D term has to be added or omitted and when the PID controller has to be turned to PID-lag, in order to cope with the existence of large zeros within the plant's transfer function.

The same results are also observed regarding the control of integrating processes which are discussed in Chap. 4. In Sect. 4.2, the conventional PID tuning method via the Symmetrical Optimum criterion is presented. There it is shown that this kind of tuning fails to tune a stable PI control action (see Sect. 4.2.2) and tunes only a PID controller which is based on pole-zero cancellation, see Sect. 4.2.3. This restriction, proves to be suboptimal in Sect. 4.4.1, and especially at cases of certain processes, the control loop proves to be even unstable. In contrast with the conventional tuning, the revised method is again proposed in Sect. 4.3 as in similar fashion with type-I control loops. The proof of the control law is presented in Sect. B.2. Once more, according to the revised method, all three PID parameters are determined in closed form expressions as a function of all time constants coming from the process. To this end, the proposed control law's proof does not involve any model reduction techniques, see Appendix B.2.

In Sect. 4.4, the conventional tuning is again compared with the revised method. This comparison focuses on the performance of the required control action, in terms of reference tracking and disturbance rejection. It is interesting to mention that the conventional tuning leads to unstable control loops in case where the complexity of the process is increased, see examples in Sects. 4.4.3–4.4.5.

The introduction of the design of type-III control loops is presented in Chap. 5 for first time within the literature, see also [2, 3]. Given the principle from pole-zero cancellation coming from the Magnitude and Symmetrical Optimum criteria, a similar methodology is presented for the design of type-III control loops, which is finally extended to the design of type-p control loops.

The proposed pole-zero cancellation method is again revised by an optimal PID control action for type-III control loops, the proof of which is presented in Appendix C.3. A performance comparison analysis in terms of simulation examples is presented in Sect. 5.2.3.

Back to type-I control loops, one interesting and attractive feature which is revealed by the conventional PID tuning procedure, is the so called preservation of the shape of the step and frequency response, thoroughly discussed in Sects. 3.2.2 and 7.2. From the conventional tuning, it is shown that when the control loop (for processes with one or two dominant time constants) is designed via the conventional way, the step and frequency response exhibit a certain performance. Moreover, it can be easily seen that from the conventional control law, the integrator's time constant can be expressed as a function of the zeros of the PID controller. To this end, by tuning the zeros of the PID type controller, the integrator's time constant is tuned automatically. The way how the zeros of the controller are tuned is driven by the aforementioned performance already observed in the direct tuning, see Sects. 7.2.5 and 7.4.

Taking into account these two features, a methodology for the automatic tuning of the PID regulator is presented in Part III. There, and given little information about the process (open loop experiment) and having access only to its output and not to its states, an automatic tuning algorithm is presented for type-I control loops. The proposed method is also extended for the control of processes with conjugate complex poles. The performance of the proposed automatic tuning algorithm is also compared with the explicit solution in Sect. 7.3.

For the aforementioned design presented in Part II (analog design of type-I, type-II, type-III control loops), the proposed theory covers the design of the same control loops but also when the PID controller is implemented digitally. To do this, the sampling time T_s of the PID controller is introduced in the analysis, and the optimal control action is proved in Appendix C. Again the optimal control law for digital PID controllers, consists of closed form expressions which at this case involve apart from the time constants of the process, the sampling time T_s of the controller.

The introduction of the sampling time to the PID control law gives the benefit to control engineers to investigate the effect of the choice of the sampling time to the control loop's performance. Such an investigation is presented in Sects. 6.1.1, 6.2.1, 6.3.1 and 6.3.2 where useful results are obtained. A basic result which comes out of this investigation is the fact that the sampling time of the controller cannot be chosen small enough compared to the plant's dominant time constant. A small enough sampling time proved to reduce the bandwidth of the closed loop transfer function for which $|T(j\omega)| \approx 1$, something which is in contradiction with the Magnitude Optimum criterion. For that reason, the choice of the sampling time has to be such so that all requirements the Magnitude Optimum criterion introduces are satisfied in the time and frequency domain.

8.2 Open Issues and Future Work

As it has been clearly stated in Chap. 1, the proposed theory is dedicated to single-input single-output systems. For that reason, one open issue that has to be fulfilled is to follow the same approach for the design of multiple-input multiple-output control systems by incorporating the theory for multivariable systems, see [4].

The proposed explicit solution for type-I, type-II, type-III control loops requires the involvement of a system identification method for the modelling of the process to be controlled. The big advantage of the proposed automatic tuning method is the fact that such an identification method is not required for tuning the PID controller. In Sect. 7.3, it was shown that by tuning the PID controller based on a certain overshoot in the output of the control loop (i.e., 4.4 %) leads often to suboptimal performance, especially in cases where the process involves more than two dominant time constants, long time delay or right half plane zeros. For that reason, the automatic tuning method has to be improved and one idea to do this is to estimate the optimal overshoot based on which the tuning of the PID parameters takes place. The tools to achieve this goal are available, see [1, 5–7], and the whole problem is under investigation.

Finally, the challenging target for defining explicit closed form expressions for the PID controller and higher-order type control loops (type-IV, type-V) is always the case, since such kind of loops are able to track fast reference signals.

References

1. Jang JSR (1993) ANFIS: adaptive-network-based fuzzy inference systems. IEEE Trans Syst Man Cybern B, Cybern 23(3):665–685
2. Papadopoulos KG, Margaris NI (2012) Extending the symmetrical optimum criterion to the design of PID type-p control loops. J Process Control 12(1):11–25
3. Papadopoulos KG, Papastefanaki EN, Margaris NI (2013) Explicit analytical PID tuning rules for the design of type-III control loops. IEEE Trans Ind Electron 60(10):4650–4664
4. Skogestad S, Postlethwaite I (2005) Multivariable feedback control: analysis and design. Wiley, New York
5. Takagi T, Sugeno M (1985) Fuzzy identification of systems and its applications to modeling and control. IEEE Trans Syst Man Cybern 15(1):1–13
6. Zadeh LA (1965) Fuzzy sets. Inf Control 8(3):338–353
7. Zadeh LA (1973) Outline of a new approach to the analysis of complex systems and decision processes. IEEE Trans Syst Man Cybern 3(1):28–44

Appendix A
The Magnitude Optimum Criterion

Abstract In this chapter, an optimization process is presented for forcing the magnitude $|T(j\omega)| \approx 1$. For achieving this goal, a general transfer function $T(s)$ is employed. At the end of the chapter, certain optimization conditions are presented which serve for proving each time the proposed optimal control law for type-I, type-II, type-III control loops presented in Sects. B.1–B.3.

A.1 Optimization Conditions

In principle, let the closed loop transfer function be defined by

$$T(s) = \frac{s^m b_m + s^{m-1} b_{m-1} + \cdots + s^2 b_2 + s b_1 + b_0}{s^n a_n + s^{n-1} a_{n-1} + \cdots + s^2 a_2 + s a_1 + a_0} = \frac{N(s)}{D(s)} \tag{A.1}$$

where $m \leq n$. The Magnitude Optimum criterion requires to force $|T(j\omega)| \simeq 1$ in the wider possible frequency range starting from the lower frequency region. Thus, by setting $s = j\omega$ into (A.1) and squaring $|T(j\omega)|$, results in

$$|T(j\omega)|^2 = \frac{|N(j\omega)|^2}{|D(j\omega)|^2} \tag{A.2}$$

or

$$T(j\omega) = \frac{N(j\omega)}{D(j\omega)} = \frac{(j\omega)^m b_m + \cdots + (j\omega)^2 b_2 + (j\omega) b_1 + b_0}{(j\omega)^n a_n + \cdots + (j\omega)^2 a_2 + (j\omega) a_1 + a_0}. \tag{A.3}$$

Separating the real from the imaginary part in (A.3), polynomials $N(j\omega)$ and $D(j\omega)$ are rewritten as follows:

$$\begin{aligned} N(j\omega) \simeq & \cdots + b_8\omega^8 - b_6\omega^6 + b_4\omega^4 - b_2\omega^2 + b_0 \\ & + j\left(\cdots - b_7\omega^7 + b_5\omega^5 - b_3\omega^3 + b_1\omega\right) \end{aligned} \tag{A.4}$$

K.G. Papadopoulos, *PID Controller Tuning Using the Magnitude Optimum Criterion*,
DOI 10.1007/978-3-319-07263-0

and

$$D(j\omega) \simeq \cdots + a_8\omega^8 - a_6\omega^6 + a_4\omega^4 - a_2\omega^2 + a_0 \\ + j\left(\cdots - a_7\omega^7 + a_5\omega^5 - a_3\omega^3 + a_1\omega\right) \tag{A.5}$$

or

$$|D(j\omega)|^2 \simeq \cdots + a_8^2\omega^{16} + (a_7^2 - a_8a_6)\omega^{14} + (a_6^2 + 2a_4a_8 - 2a_5a_7)\omega^{12} \\ + (a_5^2 + 2a_3a_7 - 2a_2a_8 - 2a_4a_6)\omega^{10} \\ + (a_4^2 + 2a_0a_8 + 2a_2a_6 - 2a_1a_7 - 2a_3a_5)\omega^8 \\ + (a_3^2 + 2a_1a_5 - 2a_6a_0 - 2a_2a_4)\omega^6 + (a_2^2 + 2a_0a_4 - 2a_1a_3)\omega^4 \\ + (a_1^2 - 2a_0a_2)\omega^2 + a_0\omega^0 \tag{A.6}$$

and

$$|N(j\omega)|^2 \simeq \cdots + b_8^2\omega^{16} + (b_7^2 - b_8b_6)\omega^{14} + (b_6^2 + 2b_4b_8 - 2b_5b_7)\omega^{12} \\ + (b_5^2 + 2b_3b_7 - 2b_2b_8 - 2b_4b_6)\omega^{10} \\ + (b_4^2 + 2b_0b_8 + 2b_2b_6 - 2b_1b_7 - 2b_3b_5)\omega^8 \\ + (b_3^2 + 2b_1b_5 - 2b_6b_0 - 2b_2b_4)\omega^6 \\ + (b_2^2 + 2b_0b_4 - 2b_1b_3)\omega^4 \\ + (b_1^2 - 2b_0b_2)\omega^2 + b_0\omega^0 \tag{A.7}$$

Finally $|T(j\omega)|^2$ is equal to

$$|T(j\omega)|^2 = \frac{|N(j\omega)|^2}{|D(j\omega)|^2} = \frac{\cdots + B_4\omega^8 + B_3\omega^6 + B_2\omega^4 + B_1\omega^2 + B_0}{\cdots + A_4\omega^8 + A_3\omega^6 + A_2\omega^4 + A_1\omega^2 + A_0} \tag{A.8}$$

where $A_0 = a_0$, $A_1 = a_1^2 - 2a_2a_0$, $A_2 = a_2^2 - 2a_3a_1 + 2a_4a_0$ … and $B_0 = b_0$, $B_1 = b_1^2 - 2b_2b_0$, $B_2 = b_2^2 - 2b_3b_1 + 2b_4b_0$…. By making equal the terms of ω^j $(j = 1, 2, \ldots, n)$ in polynomials $|D(j\omega)|^2$, $|N(j\omega)|^2$ $(A_j = B_j,\ j = 0, 1, 2, \ldots)$ it is easily proved that

$$a_0 = b_0 \tag{A.9}$$

$$a_1^2 - 2a_2a_0 = b_1^2 - 2b_2b_0 \tag{A.10}$$

$$a_2^2 - 2a_3a_1 + 2a_4a_0 = b_2^2 - 2b_3b_1 + 2b_4b_0 \tag{A.11}$$

$$a_3^2 + 2a_1a_5 - 2a_6a_0 - 2a_4a_2 = b_3^2 + 2b_1b_5 - 2b_6b_0 - 2b_4b_2 \tag{A.12}$$

$$\left(a_4^2 + 2a_0a_8 + 2a_6a_2 - 2a_1a_7 - 2a_3a_5\right) = \left(b_4^2 + 2b_0b_8 + 2b_6b_2 - 2b_1b_7 - 2b_3b_5\right) \tag{A.13}$$

$$\cdots = \cdots$$

Equations (A.9)–(A.13) are the basis for proving the optimal control law for type-I, type-II, type-III control loops, which is presented in Appendix.

Appendix B
Analog Design-Proof of the Optimal Control Law

Abstract In this chapter, the proof of the optimal PID control law for type-I, type-II, type-III control loops is presented. Basis of the design of the control law are the optimization conditions (A.9)–(A.13) of the magnitude $|T(j\omega)|$ of the closed loop transfer function presented in Sect. A.1.

B.1 Type-I Control Loops

For deriving the revised PID type control law, a general transfer function of the process model consisting of $(n-1)$ poles, m zeros plus a time delay constant T_{d} is adopted, see (B.1). Zeros of the plant may lie both in the left or right imaginary half plane. The plant transfer function may also contain second-order oscillatory terms in the denominator, described by polynomials of the form $1+2\zeta Ts+s^2T^2$, where $\zeta \in (0,1], \in \Re$ and $T>0, \in \Re$. Hence, the plant transfer function can be described in general by

$$G(s)=\frac{s^m\beta_m+s^{m-1}\beta_{m-1}+\cdots+s^2\beta_2+s\beta_1+1}{s^{n-1}\alpha_{n-1}+\cdots+s^3\alpha_3+s^2\alpha_2+s\alpha_1+1}\mathrm{e}^{-sT_{\mathrm{d}}} \tag{B.1}$$

where $n-1>m$. The proposed PID-type controller is given by the flexible form

$$C(s)=\frac{1+sX+s^2Y}{sT_{\mathrm{i}}(1+sT_{\mathrm{p_n}})} \tag{B.2}$$

allowing its zeros to become conjugate complex. Time constant $T_{\mathrm{p_n}}$ stands for the unmodelled controller dynamics coming from the controller's implementation.

According to Fig. 3.1, the closed loop transfer function $T(s)$ is given by

$$T(s)=\frac{k_{\mathrm{p}}C(s)G_{\mathrm{p}}(s)}{1+k_{\mathrm{h}}k_{\mathrm{p}}C(s)G_{\mathrm{p}}(s)}=\frac{N(s)}{D(s)}=\frac{N(s)}{D_1(s)+k_{\mathrm{h}}N(s)}. \tag{B.3}$$

K.G. Papadopoulos, *PID Controller Tuning Using the Magnitude Optimum Criterion*, DOI 10.1007/978-3-319-07263-0

Polynomials $N(s),\ D_1(s)$ are equal to

$$N(s) = k_\mathrm{p}(1 + sX + s^2Y)\sum_{i=0}^{m}(s^i\beta_i), \tag{B.4}$$

$$D_1(s) = sT_\mathrm{i}\mathrm{e}^{sT_\mathrm{d}}\sum_{j=0}^{n}(s^j\alpha_j) \tag{B.5}$$

where $\alpha_0 = \beta_0 = 1$ according to (B.1). Normalizing $N(s)$, $D_1(s)$ by making the substitution $s' = sc_1$ results in

$$N(s') = k_\mathrm{p}(1 + s'x + s'^2y)\sum_{i=0}^{m}(s'^iz_i) \tag{B.6}$$

$$D_1(s') = s't_\mathrm{i}\mathrm{e}^{s'd}\sum_{j=0}^{n}(s'^jr_j) \tag{B.7}$$

respectively. The corresponding normalized terms involved in the control loop are given by

$$x = \frac{X}{c_1}, \quad y = \frac{Y}{c_1^2}, \quad t_\mathrm{i} = \frac{T_\mathrm{i}}{c_1}, \quad d = \frac{T_\mathrm{d}}{c_1},$$

$$r_i = \frac{\alpha_i}{c_1^i}, \quad \forall i = 1, 2, \ldots, n, \quad z_j = \frac{\beta_j}{c_1^j}, \quad \forall j = 1, 2, \ldots, m.$$

The normalized time delay constant d is substituted with the "all pole" series approximation

$$\begin{aligned}\mathrm{e}^{s'd} &= \sum_{k=0}^{n}\frac{1}{k!}s'^kd^k = 1 + s'd + \frac{1}{2!}s'^2d^2 + \frac{1}{3!}s'^3d^3 \\ &+ \frac{1}{4!}s'^4d^4 + \frac{1}{5!}s'^5d^5 + \cdots\end{aligned} \tag{B.8}$$

By substituting (B.2) into (B.7), $D_1(s')$ becomes

$$D_1\left(s'\right) = \sum_{i=1}^{k}(t_\mathrm{i}s'^iq_{(i-1)}), \quad q_0 = 1, \tag{B.9}$$

where

$$q_k = \sum_{i=0}^{k}r_{(k-i)}\left(\frac{1}{i!}d^i\right), \quad k = 0, 1, 2, \ldots n, \quad r_0 = 1 \tag{B.10}$$

or

$$q_0 = 1 \tag{B.11}$$

$$q_1 = r_1 + d \tag{B.12}$$

$$q_2 = r_2 + r_1 d + \frac{1}{2!}d^2 \tag{B.13}$$

$$q_3 = r_3 + r_2 d + \frac{1}{2!}d^2 r_1 + \frac{1}{3!}d^3 \tag{B.14}$$

$$q_4 = r_4 + r_3 d + \frac{1}{2!}d^2 r_2 + \frac{1}{3!}d^3 r_1 + \frac{1}{4!}d^4 \tag{B.15}$$

$$q_5 = r_5 + r_4 d + \frac{1}{2!}d^2 r_3 + \frac{1}{3!}d^3 r_2 + \frac{1}{4!}d^4 r_1 + \frac{1}{5!}d^5 \tag{B.16}$$

Polynomials $N(s')$, $D(s') = N(s') + k_{\text{h}} D_1(s')$ are then finally defined by

$$N(s') = \sum_{i=0}^{n} \left[s'^i k_{\text{p}} (z_{(i)} + z_{(i-1)} x + z_{(i-2)} y) \right], \tag{B.17}$$

or in an expanded form by

$$\begin{aligned} N(s') = \cdots &+ s'^6 \underbrace{k_{\text{p}} (y z_4 + x z_5)}_{b_6} + s'^5 \underbrace{k_{\text{p}} (y z_3 + x z_4 + z_5)}_{b_5} \\ &+ s'^4 \underbrace{k_{\text{p}} (y z_2 + x z_3 + z_4)}_{b_4} + s'^3 \underbrace{k_{\text{p}} (y z_1 + x z_2 + z_3)}_{b_3}. \\ &+ s'^2 \underbrace{k_{\text{p}} (y + x z_1 + z_2)}_{b_2} + s' \underbrace{k_{\text{p}} (x + z_1)}_{b_1} + \underbrace{k_{\text{p}}}_{b_0} \end{aligned} \tag{B.18}$$

In similar fashion polynomial $D(s')$ is defined by

$$D(s') = \sum_{j=0}^{k} s'^j \left[t_{\text{i}} q_{(j-1)} + \left(k_{\text{p}} k_{\text{h}} \left(z_{(j)} + z_{(j-1)} x + z_{(j-2)} y \right) \right) \right] \tag{B.19}$$

or in an expanded form by

$$\begin{aligned} D(s') &= D_1(s') + k_{\text{h}} N(s') \\ &= \cdots + s'^4 \underbrace{\left[t_{\text{i}} q_3 + k_{\text{h}} k_{\text{p}} (z_2 y + z_3 x + z_4) \right]}_{a_4} \\ &\quad + s'^3 \underbrace{\left[t_{\text{i}} q_2 + k_{\text{h}} k_{\text{p}} (z_1 y + z_2 x + z_3) \right]}_{a_3} \\ &\quad + s'^2 \underbrace{\left[t_{\text{i}} q_1 + k_{\text{h}} k_{\text{p}} (y + z_1 x + z_2) \right]}_{a_2} + s' \underbrace{\left[t_{\text{i}} + k_{\text{h}} k_{\text{p}} (x + z_1) \right]}_{a_1} + \underbrace{k_{\text{p}} k_{\text{h}}}_{a_0}, \end{aligned} \tag{B.20}$$

where $z_{(-2)} = z_{(-1)} = 0, z_0 = 1$. Substituting polynomials (B.18) and (B.20) within the closed loop transfer function, it is easily shown that $T(s')$ is given by

$$T(s') = \frac{N(s')}{D(s')} = \frac{\sum_{i=0}^{n}\left[s'^{i} k_\mathrm{p}\left(z_{(i)} + z_{(i-1)}x + z_{(i-2)}y\right)\right]}{\sum_{j=0}^{k} s'^{j}\left[t_\mathrm{i} q_{(j-1)} + \left(k_\mathrm{p}k_\mathrm{h}\left(z_{(j)} + z_{(j-1)}x + z_{(j-2)}y\right)\right)\right]}. \tag{B.21}$$

Optimization Condition: $a_0 = b_0$.

From the application of (A.9)–(B.21) it is obtained

$$k_\mathrm{h} = 1. \tag{B.22}$$

Condition (B.22) renders the zero order terms of the numerator and denominator polynomial of the closed loop transfer function equal, which means that the closed loop system has zero steady state position error (type-I control loops). Note that if $k_\mathrm{h} = 1$ then $N(s') = \cdots + k_\mathrm{p}$ and $D(s') = \cdots + k_\mathrm{p}k_\mathrm{h}$.

Optimization Condition: $a_1^2 - 2a_2a_0 = b_1^2 - 2b_2b_0$.

The application of (A.10) to (B.21) results in

$$t_\mathrm{i} = 2k_\mathrm{p}k_\mathrm{h}(q_1 - z_1 - x) \tag{B.23a}$$

$$= 2k_\mathrm{p}k_\mathrm{h}\left(\sum_{i=0}^{1}\left(r_{(1-i)}\frac{1}{i!}d^{i}\right) - z_1 - x\right) \tag{B.23b}$$

or

$$t_\mathrm{i} = 2k_\mathrm{p}k_\mathrm{h}\,(r_1 + d - z_1 - x) \tag{B.24a}$$

$$= 2k_\mathrm{p}k_\mathrm{h}\left(\frac{\alpha_1}{c_1^1} + \frac{T_\mathrm{d}}{c_1} - \frac{b_1}{c_1^1} - \frac{X}{c_1}\right) \tag{B.24b}$$

or $c_1t_\mathrm{i} = 2k_\mathrm{p}k_\mathrm{h}(\alpha_1 + T_\mathrm{d} - b_1 - X)$. If the process consists of stable real poles then $\alpha_1 = \sum_{i=1}^{n} T_{p_i}$. Accordingly, the sum of the plant's zeros is given by $b_1 = \sum_{i=1}^{m} T_{z_i}$. Finally, the integral gain is defined by

$$c_1t_\mathrm{i} = 2k_\mathrm{p}k_\mathrm{h}\left(\sum_{i=1}^{n} T_{p_i} + T_\mathrm{d} - \sum_{i=1}^{m} T_{z_i} - X\right), \tag{B.25}$$

or

$$T_{\mathrm{i}} = 2k_{\mathrm{p}}k_{\mathrm{h}}\left(\underbrace{\sum_{i=1}^{n} T_{p_i} + T_{\mathrm{d}}}-\underbrace{\sum_{i=1}^{m} T_{z_i} - X}\right). \tag{B.26}$$

It is critical to point out that in comparison to the conventional definition of T_{i}, the new definition of the integral gain contains all the dynamics involved in the closed loop.

Optimization Condition: $a_2^2 - 2a_3a_1 + 2a_4a_0 = b_2^2 - 2b_3b_1 + 2b_4b_0$.

The application of (A.10) to (B.21) results in

$$x - a_{12}y = b_{11} \tag{B.27}$$

where

$$a_{12} = \frac{q_1 - z_1}{(q_1 - z_1)q_1 - (q_2 - z_2)}, \tag{B.28}$$

$$b_{11} = \frac{(q_1^2 - 2q_2)(q_1 - z_1) + q_1z_2 - q_2z_1 + q_3 - z_3}{(q_1 - z_1)q_1 - (q_2 - z_2)}. \tag{B.29}$$

Note that a_{12}, b_1 depend explicitly on process parameters.

Optimization Condition: $a_3^2 + 2a_1a_5 - 2a_6a_0 - 2a_4a_2 = b_3^2 + 2b_1b_5 - 2b_6b_0 - 2b_4b_2$.

In similar fashion, the application of (A.11) to (B.21) results in

$$x + a_{22}y = b_{22} \tag{B.30}$$

where

$$a_{22} = \frac{q_1z_2 - q_2z_1 + q_3 - z_3}{q_2^2 - 2q_1q_3 - q_2z_2 + q_1z_3 + q_3z_1 + q_4 - z_4} \tag{B.31}$$

and

$$b_{22} = \frac{Q_0Q_1 + Q_2}{Q_3} \tag{B.32}$$

where

$$Q_0 = q_2^2 - 2q_1q_3 + 2q_4 \tag{B.33}$$

$$Q_1 = q_1 - z_1 \tag{B.34}$$

$$Q_2 = q_2z_3 - q_3z_2 - q_1z_4 + q_4z_1 - q_5 + z_5 \tag{B.35}$$

$$Q_3 = q_2^2 - 2q_1q_3 - q_2z_2 + q_1z_3 + q_3z_1 + q_4 - z_4. \tag{B.36}$$

In compact form, the final optimal control law is defined by

$$\begin{bmatrix} 1 & 2k_{\mathrm{p}}k_{\mathrm{h}} & 0 \\ 0 & 1 & -a_{12} \\ 0 & 1 & a_{22} \\ 0 & 0 & 1 \end{bmatrix} \begin{bmatrix} t_{\mathrm{i}} \\ x \\ y \\ k_{\mathrm{h}} \end{bmatrix} = \begin{bmatrix} 2k_{\mathrm{p}}k_{\mathrm{h}}\,(q_1 - z_1) \\ b_1 \\ b_2 \\ 1 \end{bmatrix} \tag{B.37}$$

or finally by

$$\begin{bmatrix} t_{\mathrm{i}} \\ x \\ y \\ k_{\mathrm{h}} \end{bmatrix} = \begin{bmatrix} 1 & 2k_{\mathrm{p}}k_{\mathrm{h}} & 0 \\ 0 & 1 & -a_{12} \\ 0 & 1 & a_{22} \\ 0 & 0 & 1 \end{bmatrix}^{-1} \begin{bmatrix} 2k_{\mathrm{p}}k_{\mathrm{h}}\,(q_1 - z_1) \\ b_1 \\ b_2 \\ 1 \end{bmatrix}. \tag{B.38}$$

B.2 Type-II Control Loops

Since the closed loop control system is to be of type-II the number of pure integrators involved within the open loop transfer function $F_{\mathrm{ol}}(s)$ must be equal to 2, according to 2.6. For that reason, if the process is defined by (B.1), one more pure integrator has to be added in the proposed control law. If the process exhibits integrating behavior, the proposed controller has be of the form of (B.2). Within this section the controlled process is defined now by

$$G(s) = \frac{s^m\beta_m + s^{m-1}\beta_{m-1} + \cdots + s\beta_1 + 1}{s(s^{n-1}a_{n-1} + \cdots + s^3a_3 + sa_1 + 1)}\mathrm{e}^{-sT_{\mathrm{d}}} \tag{B.39}$$

where $n - 1 > m$. The proposed PID controller is again given by

$$C(s) = \frac{1 + sX + s^2Y}{sT_{\mathrm{i}}^2(1 + sT_{\mathrm{p_n}})} \tag{B.40}$$

where parameter $T_{\mathrm{p_n}}$ stands for the parasitic controller's time constant and is considered known from the controller's implementation. Purpose of the following analysis is to determine analytically controller parameters as a function of all modeled time constants within the control loop, $X = f_1(\beta_i, a_j, T_{\mathrm{d}})$, $Y = f_2(\beta_i, a_j, T_{\mathrm{d}})$, $T_{\mathrm{i}} = f_3(\beta_i, a_j, T_{\mathrm{d}})$. According to (B.39), (B.40) the product $C(s)G(s)$ is defined by

$$C(s)G(s) = \frac{(1 + sX + s^2Y)\sum_{j=0}^{m}(s^j\beta_j)}{s^2T_{\mathrm{i}}^2\mathrm{e}^{sT_{\mathrm{d}}}\sum_{i=0}^{n}(s^ip_i)} \tag{B.41}$$

where

$$\sum_{i=0}^{n}(s^ip_i) = (1 + sT_{\mathrm{p_n}})\sum_{j=0}^{n-1}(s^ja_j). \tag{B.42}$$

According to Fig. 4.1, the closed loop transfer function is given by

$$T(s) = \frac{F_{\mathrm{fp}}(s)}{1 + F_{\mathrm{ol}}(s)} = \frac{k_{\mathrm{p}} C(s) G(s)}{1 + k_{\mathrm{p}} k_{\mathrm{h}} C(s) G(s)} \tag{B.43}$$

where $F_{\mathrm{fp}}(s)$, $F_{\mathrm{ol}}(s)$ stand for the forward path and the open loop transfer function, respectively. Along with the aid of (B.41) $T(s)$ becomes equal to

$$T(s) = \frac{k_{\mathrm{p}}(1 + sX + s^2 Y) \sum_{j=0}^{m} (s^j \beta_j)}{s^2 T_{\mathrm{i}}^2 \mathrm{e}^{sT_{\mathrm{d}}} \sum_{i=0}^{n} (s^i p_i) + k_{\mathrm{p}} k_{\mathrm{h}} (1 + sX + s^2 Y) \sum_{j=0}^{m} (s^j \beta_j)}. \tag{B.44}$$

In the sequel, a general purpose time constant c_1 is considered for normalizing all time constants within the control loop. Therefore, frequency is normalized by setting $s' = sc_1$ and the following substitutions

$$x = \frac{X}{c_1}, \quad y = \frac{Y}{c_1^2}, \quad t_{\mathrm{i}} = \frac{t_{\mathrm{i}}}{c_1}, \; d = \frac{T_{\mathrm{d}}}{c_1} \tag{B.45}$$

$$r_i = \frac{p_i}{c_1^i}, \; \forall i = 1, \ldots, n, \quad z_j = \frac{\beta_j}{c_1^j}, \; \forall j = 1, \ldots, m \tag{B.46}$$

are considered.

The time delay constant is approximated by the series

$$\mathrm{e}^{s'd} = \sum_{k=0}^{\infty} \frac{1}{k!} s^k d^k. \tag{B.47}$$

Substituting the normalized parameters along with the approximation of $\mathrm{e}^{s'd}$ into (B.44) results in

$$T(s') = \frac{k_{\mathrm{p}}(1 + s'x + s'^2 y) \sum_{j=0}^{m} \left(s'^j z_j\right)}{s'^2 t_{\mathrm{i}}^2 \mathrm{e}^{s'd} \sum_{i=0}^{n} \left(s'^i r_i\right) + k_{\mathrm{p}} k_{\mathrm{h}} \left(1 + s'x + s'^2 y\right) \sum_{j=0}^{m} \left(s'^j z_j\right)} \tag{B.48}$$

or in a more compact form

$$T(s') = \frac{N(s')}{D_1(s') + k_{\mathrm{h}} N(s')} = \frac{N(s')}{D(s')}, \tag{B.49}$$

where

$$N(s') = k_{\mathrm{p}}(1 + s'x + s'^2 y) \sum_{j=0}^{m} (s'^j z_j) \tag{B.50}$$

and

$$D_1(s') = s'^2 t_i^2 \left(\sum_{k=0}^{7} \frac{1}{k!}(s'^k) d^k \right) \sum_{i=0}^{n} (s'^i r_i). \tag{B.51}$$

If (B.51) is expanded, results in

$$\begin{aligned} D_1(s') = {} & s'^2 t_i^2 + s'^3 t_i^2 (r_1 + d) + s'^4 t_i^2 \left(r_2 + r_1 d + \frac{1}{2!} d^2 \right) \\ & + s'^5 t_i^2 \left(r_3 + r_2 d + \frac{1}{2!} d^2 r_1 + \frac{1}{3!} d^3 \right) \\ & + s'^6 t_i^2 \left(r_4 + r_3 d + \frac{1}{2!} d^2 r_2 + \frac{1}{3!} d^3 r_1 + \frac{1}{4!} d^4 \right) + \cdots . \end{aligned} \tag{B.52}$$

Substituting the constant terms of (B.52) with

$$q_0 = 1, \tag{B.53}$$

$$q_1 = r_1 + d, \tag{B.54}$$

$$q_2 = r_2 + r_1 d + \frac{1}{2!} d^2, \tag{B.55}$$

$$q_3 = r_3 + r_2 d + \frac{1}{2!} d^2 r_1 + \frac{1}{3!} d^3, \tag{B.56}$$

$$q_4 = r_4 + r_3 d + \frac{1}{2!} d^2 r_2 + \frac{1}{3!} d^3 r_1 + \frac{1}{4!} d^4 \tag{B.57}$$

results in

$$\begin{aligned} D_1(s') = {} & \cdots + s'^8 t_i^2 q_6 + s'^7 t_i^2 q_5 + s'^6 t_i^2 q_4 \\ & + s'^5 t_i^2 q_3 + s'^4 t_i^2 q_2 + s'^3 t_i^2 q_1 + s'^2 t_i^2 q_0 \end{aligned} \tag{B.58}$$

where $q_{(-2)} = q_{(-1)} = 0$. From (B.50) it is also apparent that

$$N(s') = k_p \sum_{r=0}^{p} (s'^r)(y z_{r-2} + x z_{r-1} + z_r) \tag{B.59}$$

where $z_r = 0$, if $r < 0$, and $z_0 = 1$. In an expanded form (B.59) is rewritten in the form of

$$N(s') = \cdots + s'^6 \underbrace{k_p (y z_4 + x z_5)}_{b_6} + s'^5 \underbrace{k_p (y z_3 + x z_4 + z_5)}_{b_5}$$

$$+s'^4 \underbrace{k_p\left(yz_2+xz_3+z_4\right)}_{b_4} + s'^3 \underbrace{k_p\left(yz_1+xz_2+z_3\right)}_{b_3}$$
$$+s'^2 \underbrace{k_p\left(y+xz_1+z_2\right)}_{b_2} + s' \underbrace{k_p\left(x+z_1\right)}_{b_1} + \underbrace{k_p}_{b_0} \tag{B.60}$$

As a result, the final polynomial $D(s')$ of the closed loop transfer function is defined by

$$\begin{aligned} D(s') &= D_1(s') + k_h N(s') \\ &= \sum_{j=0}^{k} (t_i^2 q_j)(s')^{(j+2)} + k_h k_p \sum_{r=0}^{p} (s'^r)\left(yz_{r-2} + xz_{r-1} + z_r\right) \end{aligned} \tag{B.61}$$

or in an expanded form

$$D(s') = \cdots + s'^7 \underbrace{\left(t_i^2 q_5 + k_h k_p z_5 y\right)}_{a_7} + s'^6 \underbrace{\left[t_i^2 q_4 + k_h k_p\left(z_4 y + z_5 x\right)\right]}_{a_6}$$
$$+ s'^5 \underbrace{\left[t_i^2 q_3 + k_h k_p\left(z_3 y + z_4 x + z_5\right)\right]}_{a_5}$$
$$+ s'^4 \underbrace{\left[t_i^2 q_2 + k_h k_p\left(z_2 y + z_3 x + z_4\right)\right]}_{a_4}.$$
$$+ s'^3 \underbrace{\left[t_i^2 q_1 + k_h k_p (z_1 y + z_2 x + z_3)\right]}_{a_3} + s'^2 \underbrace{\left[t_i^2 + k_h k_p (y + z_1 x + z_2)\right]}_{a_2}$$
$$+ s' \underbrace{k_p\left(x + z_1\right)}_{a_1} + \underbrace{k_h k_p}_{a_0} \tag{B.62}$$

According to (B.49), (B.59) and (B.60), the resulting closed loop transfer function is given by

$$T(s') = \frac{k_p \sum_{r=0}^{p} (s'^r)(yz_{r-2} + xz_{r-1} + z_r)}{\sum_{j=0}^{k} (t_i^2 q_j)(s')^{(j+2)} + k_p k_h \sum_{r=0}^{p} (s'^r)\left(yz_{r-2} + xz_{r-1} + z_r\right)}. \tag{B.63}$$

Since (B.61) is now written in the same form of (A.1), for determining the optimal control law we can make use of the optimization conditions proved in Sect. A.1. Equations (A.9)–(A.12) are used for the derivation of the optimal control law.

Therefore, the problem to be solved is formulated as follows: given known the parameters of the process, calculate explicitly the PID control action $x,\ y,\ t_i$.

Optimization Condition 1: $a_0 = b_0$.

The application of (A.9) to (B.63) results in

$$k_{\mathrm{h}} = 1 \tag{B.64}$$

which implies that the final closed loop control system exhibits zero steady state position and velocity error. From (B.63) it is apparent that if $k_{\mathrm{h}} = 1$, then $N(s') = \cdots + s'k_{\mathrm{p}}(x + z_1) + k_{\mathrm{p}}$ and $D(s') = \cdots + s'k_{\mathrm{p}}k_{\mathrm{h}}(x + z_1) + k_{\mathrm{p}}k_{\mathrm{h}}$, respectively. According to the analysis presented Sect. 2.5, the closed loop system is of type-II.

Optimization Condition 2: $a_1^2 - 2a_2a_0 = 0$.

By making use of $a_1^2 - 2a_2a_0 = b_1^2 - 2b_2b_0$ we end up with $t_{\mathrm{i}} = 0$. For that reason, we set $a_1^2 - 2a_2a_0 = 0$ as another means of optimizing the magnitude of (B.63). This results in,

$$t_{\mathrm{i}}^2 = \frac{1}{2}k_{\mathrm{p}}k_{\mathrm{h}}\left(x^2 - 2y + z_1^2 - 2z_2\right). \tag{B.65}$$

Let it be noted, that in cases where no zeros exist in the plant transfer function $z_i = 0,\ i = 1, \ldots, m$, the integral gain is equal to

$$t_{\mathrm{i}}^2 = \frac{1}{2}k_{\mathrm{p}}k_{\mathrm{h}}(x^2 - 2y). \tag{B.66}$$

Optimization Condition 3: $a_2^2 - 2a_3a_1 + 2a_4a_0 = b_2^2 - 2b_3b_1 + 2b_4b_0$.

The application of (A.11) to (B.63) leads to

$$\begin{aligned}&\left(\left[t_{\mathrm{i}} + k_{\mathrm{p}}k_{\mathrm{h}}\left(y + z_1x + z_2\right)\right]^2 - 2k_{\mathrm{p}}k_{\mathrm{h}}q_1\left(x + z_1\right)t_{\mathrm{i}} + 2k_{\mathrm{p}}k_{\mathrm{h}}q_2t_{\mathrm{i}}\right)\\&\quad = k_{\mathrm{p}}^2(y + xz_1 + z_2)^2.\end{aligned} \tag{B.67}$$

By substituting (B.64), (B.65) into (B.67), it is easily found that

$$x^2 + 4\left(z_1 - q_1\right)x + 2y + z_1^2 + 2z_2 + 4q_2 - 4q_1z_1 = 0 \tag{B.68}$$

and in cases where no zeros exist, ($z_i = 0, i = 1, \ldots, m$), (B.68) becomes equal to

$$x^2 - 4q_1x + 2y + 4q_2 = 0. \tag{B.69}$$

Optimization Condition 4: $a_3^2 + 2a_1a_5 - 2a_6a_0 - 2a_4a_2 = b_3^2 + 2b_1b_5 - 2b_6b_0 - 2b_4b_2$.

The application of (A.12) to (B.63), along with the use of (B.64), (B.65) leads to

$$\left(q_1^2 - 2q_2\right)x^2 + 4\left(q_1z_2 - q_2z_1 + q_3 - z_3\right)x$$

$$-2\left(q_1^2-2q_1z_1+2z_2\right)y+\left(q_1^2-2q_2\right)\left(z_1^2-2z_2\right)$$
$$+4\left(q_1z_3+q_3z_1-q_4-z_4-q_2z_2\right)=0. \qquad \text{(B.70)}$$

From (B.68) to (B.70), we finally end up with the control law given by,

$$k_{\mathrm{h}}=1, \qquad \text{(B.71)}$$

$$t_{\mathrm{i}}^2=\frac{1}{2}k_{\mathrm{p}}k_{\mathrm{h}}(x^2-2y+z_1^2-2z_2), \qquad \text{(B.72)}$$

$$x^2 2\left[q_1\left(q_1-z_1\right)-q_2+z_2\right]$$
$$-4x\left(q_1^3-3q_1^2z_1+2q_1z_1^2+q_1z_2+q_2z_1-q_3+z_3-2z_1z_2\right)$$
$$+\Big[\left(q_1^2-2q_1z_1+2z_2\right)\left(z_1^2+2z_2+4q_2-4q_1z_1\right)+\left(q_1^2-2q_2\right)$$
$$\times\left(z_1^2-2z_2\right)+4\left(q_1z_3+q_3z_1-q_4-z_4-q_2z_2\right)\Big]=0, \qquad \text{(B.73)}$$

$$y=-\frac{1}{2}x^2+2(q_1-z_1)x-\frac{1}{2}(z_1^2+2z_2+4q_2-4q_1z_1). \qquad \text{(B.74)}$$

B.3 Type-III Control Loops

From the analysis presented in Sect. 5.2 the closed loop transfer function is now in the form of (A.1) or

$$T(s')=\frac{k_{\mathrm{p}}\sum_{r=0}^{p}s'^{(r)}\left(yz_{(r-2)}+xz_{(r-1)}+z_{(r)}\right)}{\sum_{j=0}^{k}(t_{\mathrm{i}}^3q_j)s'^{(j+3)}+k_{\mathrm{h}}k_{\mathrm{p}}\sum_{r=0}^{p}s'^{(r)}\left(yz_{(r-2)}+xz_{(r-1)}+z_{(r)}\right)}. \qquad \text{(B.75)}$$

Therefore, for determining the optimal control law according to the Magnitude Optimum criterion, optimization conditions (A.9)–(A.12) can be applied in (B.75).

According to the proposed PID control action proposed in (5.15), given known the plant transfer function (parameters $d,\ r_j,\ z_i,\ \forall j=1,\ldots n,\ \forall i=1,\ldots m$) in (5.14) our goal is to determine explicitly parameters $t_{\mathrm{i}},\ x,\ y$ plus the feedback k_{h} as a function of the plant's parameters. The proof takes place on the normalized closed loop transfer function $T(s')$ for which $s'=sc_1$ has been set, where c_1 is a general purpose normalizing time constant. From the first optimization condition (A.9) it is apparent that

Optimization Condition 1: $a_0 = b_0$

The application of (A.9) to (B.75) leads to

$$k_h = 1 \tag{B.76}$$

which implies that the final closed loop control system exhibits steady state position, velocity, and acceleration error. From (B.75) it is apparent that if $k_h = 1$, then numerator's polynomial

$$N(s') = \cdots + s'^2 \underbrace{k_p\,(y + xz_1 + z_2)}_{b_2} + s' \underbrace{k_p\,(x + z_1)}_{b_1} + \underbrace{k_p}_{b_0} \tag{B.77}$$

and denominator's

$$D(s') = \cdots + s'^2 \underbrace{k_p k_h\,(y + xz_1 + z_2)}_{a_2} + s' \underbrace{k_p k_h\,(x + z_1)}_{a_1} + \underbrace{k_p k_h}_{a_0} \tag{B.78}$$

are resulted. According to the definition regarding the type of the control loop, in 2.5 the closed loop control system is said to be of type-III.

Optimization Condition 2: $a_1^2 - 2a_2a_0 = 0$.

By making use of $a_1^2 - 2a_2a_0 = b_1^2 - 2b_2b_0$ results in $k_p = 0$ and $x = -z_1$ which does not lead to a feasible control law. For that reason, $a_1^2 - 2a_2a_0 = 0$ is set, as another means of optimizing the magnitude of (A.1). This results in

$$y = \frac{1}{2}x^2 + \frac{1}{2}(z_1^2 - 2z_2). \tag{B.79}$$

Optimization Condition 3: $a_2^2 - 2a_3a_1 + 2a_4a_0 = b_2^2 - 2b_3b_1 + 2b_4b_0$.

The application of (A.11) to (B.75) leads to $t_i = 0$. For that reason, the same line is followed as it was done in the previous step, by setting the second part of (A.11) equal to zero. By making use of $a_2^2 - 2a_3a_1 + 2a_4a_0 = 0$ into (B.75) results in

$$t_i^3 = \frac{1}{2}\left(k_h k_p\right) \frac{\left[y^2 - \left(z_1^2 - 2z_2\right)^2 + \left(z_2^2 - 2z_1z_3 + 2z_4\right)\right]}{x + (z_1 - q_1)}. \tag{B.80}$$

Therefore, substituting (B.79) into (B.80) results in

$$t_i^3 = \frac{k_h k_p}{8} \frac{\left[x^4 + 2\left(z_1^2 - 2z_2\right)x^2 - 3\left(z_1^2 - 2z_2\right)^2 + 4\left(z_2^2 - 2z_1z_3 + 2z_4\right)\right]}{x + (z_1 - q_1)}. \tag{B.81}$$

Optimization Condition 4: $a_3^2 + 2a_1a_5 - 2a_6a_0 - 2a_4a_2 = b_3^2 + 2b_1b_5 - 2b_6b_0 - 2b_4b_2$.
The application of (A.12) to (B.75), along with the use of (B.76), (B.79) leads to

$$\begin{aligned}&\left[t_{\mathrm{i}}^3 + k_{\mathrm{p}}\left(yz_1 + xz_2 + z_3\right)\right]^2 + 2k_{\mathrm{p}}\left(x + z_1\right)\left[t_{\mathrm{i}}^3 q_2 + k_{\mathrm{p}}\left(yz_3 + xz_4 + z_5\right)\right]\\&\quad -2k_{\mathrm{p}}\left[t_{\mathrm{i}}^3 q_3 + k_{\mathrm{p}}\left(y_3z_4 + xz_5\right)\right] - 2k_{\mathrm{p}}\left(y + xz_1 + z_2\right)\\&\quad \left[t_{\mathrm{i}}^3 q_1 + k_{\mathrm{p}}\left(yz_2 + xz_3 + z_4\right)\right] = 0\end{aligned} \tag{B.82}$$

which finally yields

$$\begin{aligned}&t_{\mathrm{i}}^6 + k_{\mathrm{p}}t_{\mathrm{i}}^3\left[2y\left(z_1 - q_1\right) + 2x\left(z_2 + q_2 - q_1z_1\right) + 2\left(z_3 - q_3 + q_2z_1 - q_1z_2\right)\right]\\&\quad + k_{\mathrm{p}}^2\left[y^2\left(z_1^2 - 2z_2\right) + y\left(4z_1z_3 - 4z_4 - 2z_2^2\right)\right.\\&\qquad \left. + x^2\left(z_2^2 + 2z_4 - 2z_1z_3\right) + \left(z_3^2 + 2z_1z_5 - 2z_6 - 2z_2z_4\right)\right] = 0.\end{aligned} \tag{B.83}$$

After considering the following substitutions,

$$\begin{bmatrix} A \\ B \\ C \end{bmatrix} = \begin{bmatrix} 2y\left(z_1 - q_1\right) \\ y^2\left(z_1^2 - 2z_2\right) \\ y\left(4z_1z_3 - 4z_4 - 2z_2^2\right) \end{bmatrix} \tag{B.84}$$

and making use of (B.79) it is obtained

$$A = \left(z_1 - q_1\right)x^2 + \left(z_1^3 - q_1z_1^2 - 2z_1z_2 + 2q_1z_2\right), \tag{B.85}$$

$$\begin{aligned}B = \frac{1}{4}&\left[\left(z_1^2 - 2z_2\right)x^4 + \left(2z_1^4 - 8z_1^2z_2 + 8z_2^2\right)x^2\right.\\&\left. + \left(z_1^6 - 6z_1^4z_2 + 12z_1^2z_2^2 - 8z_2^3\right)\right],\end{aligned} \tag{B.86}$$

$$\begin{aligned}C = \frac{1}{2}&\left[\left(4z_1z_3 - 4z_4 - 2z_2^2\right)x^2\right.\\&\left. + \left(4z_1^3z_3 - 4z_1^2z_4 - 2z_1^2z_2^2 - 8z_1z_2z_3 + 8z_2z_4 + 4z_2^3\right)\right].\end{aligned} \tag{B.87}$$

By substituting (B.85)–(B.87) into (B.83) results in

$$t_i^6 + k_pt_i^3\begin{bmatrix}\left(z_1 - q_1\right)x^2 + \left(2z_2 + 2q_2 - 2q_1z_1\right)x \\ + \left(2z_3 - 2q_3 + 2q_2z_1 + z_1^3 - q_1z_1^2 - 2z_1z_2\right)\end{bmatrix}$$

$$+k_p^2\left[\begin{array}{l}\frac{1}{4}\left(z_1^2-2z_2\right)x^4+\left(\frac{z_1^4}{2}+2z_2^2-2z_1^2z_2\right)x^2\\+\left(\begin{array}{l}z_3^2+2z_1z_5+\frac{z_1^6}{4}-\frac{3}{2}z_1^4z_2+2z_1^2z_2^2+2z_1^3z_3\\-2z_1^2z_4-4z_1z_2z_3+2z_2z_4\end{array}\right)\end{array}\right]=0 \tag{B.88}$$

or finally

$$\begin{aligned}&t_i^6-k_pt_i^3\left[\begin{array}{l}(q_1-z_1)\,x^2+2\,(q_1z_1-q_2-z_2)\,x\\+(q_1-z_1)\left(z_1^2-2z_2\right)+2\,(q_1z_2-q_2z_1+q_3-z_3)\end{array}\right]\\&\quad+\frac{1}{4}k_p^2\left[\begin{array}{l}\left(z_1^2-2z_2\right)x^4+2\left(z_1^2-2z_2\right)^2x^2+\left(z_1^2-2z_2\right)^3\\-4\left[\left(z_1^2-2z_2\right)\left(z_2^2+2z_4-2z_1z_3\right)-\left(z_3^2-2z_2z_4+2z_1z_5\right)\right]\end{array}\right]=0.\end{aligned} \tag{B.89}$$

Finally, it is set

$$\begin{bmatrix}Q_1\\Q_2\\Q_3\end{bmatrix}=\begin{bmatrix}q_1-z_1\\q_1z_1-q_2-z_2\\q_1z_2-q_2z_1+q_3-z_3\end{bmatrix} \tag{B.90}$$

where $Q_0=Q_1Z_1+2Q_3$. Moreover, the following substitutions are made

$$\begin{bmatrix}Z_1\\Z_2\\Z_3\end{bmatrix}=\begin{bmatrix}z_1 & z_2 & z_3 & z_4\end{bmatrix}\begin{bmatrix}z_1 & -2z_3 & 2z_5\\-2 & z_2 & -2z_4\\0 & 0 & z_3\\0 & 2 & 0\end{bmatrix} \tag{B.91}$$

and

$$Z_0=Z_1^3-4Z_1Z_2+4Z_3, \tag{B.92}$$
$$Z_4=3Z_1^2-4Z_2. \tag{B.93}$$

Combining (B.89) with (B.81) the optimal control law is finally derived

$$k_h=1 \tag{B.94}$$

$$y=\frac{x^2+z_1^2-2z_2}{2} \tag{B.95}$$

$$t_i^3=\frac{k_hk_p}{2}\frac{\left[y^2-\left(z_1^2-2z_2\right)^2+z_2^2-2z_1z_3+2z_4\right]}{x+z_1-q_1} \tag{B.96}$$

$$\sum_{j=0}^{8} C_j x^j = 0 \tag{B.97}$$

determining explicitly PID controller's parameters as a function of the process model. Coefficients $C_0,\ C_1$ are given by,

$$C_0 = Z_4^2 - 8Q_0Q_1Z_4 + 16Q_1^2Z_0, \tag{B.98}$$

$$C_1 = 8(-4Q_1Z_0 - 2Q_1Q_2Z_4 + Q_0Z_4) \tag{B.99}$$

respectively. $C_2,\ C_3$ are defined by

$$C_2 = 4\left(4Z_0 + 8Q_1^2Z_1^2 - Z_1Z_4 - 2Q_1^2Z_4 + 4Q_2Z_4 + 4Q_0Q_1Z_1\right), \tag{B.100}$$

$$C_3 = 8\left(Q_1Z_4 + 4Q_1Q_2Z_1 - 2Q_0Z_1 - 8Q_1Z_1^2\right). \tag{B.101}$$

Parameters $C_4,\ C_5$ are given by

$$C_4 = 2\left[2Z_1^2 - Z_4 + 4\left(2Q_1^2Z_1 - 4Z_1Q_2 + Q_0Q_1\right) + 8\left(2Z_1^2 + Q_1^2Z_1\right)\right] \tag{B.102}$$

$$C_5 = 8(-6Q_1Z_1 + 2Q_1Q_2 - Q_0) \tag{B.103}$$

where finally coefficients $C_6,\ C_7,\ C_8$ are defined by

$$C_6 = 4\left(2Q_1^2 - 4Q_2 + 5Z_1\right) \tag{B.104}$$

$$C_7 = -8Q_1 \tag{B.105}$$

$$C_8 = 1 \tag{B.106}$$

For determining parameter x the real maximum positive value of the eighth order polynomial solution of (B.97) is always adopted. Therefore, $x = \max\{x_j\}$, $x_j > 0,\ x \in \Re$.

Appendix C
Digital Design-Proof of the Optimal Control Law

Abstract In this chapter the proof of the optimal PID control law for type-I, type-II, type-III control loops is presented. Basis of the design of the control law are the optimization conditions (A.9)–(A.13) of the magnitude $|T(j\omega)|$ of the closed loop transfer function presented in Sect. A.1. Controller parameters are determined explicitly as a function of the process parameters and the sampling time of the controller T_s. For developing the proposed theory a generalized single-input single-output stable process model is employed consisting of n-poles, m-zeros plus unknown time delay-d.

C.1 Type-I Control Loops

In that section the analytic tuning rules for digital PID–type controllers are proved. The plant transfer function consists of n-poles, m-zeros plus time delay d. Zeros of the plant may lie both in the left or right imaginary half plane. The plant transfer function may contain second-order oscillatory terms in the denominator, described by polynomials of the form $1 + 2\zeta T + s^2T^2$, where $z \in (0, 1], \Re$ and $T > 0, \in \Re$. In that, the plant transfer function is defined by

$$G(s) = k_p \frac{s^m \beta_m + \cdots + s^4\beta_4 + s^3\beta_3 + s^2\beta_2 + s\beta_1 + 1}{\begin{pmatrix} s^n p_n + s^{n-1} p_{n-1} + \cdots + s^5 p_5 + s^4 p_4 + s^3 p_3 \\ +s^2 p_2 + s p_1 + 1 \end{pmatrix}} e^{-sT_d}, \quad n > m \tag{C.1}$$

The proposed PID type controller is given by

$$C(s) = C^*(s)C_{ZOH}(s) = \left(\frac{1 + sX + s^2Y}{sT_i}\right)^* \left(\frac{1 - e^{-sT_s}}{sT_s}\right) \tag{C.2}$$

where the $C^*(s)$ controller stands for the digital representation of the PID control law. $C_{ZOH}(s)$ stands for the zero order hold module and T_s stands for the controller sampling period.

K.G. Papadopoulos, *PID Controller Tuning Using the Magnitude Optimum Criterion*,
DOI 10.1007/978-3-319-07263-0

The analysis proceeds by normalizing all time constants in the frequency domain with the sampling period T_s of the zero order hold. In that,

$$s' = sT_s \tag{C.3}$$

is set and the resulting expressions (C.1) and (C.2) take the form

$$G(s') = k_p \frac{\left(s'^m z_m + \cdots + s'^4 z_4 + s'^3 z_3 + s'^2 z_2 + s' z_1 + 1\right)}{\left(s'^n r_n + s'^{n-1} r_{n-1} + \cdots + s'^5 r_5 + s'^4 r_4 + s'^3 r_3 + s'^2 r_2 + s' r_1 + 1\right)} e^{-s'd} \tag{C.4}$$

and

$$C(s') = C^*(s') C_{ZOH}(s') = \left(\frac{1 + s'x + s'^2 y}{s' t_i}\right)^* \left(\frac{1 - e^{-s'}}{s'}\right) \tag{C.5}$$

$$x = \frac{X}{T_s}, \quad y = \frac{Y}{T_s^2}, \quad t_i = \frac{T_i}{T_s}, \quad d = \frac{T_d}{T_s}, \tag{C.6}$$

$$r_j = \frac{p_j}{T_s^i}, \quad \forall j = 1, \ldots n, \quad z_i = \frac{\beta_i}{T_s^i}, \quad \forall i = 1, \ldots m. \tag{C.7}$$

The transition from the $\mathcal{L}\{.\}$ to the $\mathcal{Z}\{.\}$ domain takes place by making the transformation

$$s' = \frac{z - 1}{z} = \frac{e^{s'} - 1}{e^{s'}}. \tag{C.8}$$

Since $z = e^{s'}$, the digital PID type controller takes the form

$$\begin{aligned} C(s') &= C^*(s') C_{ZOH}(s') \\ &= \frac{1}{t_i} \frac{(1 + x + y)e^{2s'} - (x + 2y)e^{s'} + y}{e^{s'}(e^{s'} - 1)}. \end{aligned} \tag{C.9}$$

By setting

$$\hat{x} = x + 2y \quad \text{and} \quad \hat{y} = 1 + x + y \tag{C.10}$$

results in

$$x = 2\hat{y} - \hat{x} - 2 \quad \text{and} \quad y = \hat{x} - \hat{y} + 1. \tag{C.11}$$

By substituting Eqs. (C.11) into (C.9) results in

$$C(s') = C^*(s')C_{\mathrm{ZOH}}(s') = \frac{1}{t_\mathrm{i}}\frac{(1-\mathrm{e}^{s'})\hat{x}+(\mathrm{e}^{2s'}-1)\hat{y}+1}{\mathrm{e}^{s'}(\mathrm{e}^{s'}-1)}. \tag{C.12}$$

In addition, the respective open and closed loop transfer functions become

$$F_{\mathrm{ol}}(s') = k_\mathrm{h}C(s')G(s') \tag{C.13}$$

or

$$F_{\mathrm{ol}}(s') = k_\mathrm{h}\frac{k_\mathrm{p}}{t_\mathrm{i}}\frac{\left[\left(s'^m z_m + \cdots + s'^3 z_3 + s'^2 z_2 + s' z_1 + 1\right)\left[(1-\mathrm{e}^{s'})\hat{x} + (\mathrm{e}^{2s'}-1)\hat{y}+1\right]\right]}{\left[\left(s'^n r_n + \cdots + s'^3 r_3 + s'^2 r_2 + s' r_1 + 1\right)\mathrm{e}^{s'(d+1)}(\mathrm{e}^{s'}-1)\right]} \tag{C.14}$$

and

$$T(s') = \frac{C(s')G(s')}{1+k_\mathrm{h}C(s')G(s')} = \frac{N(s')}{D(s')} = \frac{N(s')}{D_1(s')+k_\mathrm{h}N(s')} \tag{C.15}$$

or

$$T(s') = \frac{k_\mathrm{p}\begin{pmatrix} s'^m z_m + \cdots + s'^3 z_3 \\ +s'^2 z_2 + s' z_1 + 1\end{pmatrix}\left[(1-\mathrm{e}^{s'})\hat{x} + (\mathrm{e}^{2s'}-1)\hat{y}+1\right]}{\begin{bmatrix} t_\mathrm{i}\begin{pmatrix} s'^n r_n + \cdots + s'^3 r_3 \\ +s'^2 r_2 + s' r_1 + 1\end{pmatrix}\mathrm{e}^{s'(d+1)}(\mathrm{e}^{s'}-1) \\ +k_\mathrm{h}k'_p\begin{pmatrix} s'^m z_m + \cdots + s'^3 z_3 \\ +s'^2 z_2 + s' z_1 + 1\end{pmatrix}\left[(1-\mathrm{e}^{s'})\hat{x} + (\mathrm{e}^{2s'}-1)\hat{y}+1\right]\end{bmatrix}}. \tag{C.16}$$

Substituting the time delay constant by the "all pole" series approximation

$$\mathrm{e}^{s'} = 1 + s' + \frac{1}{2!}s'^2 + \frac{1}{3!}s'^3 + \frac{1}{4!}s'^4 + \frac{1}{5!}s'^5 + \frac{1}{6!}s'^6 + \cdots \tag{C.17}$$

results in

$$\begin{aligned}\mathrm{e}^{s'(1+d)} &= 1 + d's' + \frac{1}{2!}d'^2 s'^2 + \frac{1}{3!}d'^3 s'^3 + \frac{1}{4!}d'^4 s'^4 \\ &\quad + \frac{1}{5!}d'^5 s'^5 + \frac{1}{6!}d'^6 s'^6 + \cdots\end{aligned} \tag{C.18}$$

where

$$d' = 1 + d. \tag{C.19}$$

Additionally,

$$\begin{aligned} e^{s'(1+d)}(e^{s'} - 1) = d_1 s' + d_2 s'^2 + d_3 s'^3 + d_4 s'^4 \\ + d_5 s'^5 + d_6 s'^6 + d_7 s'^7 + \cdots \end{aligned} \tag{C.20}$$

where

$$\begin{bmatrix} d_1 \\ d_2 \\ d_3 \\ d_4 \\ d_5 \\ \vdots \end{bmatrix} = \begin{bmatrix} 1 \\ \frac{1}{2!} + d' \\ \frac{1}{3!} + \frac{1}{2!}d' + \frac{1}{2!}d'^2 \\ \frac{1}{4!} + \frac{1}{3!}d' + \frac{1}{2!}\frac{1}{2!}d'^2 + \frac{1}{3!}d'^3 \\ \frac{1}{5!} + \frac{1}{4!}d' + \frac{1}{2!}\frac{1}{3!}d'^2 + \frac{1}{2!}\frac{1}{3!}d'^3 + \frac{1}{4!}d'^4 \\ \vdots \end{bmatrix} \tag{C.21}$$

and

$$\begin{bmatrix} q_1 \\ q_2 \\ q_3 \\ q_4 \\ q_5 \\ \vdots \end{bmatrix} = \begin{bmatrix} 1 \\ r_1 + d_2 \\ r_2 + r_1 d_2 + d_3 \\ r_3 + r_2 d_2 + r_1 d_3 + d_4 \\ r_4 + r_3 d_2 + r_2 d_3 + r_1 d_4 + d_5 \\ \vdots \end{bmatrix}. \tag{C.22}$$

For that reason, polynomial $D_1(s')$ can be rewritten in the form of

$$D_1(s') = t_i(\cdots + q_7 s'^7 + q_6 s'^6 + q_5 s'^5 + q_4 s'^4 + q_3 s'^3 + q_2 s'^2 + s') \tag{C.23}$$

The numerator of $C(s')$ is then equal to

$$\begin{aligned} &(1 - e^{s'})\hat{x} + (e^{2s'} - 1)\hat{y} + 1 \\ &\quad = 1 + (2\hat{y} - \hat{x})s' \\ &\qquad + \frac{1}{2!}(4\hat{y} - \hat{x})s'^2 + \frac{1}{3!}(8\hat{y} - \hat{x})s'^3 + \frac{1}{4!}(16\hat{y} - \hat{x})s'^4. \\ &\qquad + \frac{1}{5!}(32\hat{y} - \hat{x})s'^5 + \frac{1}{6!}(64\hat{y} - \hat{x})s'^6 + \cdots \end{aligned} \tag{C.24}$$

The numerator of the closed loop transfer function proves to be equal to

$$N(s') = k_{\mathrm{p}} \begin{bmatrix} \cdots + (z_6 + y_6\hat{y} - x_6\hat{x})s'^6 \\ + (z_5 + y_5\hat{y} - x_5\hat{x})s'^5 + (z_4 + y_4\hat{y} - x_4\hat{x})s'^4 \\ + (z_3 + y_3\hat{y} - x_3\hat{x})s'^3 + (z_2 + y_2\hat{y} - x_2\hat{x})s'^2 \\ + (z_1 + 2\hat{y} - \hat{x})s' + 1 \end{bmatrix} \tag{C.25}$$

where

$$x_k = \sum_{j=1}^{k} \left(z_{(j-1)} \frac{1}{[k-(j-1)]!} \right) \tag{C.26}$$

and

$$y_k = 2\sum_{j=1}^{k} \left(\frac{2^{j-1}}{j!} \right) z_{k-j} \tag{C.27}$$

or in an expanded form

$$\begin{bmatrix} x_1 \\ x_2 \\ x_3 \\ x_4 \\ x_5 \\ x_6 \\ \vdots \end{bmatrix} = \begin{bmatrix} 1 \\ \frac{1}{2!} + z_1 \\ \frac{1}{3!} + \frac{1}{2!}z_1 + z_2 \\ \frac{1}{4!} + \frac{1}{3!}z_1 + \frac{1}{2!}z_2 + z_3 \\ \frac{1}{5!} + \frac{1}{4!}z_1 + \frac{1}{3!}z_2 + \frac{1}{2!}z_3 + z_4 \\ \frac{1}{6!} + \frac{1}{5!}z_1 + \frac{1}{4!}z_2 + \frac{1}{3!}z_3 + \frac{1}{2!}z_4 + z_5 \\ \vdots \end{bmatrix}, \tag{C.28}$$

and

$$\begin{bmatrix} y_2 \\ y_3 \\ y_4 \\ y_5 \\ y_6 \\ \vdots \end{bmatrix} = \begin{bmatrix} 2z_1 + \frac{4}{2!} \\ 2z_2 + \frac{4}{2!}z_1 + \frac{8}{3!} \\ 2z_3 + \frac{4}{2!}z_2 + \frac{8}{3!}z_1 + \frac{16}{4!} \\ 2z_4 + \frac{4}{2!}z_3 + \frac{8}{3!}z_2 + \frac{16}{4!}z_1 + \frac{32}{5!} \\ 2z_5 + \frac{4}{2!}z_4 + \frac{8}{3!}z_3 + \frac{16}{4!}z_2 + \frac{32}{5!}z_1 + \frac{64}{6!} \\ \vdots \end{bmatrix}. \tag{C.29}$$

Finally, the corresponding polynomials for both the numerator and denominator of the closed loop transfer function are given by

$$\begin{aligned} N(s') = \cdots &+ \underbrace{k_p(z_6 + y_6\hat{y} - x_6\hat{x})}_{b_6} s'^6 \\ &+ \underbrace{k_p(z_5 + y_5\hat{y} - x_5\hat{x})}_{b_5} s'^5 + \underbrace{k_p(z_4 + y_4\hat{y} - x_4\hat{x})}_{b_4} s'^4 \\ &+ \underbrace{k_p(z_3 + y_3\hat{y} - x_3\hat{x})}_{b_3} s'^3 + \underbrace{k_p(z_2 + y_2\hat{y} - x_2\hat{x})}_{b_2} s'^2 \\ &+ \underbrace{k_p(z_1 + 2\hat{y} - \hat{x})}_{b_1} s' + \underbrace{k_p}_{b_0} \end{aligned} \tag{C.30}$$

and

$$\begin{aligned} D(s') = D_1(s') + k_h N(s') = \cdots &+ \underbrace{\left[t_i q_6 + k_h k_p(z_6 + y_6\hat{y} - x_6\hat{x})\right]}_{a_6} s'^6 \\ &+ \underbrace{\left[t_i q_5 s'^5 + k_h k_p(z_5 + y_5\hat{y} - x_5\hat{x})\right]}_{a_5} s'^5 \\ &+ \underbrace{\left[t_i q_4 + k_h k_p(z_4 + y_4\hat{y} - x_4\hat{x})\right]}_{a_4} s'^4 \\ &+ \underbrace{\left[t_i q_3 + k_h k_p(z_3 + y_3\hat{y} - x_3\hat{x})\right]}_{a_3} s'^3 + \underbrace{\left[t_i q_2 + k_h k_p(z_2 + y_2\hat{y} - x_2\hat{x})\right]}_{a_2} s'^2 \\ &+ \underbrace{\left[t_i + k_h k_p(z_1 + 2\hat{y} - \hat{x})\right]}_{a_1} s' + \underbrace{k_h k_p}_{a_0} . \end{aligned} \tag{C.31}$$

For determining the optimal PID controller's parameters, Eqs. (A.9)–(A.13) are applied to (C.16). For that reason, from the application of

Optimization Condition: $a_0 = b_0$.

To the closed loop transfer function (C.15) and since $a_0 = k_p k_h$, $b_0 = k_p$ and within (C.30), (C.31) results in

$$k_h = 1, \tag{C.32}$$

which implies that the final closed loop control system exhibits zero steady position error if $k_h = 1$.

Optimization Condition: $a_1^2 - 2a_2a_0 = b_1^2 - 2b_2b_0$.

From (C.30) and (C.31) it is apparent that

$$a_0 = k_\mathrm{h}k_\mathrm{p}, \tag{C.33a}$$

$$a_1 = t_\mathrm{i} + k_\mathrm{h}k_\mathrm{p}(z_1 + 2\hat{y} - \hat{x}), \tag{C.33b}$$

$$a_2 = t_\mathrm{i}q_2 + k_\mathrm{h}k_\mathrm{p}(z_2 + y_2\hat{y} - x_2\hat{x}) \tag{C.33c}$$

and

$$b_0 = k_\mathrm{p}, \tag{C.34a}$$

$$b_1 = k_\mathrm{p}(z_1 + 2\hat{y} - \hat{x}), \tag{C.34b}$$

$$b_2 = k_\mathrm{p}(z_2 + y_2\hat{y} - x_2\hat{x}). \tag{C.34c}$$

Substituting into the second optimization condition results in

$$t_\mathrm{i} = 2k_\mathrm{h}k_\mathrm{p}(r_1 + d_2 + \hat{x} - 2\hat{y} - z_1), \tag{C.35}$$

or according to (C.10) and (C.11)

$$t_\mathrm{i} = 2k_\mathrm{h}k_\mathrm{p}\left(r_1 + d - z_1 - x - \frac{1}{2}\right). \tag{C.36}$$

Optimization Condition: $a_2^2 - 2a_3a_1 + 2a_4a_0 = b_2^2 - 2b_3b_1 + 2b_4b_0$.

The application of the third optimization condition (A.12) to the closed loop transfer function and after taking into account that

$$a_3 = t_\mathrm{i}q_3 + k_\mathrm{h}k_\mathrm{p}(z_3 + y_3\hat{y} - x_3\hat{x}), \tag{C.37}$$

$$a_4 = t_\mathrm{i}q_4 + k_\mathrm{h}k_\mathrm{p}(z_4 + y_4\hat{y} - x_4\hat{x}), \tag{C.38}$$

and

$$b_3 = k_\mathrm{p}(z_3 + y_3\hat{y} - x_3\hat{x}), \tag{C.39}$$

$$b_4 = k_\mathrm{p}(z_4 + y_4\hat{y} - x_4\hat{x}) \tag{C.40}$$

results in

$$\begin{aligned}&\left[(q_2^2 - q_3) - (q_2x_2 - x_3)\right]\hat{x} - \left[2(q_2^2 - q_3) - (q_2y_2 - y_3)\right]\hat{y}\\&= (q_3z_1 - q_2z_2 + z_3 - q_4) - (q_2^2 - 2q_3)(q_2 - z_1).\end{aligned} \tag{C.41}$$

Optimization Condition: $a_3^2 + 2a_1a_5 - 2a_6a_0 - 2a_4a_2 = b_3^2 + 2b_1b_5 - 2b_6b_0 - 2b_4b_2$.

Finally, the application of the fourth optimization condition to the closed loop transfer function taking into account also that

$$a_5 = t_i q_5 + k_h k_p (z_5 + y_5\hat{y} - x_5\hat{x}), \tag{C.42}$$
$$a_6 = t_i q_6 + k_h k_p (z_6 + y_6\hat{y} - x_6\hat{x}), \tag{C.43}$$

and

$$b_5 = k_p (z_5 + y_5\hat{y} - x_5\hat{x}), \tag{C.44}$$
$$b_6 = k_p (z_6 + y_6\hat{y} - x_6\hat{x}) \tag{C.45}$$

leads to

$$\begin{aligned} &[(q_3 - x_3)q_3 - (q_4 - x_4)q_2 - (q_2 - x_2)q_4 + q_5 - x_5]\,\hat{x} \\ &\quad - \left[2q_3^2 - 4q_2q_4 + q_2y_4 - q_3y_3 - y_5 + 2q_5 + q_4y_2\right]\hat{y} \\ &= -(q_3^2 - 2q_2q_4 + 2q_5)(q_2 - z_1) \\ &\quad + (q_2z_4 - q_3z_3 + q_4z_2 - q_5z_1 - z_5 + q_6) \end{aligned} \tag{C.46}$$

To that end, the optimal PID controller's parameters are given by

$$k_h = 1 \tag{C.47}$$

$$t_i = 2k_h k_p \left(r_1 + d - z_1 - x - \frac{1}{2}\right) \tag{C.48}$$

$$\hat{x} - a_1\hat{y} = b_1 \quad \text{and} \quad \hat{x} - a_2\hat{y} = b_2 \tag{C.49}$$

where

$$a_1 = \frac{2(q_2^2 - q_3) - (q_2y_2 - y_3)}{(q_2^2 - q_3) - (q_2x_2 - x_3)} \tag{C.50}$$

$$b_1 = \frac{(q_3z_1 - q_2z_2 + z_3 - q_4) - (q_2^2 - 2q_3)(q_2 - z_1)}{(q_2^2 - q_3) - (q_2x_2 - x_3)} \tag{C.51}$$

$$a_2 = \frac{2q_3^2 - 4q_2q_4 + q_2y_4 - q_3y_3 - y_5 + 2q_5 + q_4y_2}{(q_3 - x_3)q_3 - (q_4 - x_4)q_2 - (q_2 - x_2)q_4 + q_5 - x_5} \tag{C.52}$$

$$b_2 = \frac{-\left(q_3^2 - 2q_2q_4 + 2q_5\right)(q_2 - z_1) + \left(q_2z_4 - q_3z_3 + q_4z_2 - q_5z_1 - z_5 + q_6\right)}{\left[(q_3 - x_3)q_3 - (q_4 - x_4)q_2 - (q_2 - x_2)q_4 + (q_5 - x_5)\right]} \tag{C.53}$$

By solving (C.48), (C.49) parameters $\hat{x}$, $\hat{y}$ are determined by

$$\hat{x} = \frac{a_1 b_2 - a_2 b_1}{a_1 - a_2}, \quad \hat{y} = \frac{b_2 - b_1}{a_1 - a_2}. \tag{C.54}$$

From the definition of the integrator's time constant (C.48), it is critical to point out that

$$\frac{T_\mathrm{i}}{T_\mathrm{s}} = 2k_\mathrm{h}k_\mathrm{p}\left(r_1 + d - z_1 - x - \frac{1}{2}\right) \tag{C.55}$$

or according to (C.6) and (C.7)

$$\begin{aligned} T_\mathrm{i} &= 2k_h k_\mathrm{p}\left(T_\mathrm{s} r_1 + T_\mathrm{s} d - T_\mathrm{s} z_1 - T_\mathrm{s} x - \frac{1}{2}T_\mathrm{s}\right) \\ &= 2k_\mathrm{h}k_\mathrm{p}\left(p_1 + T_\mathrm{d} - \beta_1 - T_\mathrm{s} x - \frac{1}{2}T_\mathrm{s}\right) \\ &= 2k_\mathrm{h}k_\mathrm{p}\left(\sum_{i=1}^{n}(T_{p_i}) + T_\mathrm{d} - \sum_{i=1}^{m}(T_{z_i}) - X - \frac{1}{2}T_\mathrm{s}\right). \end{aligned} \tag{C.56}$$

In other words, as it was proved in (C.36), integrator's time constant is equal

$$T_{i_\mathrm{dig}} = T_{i_\mathrm{an}} - \underbrace{2k_\mathrm{h}k_\mathrm{p}\frac{1}{2}T_\mathrm{s}}, \tag{C.57}$$

where T_{i_dig} and T_{i_an} the optimal values for the integrator's time constant regarding the analog and digital design, respectively.

C.2 Type-II Control Loops

Let the stable process in Fig. 3.2 be defined by

$$G(s) = k_\mathrm{p}\frac{\left(s^m \beta_m + \cdots + s^4\beta_4 + s^3\beta_3 + s^2\beta_2 + s\beta_1 + 1\right)}{\left(s^n \mathrm{p_n} + s^{n-1}p_{n-1} + \cdots + s^5 p_5 + s^4 p_4 + s^3 p_3 + s^2 p_2 + s p_1 + 1\right)}\mathrm{e}^{-sT_\mathrm{d}} \tag{C.58}$$

where $n > m$. The proposed PID type controller is given by

$$C(s) = C^*(s)C_\mathrm{ZOH}(s) = \left(\frac{1 + sX + s^2 Y}{s^2 T_\mathrm{i}}\right)^* \left(\frac{1 - \mathrm{e}^{-sT_\mathrm{s}}}{s}\right) \tag{C.59}$$

where controller $C^*(s)$ stands for the digital representation of the analog PID control law. $C_{\mathrm{ZOH}}(s)$ stands for the zero order hold module and T_s stands for the controller sampling period. The analysis proceeds by normalizing all time constants in the frequency domain with the sampling period T_s. In that, we make the substitution

$$s' = sT_s. \tag{C.60}$$

The resulting expressions (C.58) and (C.59) take the form

$$G(s') = k_p \frac{\left(s'^m z_m + \cdots + s'^4 z_4 + s'^3 z_3 + s'^2 z_2 + s' z_1 + 1\right)}{\left(s'^n r_n + s'^{n-1} r_{n-1} + \cdots + s'^5 r_5 + s'^4 r_4 + s'^3 r_3 + s'^2 r_2 + s' r_1 + 1\right)} e^{-s'd} \tag{C.61}$$

and

$$C(s') = C^*(s')C_{\mathrm{ZOH}}(s') = T_s\left(\frac{1 + s'x + s'^2 y}{s'^2 t_i^2}\right)^* \frac{1 - e^{-s'}}{s'} \tag{C.62}$$

$$x = \frac{X}{T_s}, \quad y = \frac{Y}{T_s^2}, \quad t_i = \frac{T_i}{T_s}, \quad d = \frac{T_d}{T_s}, \tag{C.63}$$

$$r_j = \frac{p_j}{T_s^j}, \quad \forall j = 1, \ldots, n, \quad z_i = \frac{\beta_i}{T_s^i}, \quad \forall i = 1, \ldots, m. \tag{C.64}$$

The transition from the $\mathcal{L}\{.\}$ to the $\mathcal{Z}\{.\}$ domain takes place by utilizing the relation

$$s' = \frac{z - 1}{z} = \frac{e^{s'} - 1}{e^{s'}} \tag{C.65a}$$

$$\frac{1}{s'^2} = \frac{T_s z'}{(z' - 1)^2} = \frac{T_s e^{s'}}{(e^{s'} - 1)^2}. \tag{C.65b}$$

Since $z = e^{s'}$, the digital PID type controller takes the form

$$C(s') = C^*(s')C_{\mathrm{ZOH}}(s') = \frac{T_s}{t_i^2}\left(\frac{1}{s'^2} + \frac{x}{s'} + y\right)^* \left(\frac{1 - e^{-s'}}{s'}\right) \tag{C.66}$$

or

$$C(s') = \frac{T_s}{t_i^2}\left[\frac{(x + y)e^{2s'} - (x + 2y - T_s)e^{s'} + y}{(e^{s'} - 1)^2}\right] \tag{C.67}$$

or finally

$$C(s') = \frac{T_s^2}{t_i^2}\left[\frac{\left(\frac{x}{T_s}+\frac{y}{T_s}\right)e^{2s'} - \left(\frac{x}{T_s}+2\frac{y}{T_s}-1\right)e^{s'} + \frac{y}{T_s}}{(e^{s'}-1)^2}\right]. \tag{C.68}$$

The analysis proceeds by making now the transformation

$$\hat{x} = \frac{x}{T_s} + 2\frac{y}{T_s} - 1 \tag{C.69}$$

and

$$\hat{y} = \frac{x}{T_s} + \frac{y}{T_s}, \tag{C.70}$$

which finally results in

$$\frac{x}{T_s} = 2\hat{y} - \hat{x} - 1 \tag{C.71}$$

and

$$\frac{y}{T_s} = \hat{x} - \hat{y} + 1. \tag{C.72}$$

By substituting Eqs. (C.69)–(C.70), (C.68) takes the form

$$C(s') = \frac{T_s^2}{t_i^2}\left[\frac{(1-e^{s'})\hat{x} + (e^{2s'}-1)\hat{y} + 1}{(e^{s'}-1)^2}\right]. \tag{C.73}$$

With respect to the above, the corresponding open and closed loop transfer functions become

$$F_{ol}(s') = k_h C(s') G(s') \tag{C.74}$$

or

$$k_h C(s') G(s') = k_h \frac{T_s^2 k_p}{t_i^2} \times \frac{\left[\begin{array}{c}\left(s'^m z_m + \cdots + s'^3 z_3 + s'^2 z_2 + s' z_1 + 1\right)\\ \times\left[(1-e^{s'})\hat{x} + (e^{2s'}-1)\hat{y} + 1\right]\end{array}\right]}{\left(s'^n r_n + \cdots + s'^3 r_3 + s'^2 r_2 + s' r_1 + 1\right) e^{s'd}\left(e^{s'}-1\right)^2} \tag{C.75}$$

and

$$T(s') = \frac{C(s')G(s')}{1 + k_\mathrm{h}C(s')G(s')} = \frac{N(s')}{D(s')} = \frac{N(s')}{D_1(s') + k_\mathrm{h}N(s')}$$

$$= \frac{\begin{bmatrix} k'_p(s'^m z_m + \cdots + s'^3 z_3 + s'^2 z_2 + s' z_1 + 1) \\ \times\left[(1-\mathrm{e}^{s'})\hat{x} + (\mathrm{e}^{2s'}-1)\hat{y} + 1\right] \end{bmatrix}}{\begin{bmatrix} t_\mathrm{i}^2(s'^n r_n + \cdots + s'^3 r_3 + s'^2 r_2 + s' r_1 + 1)\mathrm{e}^{s'd}(\mathrm{e}^{s'}-1)^2 \\ + \begin{bmatrix} k_\mathrm{h} k'_p(s'^m z_m \cdots + s'^3 z_3 + s'^2 z_2 + s' z_1 + 1) \\ \times\left[(1-\mathrm{e}^{s'})\hat{x} + (\mathrm{e}^{2s'}-1)\hat{y} + 1\right] \end{bmatrix} \end{bmatrix}} \tag{C.76}$$

where

$$k'_p = T_\mathrm{s}^2 k_\mathrm{p}. \tag{C.77}$$

Substituting the time delay constant with the “all pole” series approximation

$$\mathrm{e}^{s'} - 1 = s' + \frac{1}{2!}s'^2 + \frac{1}{3!}s'^3 + \frac{1}{4!}s'^4 + \frac{1}{5!}s'^5 + \frac{1}{6!}s'^6 + \cdots \tag{C.78}$$

results in

$$\mathrm{e}^{2s'} - 1 = 2s' + \frac{4}{2!}s'^2 + \frac{8}{3!}s'^3 + \frac{16}{4!}s'^4 + \frac{32}{5!}s'^5 + \frac{64}{6!}s'^6 + \cdots \tag{C.79}$$

or

$$\begin{aligned}(\mathrm{e}^{s'} - 1)^2 &= (s' + \frac{1}{2!}s'^2 + \frac{1}{3!}s'^3 + \frac{1}{4!}s'^4 + \frac{1}{5!}s'^5 + \frac{1}{6!}s'^6 + \cdots)^2 \\ &= \cdots + 0.0861s'^6 + 0.25s'^5 + 0.5833s'^4 + 1s'^3 + s'^2\end{aligned} \tag{C.80}$$

and finally in

$$\begin{aligned}\mathrm{e}^{s'd} &= 1 + \mathrm{d}s' + \frac{1}{2!}d^2 s'^2 + \frac{1}{3!}d^3 s'^3 + \frac{1}{4!}d^4 s'^4 + \frac{1}{5!}d^5 s'^5 \\ &+ \frac{1}{6!}d^6 s'^6 + \cdots\end{aligned} \tag{C.81}$$

Additionally, we have

$$\mathrm{e}^{s'd}(\mathrm{e}^{s'} - 1)^2 = s'^2 + d_3 s'^3 + d_4 s'^4 + d_5 s'^5 + d_6 s'^6 + \cdots \tag{C.82}$$

where

$$\begin{bmatrix} d_3 \\ d_4 \\ d_5 \\ d_6 \end{bmatrix} = \begin{bmatrix} 1+d \\ 0.5833 + 1d + \frac{1}{2}d^2 \\ 0.25 + 0.5833d + \frac{1}{2}d^2 + \frac{1}{6}d^3 \\ 0.0861 + 0.25d + \frac{0.5833}{2}d^2 + \frac{1}{6}d^3 + \frac{1}{24}d^4 \end{bmatrix}. \tag{C.83}$$

According to the above, polynomial $D_1(s')$ can be rewritten in the form of

$$\begin{aligned} D_1(s') &= t_{\mathrm{i}}^2(s'^n r_n + \cdots + s'^3 r_3 + s'^2 r_2 + s' r_1 + 1)\mathrm{e}^{s'd}(\mathrm{e}^{s'} - 1)^2 \\ &= t_{\mathrm{i}}^2(\cdots + q_7 s'^7 + q_6 s'^6 + q_5 s'^5 + q_4 s'^4 + q_3 s'^3 + s'^2) \end{aligned} \tag{C.84}$$

where

$$\begin{bmatrix} q_3 \\ q_4 \\ q_5 \\ q_6 \\ q_7 \end{bmatrix} = \begin{bmatrix} r_1 + d_3 \\ r_2 + d_3 r_1 + d_4 \\ r_3 + d_3 r_2 + d_4 r_1 + d_5 \\ r_4 + d_3 r_3 + d_4 r_2 + d_5 r_1 + d_6 \\ r_5 + d_6 r_1 + d_3 r_4 + d_4 r_3 + d_5 r_2 + d_7 \end{bmatrix}. \tag{C.85}$$

Since $(1 - \mathrm{e}^{s'})\hat{x} + (\mathrm{e}^{2s'} - 1)\hat{y} + 1$ is equal to

$$\begin{aligned} &(1 - \mathrm{e}^{s'})\hat{x} + (\mathrm{e}^{2s'} - 1)\hat{y} + 1 \\ &\quad = 1 + (2\hat{y} - \hat{x})s' + \left(2\hat{y} - \frac{1}{2}\hat{x}\right)s'^2 + \left(\frac{4}{3}\hat{y} - \frac{1}{6}\hat{x}\right)s'^3 \\ &\qquad + \left(\frac{2}{3}\hat{y} - \frac{1}{24}\hat{x}\right)s'^4 + \left(\frac{32}{5!}\hat{y} - \frac{1}{5!}\hat{x}\right)s'^5 + \left(\frac{64}{6!}\hat{y} - \frac{1}{6!}\hat{x}\right)s'^6 + \cdots \end{aligned} \tag{C.86}$$

polynomial $k_{\mathrm{h}}N(s')$ takes the form

$$\begin{aligned} k_{\mathrm{h}}N(s') = \cdots &+ k_{\mathrm{h}}k'_p\left(y_7\hat{y} - x_7\hat{x} + z_7\right)s'^7 \\ &+ k_{\mathrm{h}}k'_p\left(y_6\hat{y} - x_6\hat{x} + z_6\right)s'^6 \\ &+ k_{\mathrm{h}}k'_p\left(y_5\hat{y} - x_5\hat{x} + z_5\right)s'^5 \\ &+ k_{\mathrm{h}}k'_p\left(y_4\hat{y} - x_4\hat{x} + z_4\right)s'^4 \\ &+ k_{\mathrm{h}}k'_p\left(y_3\hat{y} - x_3\hat{x} + z_3\right)s'^3 \\ &+ k_{\mathrm{h}}k'_p\left(y_2\hat{y} - x_2\hat{x} + z_2\right)s'^2 \\ &+ k_{\mathrm{h}}k'_p\left(2\hat{y} - \hat{x} + z_1\right)s' + k_{\mathrm{h}}k'_p \end{aligned} \tag{C.87}$$

where

$$x_k = \sum_{j=1}^{k}\left(z_{(j-1)}\frac{1}{[k-(j-1)]!}\right) \tag{C.88}$$

or in an expanded form

$$\begin{bmatrix} x_1 \\ x_2 \\ x_3 \\ x_4 \\ x_5 \\ x_6 \\ x_7 \end{bmatrix} = \begin{bmatrix} 1 \\ \frac{1}{2!}+z_1 \\ \frac{1}{3!}+\frac{1}{2}z_1+z_2 \\ \frac{1}{4!}+\frac{1}{6}z_1+\frac{1}{2}z_2+z_3 \\ \frac{1}{5!}+\frac{1}{24}z_1+\frac{1}{6}z_2+\frac{1}{2}z_3+z_4 \\ \frac{1}{6!}+\frac{1}{5!}z_1+\frac{1}{24}z_2+\frac{1}{6}z_3+\frac{1}{2}z_4+z_5 \\ \frac{1}{6!}z_1+\frac{1}{5!}z_2+\frac{1}{24}z_3+\frac{1}{6}z_4+\frac{1}{2}z_5+z_6+\frac{1}{7!} \end{bmatrix} \tag{C.89}$$

and

$$\begin{bmatrix} y_1 \\ y_2 \\ y_3 \\ y_4 \\ y_5 \\ y_6 \\ y_7 \end{bmatrix} = \begin{bmatrix} 1 \\ 2(1+z_1) \\ 2\left(z_1+z_2+\frac{2}{3}\right) \\ 2\left(z_2+z_3+\frac{2}{3}z_1+\frac{1}{3}\right) \\ 2\left(z_3+z_4+\frac{2}{3}z_2+\frac{1}{3}z_1+\frac{16}{5!}\right) \\ 2\left(z_5+z_4+\frac{2}{3}z_3+\frac{1}{3}z_2+\frac{16}{5!}z_1+\frac{32}{6!}\right) \\ 2\left(z_6+z_5+\frac{2}{3}z_4+\frac{1}{3}z_3+\frac{16}{5!}z_2+\frac{32}{6!}z_1+\frac{64}{7!}\right) \end{bmatrix}. \tag{C.90}$$

Finally, the corresponding polynomials $N(s')$, $D(s')$ for both the numerator and denominator of the closed loop transfer function are given by

$$N(s') = \sum_{j=0}^{m}\left[k'_p\left(y_j\hat{y} - x_j\hat{x} + z_j\right)s'^j\right] \tag{C.91}$$

where $y_1 = 2, x_1 = 1, z_0 = 1$ and

$$D\left(s'\right) = \sum_{i=0}^{n} \left[\left(q_i t_\mathrm{i}^2 + k_\mathrm{h} k'_p \left(y_i \hat{y} - x_i \hat{x} + z_i\right) s'^i\right)\right] \tag{C.92}$$

and in an expanded form

$$\begin{aligned} N(s') = \cdots &+ \underbrace{k'_p \left(y_7 \hat{y} - x_7 \hat{x} + z_7\right)}_{b_7} s'^7 + \underbrace{k'_p \left(y_6 \hat{y} - x_6 \hat{x} + z_6\right)}_{b_6} s'^6 \\ &+ \underbrace{k'_p \left(y_5 \hat{y} - x_5 \hat{x} + z_5\right)}_{b_5} s'^5 + \underbrace{k'_p \left(y_4 \hat{y} - x_4 \hat{x} + z_4\right)}_{b_4} s'^4 \\ &+ \underbrace{k'_p \left(y_3 \hat{y} - x_3 \hat{x} + z_3\right)}_{b_3} s'^3 + \underbrace{k'_p \left(y_2 \hat{y} - x_2 \hat{x} + z_2\right)}_{b_2} s'^2 \\ &+ \underbrace{k'_p \left(2\hat{y} - \hat{x} + z_1\right)}_{b_1} s' + \underbrace{k'_p}_{b_0} \end{aligned} \tag{C.93}$$

and

$$\begin{aligned} D(s') = D_1(s') + k_\mathrm{h} N(s') = \cdots \\ &+ \underbrace{\left[q_7 t_\mathrm{i}^2 + k_\mathrm{h} k'_p \left(y_7 \hat{y} - x_7 \hat{x} + z_7\right)\right]}_{a_7} s'^7 + \underbrace{\left[q_6 t_\mathrm{i}^2 + k_\mathrm{h} k'_p \left(y_6 \hat{y} - x_6 \hat{x} + z_6\right)\right]}_{a_6} s'^6 \\ &+ \underbrace{\left[q_5 t_\mathrm{i}^2 + k_\mathrm{h} k'_p \left(y_5 \hat{y} - x_5 \hat{x} + z_5\right)\right]}_{a_5} s'^5 + \underbrace{\left[q_4 t_\mathrm{i}^2 + k_\mathrm{h} k'_p \left(y_4 \hat{y} - x_4 \hat{x} + z_4\right)\right]}_{a_4} s'^4 \\ &+ \underbrace{\left[q_3 t_\mathrm{i}^2 + k_\mathrm{h} k'_p \left(y_3 \hat{y} - x_3 \hat{x} + z_3\right)\right]}_{a_3} s'^3 + \underbrace{\left[t_\mathrm{i}^2 + k_\mathrm{h} k'_p \left(y_2 \hat{y} - x_2 \hat{x} + z_2\right)\right]}_{a_2} s'^2 \\ &+ \underbrace{k_\mathrm{h} k'_p \left(2\hat{y} - \hat{x} + z_1\right)}_{a_1} s' + \underbrace{k_\mathrm{h} k'_p}_{a_0} \end{aligned} \tag{C.94}$$

where $q_2 = 1, q_1 = 1, y_1 = 2, x_1 = 1, q_0 = y_0 = x_0 = 0$ and $z_0 = 1$. Therefore, the resulting transfer function of the closed loop control system is given by

$$\begin{aligned} T\left(s'\right) &= \frac{N(s')}{D(s') + k_\mathrm{h} N(s')} \\ &= \frac{\sum_{j=0}^{m} \left[k'_p \left(y_j \hat{y} - x_j \hat{x} + z_j\right) s'^j\right]}{\left[\sum_{i=0}^{n} \left[\left(q_i t_\mathrm{i}^2 + k_\mathrm{h} k'_p \left(y_i \hat{y} - x_i \hat{x} + z_i\right) s'^i\right)\right] + k_\mathrm{h} \sum_{j=0}^{m} \left[k'_p \left(y_j \hat{y} - x_j \hat{x} + z_j\right) s'^j\right]\right]} \end{aligned} \tag{C.95}$$

By applying the first optimization condition

Optimization Condition: $a_0 = b_0$.

To the closed loop tranfer function results in

$$k_{\mathrm{h}} = 1 \tag{C.96}$$

which implies that the final closed loop control system exhibits steady state position and velocity error. From, (C.91), it is apparent that if $k_{\mathrm{h}} = 1$ then the respective terms s^0, s^1, of $N(s') = \cdots + k_p' \left(2\hat{y} - \hat{x} + z_1\right) s' + k_p'$ and $D(s') = \cdots + k_{\mathrm{h}} k_p' \left(2\hat{y} - \hat{x} + z_1\right) s' + k_{\mathrm{h}} k_p'$ are equal.

Optimization Condition: $a_1^2 - 2a_2a_0 = b_1^2 - 2b_2b_0$.

By making use of $a_1^2 - 2a_2a_0 = b_1^2 - 2b_2b_0$ results in $t_{\mathrm{i}} = 0$. For that reason, $a_1^2 - 2a_2a_0 = 0$ is set, as another means of optimizing the magnitude of (A.1). This results in,

$$t_{\mathrm{i}}^2 = \frac{1}{2} k_{\mathrm{h}} k_p' \left[(2\hat{y} - \hat{x})^2 - 2(y_2 - 2z_1)\hat{y} + 2(x_2 - z_1)\hat{x} + \left(z_1^2 - 2z_2\right) \right]. \tag{C.97}$$

In case where no zeros exist in the plant transfer function then,

$$t_{\mathrm{i}}^2 = \frac{1}{2} k_{\mathrm{h}} k_p' \left[(2\hat{y} - \hat{x})^2 - 2y_2\hat{y} + 2x_2\hat{x} \right]. \tag{C.98}$$

Optimization Condition: $a_2^2 - 2a_3a_1 + 2a_4a_0 = b_2^2 - 2b_3b_1 + 2b_4b_0$.

The application of (A.11) to (C.95) yields

$$t_{\mathrm{i}}^2 = 2k_{\mathrm{h}} k_p' \left[(x_2 - q_3)\hat{x} + (2q_3 - y_2)\hat{y} - z_2 + q_3z_1 - q_4 \right] \tag{C.99}$$

Optimization Condition: $a_3^2 + 2a_1a_5 - 2a_6a_0 - 2a_4a_2 = b_3^2 + 2b_1b_5 - 2b_6b_0 - 2b_4b_2$.

Finally, the application of (A.12) into (C.95) results in

$$t_{\mathrm{i}}^2 = \frac{2k_{\mathrm{h}} k_p'}{\left(q_3^2 - 2q_4\right)} \left[\begin{array}{c} \left(q_4y_2 - q_3y_3 - 2q_5 + y_4\right)\hat{y} + \begin{pmatrix} q_3x_3 - q_4x_2 \\ + q_5 - x_4 \end{pmatrix} \hat{x} \\ \left(q_4z_2 - q_5z_1 - q_3z_3 + q_6 + z_4\right) \end{array} \right] \tag{C.100}$$

For determining the optimal controller parameters relations (C.97), (C.99) and (C.99), (C.100) are manipulated together. Therefore, from (C.97), (C.99) it is apparent that

$$\begin{aligned} &(2\hat{y} - \hat{x})^2 - 2\left(4q_3 - y_2 - 2z_1\right)\hat{y} + 2\left(2q_3 - x_2 - z_1\right)\hat{x} \\ &\quad + \left(z_1^2 + 2z_2 - 4q_3z_1 + 4q_4\right) = 0. \end{aligned} \tag{C.101}$$

From (C.99), (C.100) it is found that

$$\begin{aligned} &\left[(q_3^2 - 2q_4)(x_2 - q_3) - (q_3x_3 - q_4x_2 + q_5 - x_4)\right]\hat{x} \\ &\quad + \left[(q_3^2 - 2q_4)(2q_3 - y_2) - (q_4y_2 - q_3y_3 - 2q_5 + y_4)\right]\hat{y} \\ &= (q_3^2 - 2q_4)(z_2 - q_3z_1 + q_4) + (q_4z_2 - q_5z_1 - q_3z_3 + q_6 + z_4). \end{aligned} \tag{C.102}$$

After making the following substitutions

$$A = 4q_3 - y_2 - 2z_1, \tag{C.103}$$

$$B = 2q_3 - x_2 - z_1, \tag{C.104}$$

$$C = z_1^2 + 2z_2 - 4q_3z_1 + 4q_4, \tag{C.105}$$

$$D = (q_3^2 - 2q_4)(x_2 - q_3) - (q_3x_3 - q_4x_2 + q_5 - x_4), \tag{C.106}$$

$$E = (q_3^2 - 2q_4)(2q_3 - y_2) - (q_4y_2 - q_3y_3 - 2q_5 + y_4), \tag{C.107}$$

$$Z = (q_3^2 - 2q_4)(z_2 - q_3z_1 + q_4) + \begin{pmatrix} q_4z_2 - q_5z_1 - q_3z_3 + q_6 \\ + z_4 \end{pmatrix} \tag{C.108}$$

and substituting (C.103)–(C.108) back into (C.101)–(C.102) it is obtained

$$(2\hat{y} - \hat{x})^2 - 2A\hat{y} + 2B\hat{x} + C = 0, \tag{C.109}$$

$$D\hat{x} + E\hat{y} = Z \tag{C.110}$$

respectively. From (C.110) it is found that $\hat{x}$ is equal to

$$\hat{x} = \frac{Z - E\hat{y}}{D}. \tag{C.111}$$

Substituting (C.111) into (C.109) results in

$$\hat{y}^2 - 2\frac{[(2D + E)Z + D(\mathrm{AD} + \mathrm{BE})]}{(2D + E)^2}\hat{y} + \frac{D(2\mathrm{BZ} + \mathrm{CD}) + Z^2}{(2D + E)^2} = 0. \tag{C.112}$$

As a result the final control law is given by

$$\begin{bmatrix} 1 - 2k_\mathrm{h}k_p'(x_2 - q_3) & -2k_\mathrm{h}k_p'(2q_3 - y_2) \\ 0 \quad D & E \\ 0 \quad 0 & -2\frac{[(2D+E)Z+D(\mathrm{AD}+\mathrm{BE})]}{(2D+E)^2} \end{bmatrix} \begin{bmatrix} t_\mathrm{i}^2 \\ \hat{x} \\ \hat{y} \end{bmatrix} = \begin{bmatrix} -(z_2 + q_3z_1 - q_4) \\ Z \\ -\frac{D(2\mathrm{BZ}+\mathrm{CD})+Z^2}{(2D+E)^2} - \hat{y}^2 \end{bmatrix} \tag{C.113}$$

or

$$\begin{bmatrix} t_{\mathrm{i}}^2 \\ \hat{x} \\ \hat{y} \end{bmatrix} = \begin{bmatrix} 1 & -2k_{\mathrm{h}}k'_p\,(x_2 - q_3) & -2k_{\mathrm{h}}k'_p\,(2q_3 - y_2) \\ 0 & D & E \\ 0 & 0 & -2\frac{[(2D+E)Z+D(AD+BE)]}{(2D+E)^2} \end{bmatrix}^{-1} \times \begin{bmatrix} -\left(z_2 + q_3 z_1 - q_4\right) \\ Z \\ -\frac{D(2BZ+CD)+Z^2}{(2D+E)^2} - \hat{y}^2 \end{bmatrix} \tag{C.114}$$

taking into account that $\frac{x}{T_{\mathrm{s}}} = 2\hat{y} - \hat{x} - 1$ and $\frac{y}{T_{\mathrm{s}}} = \hat{x} - \hat{y} + 1$ from (C.69), (C.70) respectively.

C.3 Type-III control loops

For presenting the proof for digital PID controller design in type-III control loops, the process is defined by

$$G(s) = \frac{1}{sT_m(1 + sT_{p_1})(1 + sT_{\Sigma_p})} \tag{C.115}$$

whereas the proposed controller is given by

$$C(s) = C^*(s)C_{\mathrm{ZOH}}(s) = \left[\frac{(1 + sT_n)(1 + sT_v)(1 + sT_x)}{s^2 T_{\mathrm{i}}(1 + sT_{\Sigma c_1})(1 + sT_{\Sigma c_2})}\right]^* \frac{(1 - \mathrm{e}^{-sT_{\mathrm{s}}})}{sT_{\mathrm{s}}}. \tag{C.116}$$

Note again that $C_{\mathrm{ZOH}}(s)$ stands for the zero order hold transfer function and T_{s} stands for the controller's sampling period. All time constants in the control loop are normalized in the frequency domain with the sampling period T_{s}. Therefore, by substituting

$$s' = sT_{\mathrm{s}}, \tag{C.117}$$

both (C.115) and (C.116) become

$$G(s') = \frac{1}{s' t_m(1 + s' t_{p_1})(1 + s' t_{\Sigma_p})} \tag{C.118}$$

and

$$C(s') = C^*(s')C_{\mathrm{ZOH}}(s') = T_{\mathrm{s}}\left[\frac{(1 + s't_n)(1 + s't_v)(1 + s't_x)}{s'^2 t_{\mathrm{i}}(1 + s't_{\Sigma c_1})(1 + s't_{\Sigma c_2})}\right]^* \frac{(1 - \mathrm{e}^{-s'})}{s'} \tag{C.119}$$

where

$$t_{\mathrm{i}} = \frac{T_{\mathrm{i}}}{T_{\mathrm{s}}}, \quad t_{\Sigma_{c_1}} = \frac{T_{\Sigma_{c_1}}}{T_{\mathrm{s}}}, \quad t_{\Sigma_{c_2}} = \frac{T_{\Sigma_{c_2}}}{T_{\mathrm{s}}}, \tag{C.120}$$

$$t_n = \frac{T_n}{T_{\mathrm{s}}}, \quad t_v = \frac{T_v}{T_{\mathrm{s}}}, \quad t_x = \frac{T_x}{T_{\mathrm{s}}}, \tag{C.121}$$

$$t_m = \frac{T_m}{T_{\mathrm{s}}}, \quad t_{p_1} = \frac{T_{p_1}}{T_{\mathrm{s}}}, \quad t_{\Sigma_P} = \frac{T_{\Sigma_p}}{T_{\mathrm{s}}}. \tag{C.122}$$

In similar fashion with the analog design procedure in section B, the open loop transfer function $F_{\mathrm{ol}}(s)$ is given by

$$\begin{aligned} F_{\mathrm{ol}}(s') &= k_{\mathrm{p}} k_{\mathrm{h}} C(s') G(s') \\ &= T_{\mathrm{s}} \left[\frac{(1 + s't_n)(1 + s't_v)(1 + s't_x)}{s'^2 t_{\mathrm{i}} (1 + s't_{\Sigma c1})(1 + s't_{\Sigma c2})} \right]^* \left[\frac{(1 - \mathrm{e}^{-s'})}{s'} \right] \\ &\times \frac{k_{\mathrm{p}} k_{\mathrm{h}}}{\left[s't_m (1 + s't_{p_1})(1 + s't_{\Sigma_p}) \right]}. \end{aligned} \tag{C.123}$$

For moving from the $\mathcal{L}\{.\}$ to the $\mathcal{Z}\{.\}$ domain, the following substitutions are considered

$$\frac{1}{s'} = \frac{z'}{z' - 1} = \frac{\mathrm{e}^{s'}}{\mathrm{e}^{s'} - 1}, \tag{C.124}$$

$$\frac{1}{s'^2} = \frac{T_{\mathrm{s}} z'}{(z' - 1)^2} = \frac{T_{\mathrm{s}} \mathrm{e}^{s'}}{(\mathrm{e}^{s'} - 1)^2}. \tag{C.125}$$

To this end, and since $z' = \mathrm{e}^{s'}$, $F_{\mathrm{ol}}(s')$ becomes equal to

$$\begin{aligned} F_{\mathrm{ol}}(s') = & \frac{k_{\mathrm{p}} k_{\mathrm{h}} T_{\mathrm{s}}}{s't_m (1 + s't_{p_1})(1 + s't_{\Sigma_p})} \left[\frac{(1 + s't_n)(1 + s't_v)(1 + s't_x)}{t_{\mathrm{i}} (1 + s't_{\Sigma c1})(1 + s't_{\Sigma c2})} \right] \\ & \times \frac{(1 - \mathrm{e}^{-s'}) \mathrm{e}^{s'} T_{\mathrm{s}} \mathrm{e}^{s'}}{(\mathrm{e}^{s'} - 1)(\mathrm{e}^{s'} - 1)^2} \end{aligned} \tag{C.126}$$

or finally

$$\begin{aligned} F_{\mathrm{ol}}(s') = & \frac{k_{\mathrm{p}} k_{\mathrm{h}} T_{\mathrm{s}}^2}{s't_m (1 + s't_{p_1})(1 + s't_{\Sigma_p})} \left[\frac{(1 + s't_n)(1 + s't_v)(1 + s't_x)}{t_{\mathrm{i}} (1 + s't_{\Sigma c1})(1 + s't_{\Sigma c2})} \right] \\ & \times \frac{\mathrm{e}^{s'}}{(\mathrm{e}^{s'} - 1)^2}. \end{aligned} \tag{C.127}$$

Assuming that the dominant time constant is accurately measured, as mentioned in Sect. 6.3, for determining parameter t_x, pole-zero cancellation takes place. Therefore

$$t_x = t_{p_1} \tag{C.128}$$

is set. This results in

$$F_{\mathrm{ol}}(s') = \frac{k_{\mathrm{p}} k_{\mathrm{h}} T_{\mathrm{s}}^2}{s' t_m (1 + s' t_{\Sigma_{\mathrm{p}}})} \left[\frac{\left(1 + s' t_n\right)\left(1 + s' t_v\right)}{t_{\mathrm{i}} \left(1 + s' t_{\Sigma c_1}\right)\left(1 + s' t_{\Sigma c_2}\right)} \right] \frac{\mathrm{e}^{s'}}{(\mathrm{e}^{s'} - 1)^2}. \tag{C.129}$$

After setting $k_{\mathrm{p}}' = k_{\mathrm{p}} T_{\mathrm{s}}^2$, $F_{\mathrm{ol}}(s')$ becomes equal to

$$F_{\mathrm{ol}}(s') = \frac{k_{\mathrm{p}}' k_{\mathrm{h}} \mathrm{e}^{s'} (1 + s' t_n)(1 + s' t_v)}{s' t_m t_{\mathrm{i}} (1 + s' t_{\Sigma_{\mathrm{p}}})(1 + s' t_{\Sigma c_1})(1 + s' t_{\Sigma c_2})(\mathrm{e}^{s'} - 1)^2}. \tag{C.130}$$

In similar fashion with the analog design it is set $t_{\Sigma_{c_1}} t_{\Sigma_{c_2}} \approx 0$ and $t_{\Sigma_c} = t_{\Sigma_{c_1}} + t_{\Sigma_{c_2}}$. This results in $(1 + s' t_{\Sigma_p})(1 + s' t_{\Sigma_{c_1}})(1 + s' t_{\Sigma_{c_2}}) = (1 + s' t_{\Sigma_p})(1 + s' t_{\Sigma_c})$. Moreover if $t_{\Sigma_c} t_{\Sigma_p} \approx 0$ and $t_\Sigma = t_{\Sigma_c} + t_{\Sigma_p}$ then (C.130) becomes equal to

$$F_{\mathrm{ol}}(s') = \frac{k_{\mathrm{p}}' k_{\mathrm{h}} \mathrm{e}^{s'} (1 + s' t_n)(1 + s' t_v)}{s' t_m t_{\mathrm{i}} (1 + s' t_\Sigma)(\mathrm{e}^{s'} - 1)^2}. \tag{C.131}$$

Finally the closed loop transfer function becomes equal to

$$\begin{aligned} T(s') &= \frac{k_{\mathrm{p}} C(s') G(s')}{1 + k_{\mathrm{p}} k_{\mathrm{h}} C(s') G(s')} = \frac{\dfrac{k_{\mathrm{p}}' \mathrm{e}^{s'} (1 + s' t_n)(1 + s' t_v)}{s' t_m t_{\mathrm{i}} (1 + s' t_\Sigma)(\mathrm{e}^{s'} - 1)^2}}{1 + k_{\mathrm{h}} \dfrac{k_{\mathrm{p}}' \mathrm{e}^{s'} (1 + s' t_n)(1 + s' t_v)}{s' t_m t_{\mathrm{i}} (1 + s' t_\Sigma)(\mathrm{e}^{s'} - 1)^2}} \\ &= \frac{k_{\mathrm{p}}' \mathrm{e}^{s'} (1 + s' t_n)(1 + s' t_v)}{s' t_m t_{\mathrm{i}} (1 + s' t_\Sigma)(\mathrm{e}^{s'} - 1)^2 + k_{\mathrm{h}} k_{\mathrm{p}}' \mathrm{e}^{s'} (1 + s' t_n)(1 + s' t_v)}. \end{aligned} \tag{C.132}$$

By approximating the time delay $\mathrm{e}^{s'}$ by the "all pole" series approximation

$$D_1(s') = \mathrm{e}^{s'} - 1 = s' + \frac{1}{2!} s'^2 + \frac{1}{3!} s'^3 + \frac{1}{4!} s'^4 + \frac{1}{5!} s'^5 + \frac{1}{6!} s'^6 + \cdots \tag{C.133}$$

and

$$\begin{aligned} D_2(s') &= (\mathrm{e}^{s'} - 1)^2 \\ &= (s' + \frac{1}{2!} s'^2 + \frac{1}{3!} s'^3 + \frac{1}{4!} s'^4 + \frac{1}{5!} s'^5 + \frac{1}{6!} s'^6 + \cdots)^2 \end{aligned}$$

$$= \cdots + Ds'^6 + Cs'^5 + Bs'^4 + As'^3 + s'^2 \tag{C.134}$$

where $A = \frac{1}{2!} + \frac{1}{2!}$, $B = \frac{1}{3!} + \frac{1}{2!}\frac{1}{2!} + \frac{1}{3!}$, $C = \frac{1}{4!} + \frac{1}{2!}\frac{1}{3!} + \frac{1}{3!}\frac{1}{2!} + \frac{1}{4!}$ and $D = \frac{1}{5!} + \frac{1}{2!}\frac{1}{4!} + \frac{1}{3!}\frac{1}{3!} + \frac{1}{4!}\frac{1}{2!} + \frac{1}{5!}$ it is concluded that (C.132) is equal to

$$\begin{aligned} T(s') &= \frac{k_p' e^{s'}\left(1 + s't_n\right)\left(1 + s't_v\right)}{s't_m t_i\left(1 + s't_\Sigma\right)\left(e^{s'} - 1\right)^2 + k_p' k_h e^{s'}\left(1 + s't_n\right)\left(1 + s't_v\right)} \\ &= \frac{\left[s'^2 k_p' t_n t_v + s' k_p'(t_n + t_v) + k_p'\right]\left(D_1(s') + 1\right)}{\begin{array}{l}(s'^2 t_i t_m t_\Sigma + s' t_m t_i) D_2(s') \\ + \left[s'^2 k_p' k_h t_n t_v + s' k_p' k_h (t_n + t_v) + k_p' k_h\right]\left(D_1(s') + 1\right)\end{array}} . \end{aligned} \tag{C.135}$$

Since (C.135) is in the form of (A.1), the optimization conditions (A.9)–(A.12) can be applied for proving the optimal digital PID control law.

For the simplification of the proof of the optimal control, the following substitutions are made. Within the numerator of (C.135) it is set

$$z_1 = k_p' + k_p'(t_n + t_v) \tag{C.136}$$

$$z_2 = \frac{1}{2!}k_p' + k_p'(t_n + t_v) + k_p' t_n t_v \tag{C.137}$$

$$z_3 = \frac{1}{3!}k_p' + \frac{1}{2!}k_p'(t_n + t_v) + k_p' t_n t_v \tag{C.138}$$

$$z_4 = \frac{1}{4!}k_p' + \frac{1}{3!}k_p'(t_n + t_v) + \frac{1}{2!}k_p' t_n t_v. \tag{C.139}$$

In similar fashion, within the denominator of (C.135) it is set

$$r_1 = k_h z_1 \tag{C.140}$$

$$r_2 = k_h z_2 \tag{C.141}$$

$$r_3 = k_h z_3 + t_m t_i \tag{C.142}$$

$$r_4 = k_h z_4 + A t_m t_i + t_m t_i t_\Sigma \tag{C.143}$$

$$r_5 = k_h z_5 + B t_m t_i + A t_m t_i t_\Sigma \tag{C.144}$$

$$r_6 = k_h z_6 + C t_m t_i + B t_m t_i t_\Sigma \tag{C.145}$$

$$r_7 = k_h z_7 + D t_m t_i + C t_m t_i t_\Sigma . \tag{C.146}$$

Since (C.135) is in the form of (A.1) we are now ready to apply the optimization conditions (A.9)–(A.12) for determining the proposed analytical control law regarding parameters t_n, t_v, t_i.

Optimization Condition: $a_0 = b_0$.

By applying the first optimization condition to the closed loop transfer function (C.135) results in

$$k_{\mathrm{h}} = 1 \tag{C.147}$$

which implies that the final closed loop control system exhibits steady state position, velocity error. From (C.135) it is apparent that if $k_{\mathrm{h}} = 1$, then

$$N(s') = k_{\mathrm{p}} t_n t_v s'^2 + s' k_{\mathrm{p}} (t_n + t_v) + k_{\mathrm{p}} \tag{C.148}$$

and

$$D(s') = \cdots + k_{\mathrm{p}} k_{\mathrm{h}} t_n t_v s'^2 + s' k_{\mathrm{p}} k_{\mathrm{h}} (t_n + t_v) + k_{\mathrm{p}} k_{\mathrm{h}} \tag{C.149}$$

respectively.

Optimization Condition: $a_1^2 - 2a_2a_0 = b_1^2 - 2b_2b_0$.

The application of (A.10) to (C.135) results in

$$t_n^2 + t_v^2 = 0, \tag{C.150}$$

which is not accepted since for both t_n, t_v, conditions $t_n > 0$, $t_v > 0$ must hold by.

Optimization Condition: $a_2^2 - 2a_3a_1 + 2a_4a_0 = b_2^2 - 2b_3b_1 + 2b_4b_0$.

In similar fashion, the application of (A.10) to (C.135) does not result in an acceptable relation and therefore the right part of (A.10) is set to zero. In that $a_2^2 - 2a_3a_1 + 2a_4a_0 = 0$ is used.

$$k_{\mathrm{p}}' k_{\mathrm{h}} t_n^2 t_v^2 - 2 t_m t_{\mathrm{i}} (t_n + t_v - t_\Sigma) = 0. \tag{C.151}$$

Optimization Condition: $a_3^2 + 2a_1a_5 - 2a_6a_0 - 2a_4a_2 = b_3^2 + 2b_1b_5 - 2b_6b_0 - 2b_4b_2$.

The integrator's time constant is calculated after the application of (A.12) to (C.135) and by setting again the right part of the (A.12) equal to zero $a_3^2 + 2a_1a_5 - 2a_6a_0 - 2a_4a_2 = 0$. This results in

$$t_{\mathrm{i}} t_m - k_{\mathrm{p}}' k_{\mathrm{h}} [2 t_\Sigma t_n t_v + (2B - 1)(t_\Sigma - t_n - t_v)] = 0 \tag{C.152}$$

or finally

$$t_{\mathrm{i}} = \frac{k_{\mathrm{p}}' k_{\mathrm{h}} [2 t_\Sigma t_n t_v + (2B - 1)(t_\Sigma - t_n - t_v)]}{t_m}. \tag{C.153}$$

Substituting (C.153) into (C.151) results in

$$4t_\Sigma t_n t_v(t_n + t_v - t_\Sigma) - 2(2B-1)(t_n + t_v - t_\Sigma)^2 = t_n^2 t_v^2. \tag{C.154}$$

At this point, the same line is adopted, as the one followed in Sect. 5.2.1, regarding the determination of parameter t_v. Therefore

$$t_v = nt_\Sigma \tag{C.155}$$

is set and is substituted into (C.154) which results in

$$t_{n_{1,2}} = \frac{-[(n-1)t_\Sigma(4nt_\Sigma^2 - 4(2B-1))] \pm \sqrt{\Delta}}{2[nt_\Sigma^2(4-n) - 2(2B-1)]} \tag{C.156}$$

where

$$\Delta = n(n-1)^2 t_\Sigma^4[16n - 8(2B-1)(n-3)]. \tag{C.157}$$

Since $t_\Sigma = \frac{T_\Sigma}{T_s}$, T_s is a design parameter, parameter t_n is calculated out of (C.156). Parameter n must be chosen such that condition

$$n(n-1)^2 t_\Sigma^4[16n - 8(2B-1)(n-3)] > 0 \tag{C.158}$$

is satisfied. To this end, $n > 0$ and $n > 1$. If n is chosen such that $n > 1$ then it is easily shown that $\forall n > 1,\ 16n - 8(2B-1)(n-3) > 0$.

Index

K.G. Papadopoulos, *PID Controller Tuning Using the Magnitude Optimum Criterion*,
DOI 10.1007/978-3-319-07263-0

MIX
Papier aus verantwortungsvollen Quellen
Paper from responsible sources
FSC® C105338

If you have any concerns about our products,
you can contact us on
ProductSafety@springernature.com

In case Publisher is established outside the EU,
the EU authorized representative is:
Springer Nature Customer Service Center GmbH
Europaplatz 3, 69115 Heidelberg, Germany

Printed by Libri Plureos GmbH
in Hamburg, Germany